Photography, Natural History and the Nineteenth-Century Muse

AF760715

The Victorian era heralded an age of transformation in which momentous changes in the field of natural history coincided with the rise of new visual technologies. Concurrently, different parts of the British Empire began to more actively claim their right to being acknowledged as indispensable contributors to knowledge and the progress of empire. This book addresses the complex relationship between natural history and photography from the 1850s to the 1880s in Britain and its colonies: Australia, New Zealand and, to a lesser extent, India. Coinciding with the rise of the modern museum, photography's arrival was timely, and it rapidly became an essential technology for recording and publicising rare objects and valuable collections. Also during this period, the medium assumed a more significant role in the professional practices and reputations of naturalists than has been previously recognized, and it figured increasingly within the expanding specialized networks that were central to the production and dissemination of new knowledge. In an interrogation that ranges from the first forays into museum photography and early attempts to document collecting expeditions to the importance of traditional and photographic portraiture for the recognition of scientific discoveries, this book not only recasts the parameters of what we actually identify as natural history photography in the Victorian era but also how we understand the very structure of empire in relation to this genre at that time.

Kathleen Davidson is an Honorary Associate in the School of Literature, Art and Media and teaches in the Department of Art History at the University of Sydney. Previously, she was Curator of International Photography at the National Gallery of Australia.

Science and the Arts since 1750
Series Editor
Barbara Larson, University of West Florida

This series of monographs and edited volumes explores the arts – painting and sculpture, drama, dance, architecture, design, photography, popular culture materials – as they intersect with emergent scientific theories, agendas, and technologies, from any geographical area from 1750 to now.

1 Art, Technology and Nature
Renaissance to Postmodernity
Edited by Camilla Skovbjerg Paldam and Jacob Wamberg

2 The Organic School of the Russian Avant-Garde
Nature's Creative Principles
Isabel Wünsche

Photography, Natural History and the Nineteenth-Century Museum

Exchanging Views of Empire

Kathleen Davidson

Routledge
Taylor & Francis Group
LONDON AND NEW YORK

First published 2017 by Routledge

2 Park Square, Milton Park, Abingdon, Oxfordshire OX14 4RN
52 Vanderbilt Avenue, New York, NY 10017

Routledge is an imprint of the Taylor & Francis Group, an informa business

First issued in paperback 2020

Copyright © 2017 Kathleen Davidson

The right of Kathleen Davidson to be identified as author of this work has been asserted by her in accordance with sections 77 and 78 of the Copyright, Designs and Patents Act 1988.

All rights reserved. No part of this book may be reprinted or reproduced or utilised in any form or by any electronic, mechanical, or other means, now known or hereafter invented, including photocopying and recording, or in any information storage or retrieval system, without permission in writing from the publishers.

Notice:
Product or corporate names may be trademarks or registered trademarks, and are used only for identification and explanation without intent to infringe.

British Library Cataloguing in Publication Data
A catalogue record for this book is available from the British Library

Library of Congress Cataloging in Publication Data
A catalog record for this title has been requested

ISBN: 978-1-4724-3129-5 (hbk)
ISBN: 978-0-367-33116-0 (pbk)

Typeset in Sabon
by Deanta Global Publishing Services, Chennai, India

Contents

List of illustrations vii
Colour plates xiii
Acknowledgements xv

Introduction 1

Rethinking the role of photography in Victorian natural history 4
Emerging sites of production: a comparative approach 6
Formative visions of empire 7
Exchanging views of empire 7
From 'immutable mobiles' to 'boundary objects' 10
Making and moving images 11
Peripatetic objects of empire and collections at the periphery 12
Navigating the archive 13

1 **Paper museums: photography and natural history at the British Museum** 21

Introduction 21
From institutional priorities to imperial interests 22
A virtual inventory of specimens: circulating new knowledge and filling in the gaps 33
Private operators and entrepreneurs 45
Conclusion 56

2 **Museum traffic: naturalist correspondents and the advent of photography in the colonial museum** 66

Introduction 66
From German settler to cosmopolitan scholar: colonial masculinity and the rise of the self-made man 69
Naturalist correspondents, photography and the reframing of natural history 76
Expeditionary photography and the 'New Traveller's Tales' 82
Conclusion 95

3 **The rhetoric of exemplarity: portraiture and the naturalist as celebrity** 108

Introduction 108
Gentleman amateurs and professional bodies: the social formation of Victorian science 111
The naturalist refashioned 120
Conclusion 146

4 **Nature as spectacle: encountering the moa from Christchurch to Madras via London and Paris** 154

Introduction 154
The origin of the natural history collections at the Canterbury Museum 159
A new era of the museum begins: the debate over order versus spectacle 170
Portraits and the press: the colonial naturalist as publicist 174
Conclusion 187

Conclusion 197

References 205

Archival sources 205
Newspapers and periodicals 208
Secondary sources 208

Index 227

List of illustrations

1.1 George Scharf, *Staircase of the Old British Museum, Montague House*, 1845, watercolour. Collection: British Museum, London (1862, 0614.631). 23

1.2 'Easter Monday – The Great Zoological Gallery, British Museum', *Illustrated London News*, 29 March 1845, p. 201. Collection: Wellcome Library, London. 25

1.3 Richard Owen, 'List of Suitable Objects to Be Photographed', British Museum: Officers Reports, 23 April 1857, f.367, pp. 1–2 of ten pages. Courtesy of the Trustees of the British Museum. 31

1.4 Roger Fenton, *Fossil fish, Holoptychius nobiliformus*, 1858, salted paper print. Collection: Victoria and Albert Museum, London. 32

1.5 Roger Fenton, *Elephantine Moa (Dinornis elephantopus), an Extinct Wingless Bird, in the Gallery of Fossils, British Museum*, c.1857, salted paper print. The J. Paul Getty Museum, Los Angeles (84.XP.452.3). Digital image courtesy of the Getty's Open Content Program. 36

1.6 A. C. Gunther, *New Zealand*, vol. 17, from the A. C. Gunther series of albums, 1870s–80s, case 22: mss. 551–2, f.28–29 (album opening showing lithographs and drawings of various skulls and jawbones, and a photograph of skeleton). Collection: Linnean Society, London. 39

1.7 Roger Fenton, *Dinornis elephantopus*, c.1855, salted paper print. Collection: Natural History Museum, London. 42

1.8 Richard Owen, *Palaeontology or, A Systematic Study of Extinct Animals and Their Geographical Relations*, A. & C. Black, Edinburgh, 1861, p. 330. Image from the Biodiversity Heritage Library. Digitized by Smithsonian Libraries. 43

1.9 Henry Barnes (attrib.), *Dromornis australis*, (photographs received from Rev. W. B. Clarke, Sydney, 1872), Richard Owen Collection of Palaeontological and Zoological Drawings, f.506. Collection: Natural History Museum, London. 44

1.10 Plates LVIII and LXI in Richard Owen, 'On Dinornis (Part XIX): containing a Description of the Femur indicative of a Large Wingless

Bird (*Dromornis australis* Owen) from a post-tertiary deposit in Queensland, Australia', *Transactions of the Zoological Society of London* vol. 8, no. 6, 1873. Courtesy of John Wiley & Sons. 45
1.11 Jacob Spoel, *Group of women looking at stereoscope photographs*, c.1868, oil on panel. Courtesy of the Rijksmuseum, Amsterdam. 46
1.12 Roger Fenton, *The Megatherium, British Museum* from *The Stereoscopic Magazine*, vol. 1, no. 14, 1859. Rare Books Collection, State Library Victoria. 49
1.13 Roger Fenton, *The Gigantic Irish Deer, British Museum* from *The Stereoscopic Magazine*, vol. 2, no. 17, 1859. Rare Books Collection, State Library Victoria. 51
1.14 Roger Fenton, *The Reptile Room, British Museum* from *The Stereoscopic Magazine*, vol. 2, no. 20, 1860. Rare Books Collection, State Library Victoria. 52
1.15 Roger Fenton, *Group of Corals, British Museum* from *The Stereoscopic Magazine*, vol. 1, no. 16, 1859. Rare Books Collection, State Library Victoria. 53
1.16 Roger Fenton, *The Assyrian Gallery, British Museum* from *The Stereoscopic Magazine*, vol. 2, no. 23, 1860. Rare Books Collection, State Library Victoria. 54
1.17 A. C. Gunther, *Europe Mammals,* vol. 8, from the A. C. Gunther series of albums, 1870s–80s, case 22: ms. 534, f.18–19, (album opening showing photographs of the British Museum natural history galleries, c.1878). Collection: Linnean Society, London. 55
2.1 Gerard Krefft, *Dasyurus Viverrinus* (Eastern Quolls), 1860, watercolour, in Gerard Krefft album of watercolour drawings, 1857–66. Collection: Mitchell Library, State Library of New South Wales (63890). 73
2.2 Henry Barnes (attrib.), *Australian Museum, Sydney New South Wales* (front and side views), June 1865, Richard Owen Collection of Palaeontological and Zoological Drawings, f.17 (albumen silver photographs sent by George Bennett to Richard Owen). Collection: Natural History Museum, London. 74
2.3 George Gordon McCrae, *A Saylore on Horse-Backe. Ye Adelaide Troopes Are Reviewed by a Prussian Colonel*, 1868, pencil and watercolour. Collection: National Library of Australia, Canberra (nla.obj-139386514). 76
2.4 Roberts Brothers, photographers, *H. R. H. the Duke of Edinburgh and Suite in Mining Costume after Descending the Band of Hope Gold Mine, Ballarat, Australia*, 10 December 1867, *carte-de-visite* photograph. Pictures Collection, State Library Victoria (H84.439). 77
2.5 Photographer unknown, *Portrait of Gerard Krefft Sent to Albert Gunther at the British Museum in 1864*, reproduction of an albumen silver photograph. Collection: Australian Museum, Sydney. 78

2.6 Roger Fenton, *Skeleton of Man and of the Male Gorilla (Troglodytes Gorilla) I*, 1854–58, salted paper print. Collection: Wellcome Library, London. 83

2.7 *Homo sapiens, human and Gorilla gorilla, western gorilla* [after Roger Fenton] in Richard Owen, 'Osteological Contributions to the Natural History of Anthropoid Apes – No. 7', *Transactions of the Zoological Society of London*, vol. 5, p. 1, 1866. ©The Trustees of the Natural History Museum, London. 85

2.8 'Photographing Gorillas' from Paul Du Chaillu, *The Country of the Dwarfs*, Harpers & Brothers, New York, 1872, p. 83. Source: University of California – courtesy of HathiTrust. 85

2.9 Major T. L. Mitchell (Day and Haghe, Lith.), 'Large Cavern at Wellington Valley' in Thomas L. Mitchell, *Three Expeditions into the Interior of Eastern Australia*, vol. 2, T & W Boone, London, 1839. Collection: National Library of Australia, Canberra (nla.obj-136107184). 89

2.10 Augustus Earle, *Mosman's Cave, Wellington Valley, New South Wales, no. 3*, c.1826, watercolour. Collection: National Library of Australia, Canberra. 90

2.11 Augustus Earle, *Mosman's Cave, Wellington Valley, New South Wales no. 5*, c.1826, watercolour. Collection: National Library of Australia, Canberra. 92

2.12 'Diprotodon!!', *Sydney Punch*, 26 June 1874, p. 209. La Trobe Collection, State Library Victoria. 97

3.1 After Antoine Claudet, *Michael Faraday*, c.1849, lithograph after a daguerreotype. Collection: Wellcome Library, London. 109

3.2 Sir Roderick Impey Murchison, p. 23 in Lovell Reeve (ed.), *Portraits of Men of Eminence in Literature, Science and Art with Biographical Memoirs. The Photographs from Life by Ernest Edwards*, Lovell Reeve & Co., London, 1863. Source: The Getty Research Institute – courtesy of HathiTrust. 115

3.3 Thomas Henry Huxley, p. 127 in Lovell Reeve (ed.), *Portraits of Men of Eminence in Literature, Science and Art with Biographical Memoirs. The Photographs from Life by Ernest Edwards*, Lovell Reeve & Co., London, 1863. Source: The Getty Research Institute – courtesy of HathiTrust. 116

3.4 Michael Faraday, p. 147 in Lovell Reeve (ed.), *Portraits of Men of Eminence in Literature, Science and Art with Biographical Memoirs. The Photographs from Life by Ernest Edwards*, Lovell Reeve & Co., London, 1863. Source: The Getty Research Institute – courtesy of HathiTrust. 117

3.5 W. O. Geller after Joshua Reynolds, *Portrait of John Hunter*, 1836, mezzotint. Collection: Wellcome Library, London. 118

3.6 'John Hunter – From a Painting by Sir Joshua Reynolds', *Illustrated London News*, 2 April 1859, p. 332. 119

3.7 Maull and Polyblank, *Charles Darwin*, c.1855, albumen silver photograph, from the *Literary and Scientific Portrait Club*. Collection: National Portrait Gallery, London. 121
3.8 Ernest Edwards, *Charles Darwin*, 1865–66, albumen silver photograph. Collection: National Portrait Gallery, London. 122
3.9 Professor Owen, c.1856, plate 1 in Maull and Polyblank, *Photographic Portraits of Living Celebrities*, 1856–59. Collection: Wellcome Library, London. 124
3.10 George Rowe, *William Buckland*, 1823, lithograph printed by Charles Joseph Hullmandel. Collection: National Portrait Gallery, London. 127
3.11 Samuel Cousins, after Thomas Phillips, *William Buckland*, 1832, mezzotint, published 1833. Collection: National Portrait Gallery, London. 128
3.12 H. W. Pickersgill, *Portrait of Richard Owen* (holding the Pearly Nautilus), c.1845, oil on canvas. Collection: National Portrait Gallery, London. 129
3.13 David Livingstone, 1857, plate 25 in Maull and Polyblank, *Photographic Portraits of Living Celebrities*, 1856–59. Collection: National Portrait Gallery, London. 132
3.14 Walker and Boutall after unknown photographer, *Portrait of Richard Owen*, c.1845, photo engraving after a daguerreotype in Owen, Rev Richard, *The Life of Richard Owen*, vol. II, John Murray, London, 1894. 136
3.15 Photographer unknown, *Portrait of Richard Owen*, c.1877, albumen silver photograph, plate 97 in Richard Owen, *Memoirs of the Extinct Wingless Birds of New Zealand*, John Van Voorst, London, 1879. Courtesy of the University of Texas Libraries, University of Texas at Austin. 137
3.16 Maull and Polyblank, Dr. Thomas Thomson, c.1854, albumen silver photograph in *R. C. A. Prior Album* (assembled from 1854). Collection: Linnean Society, London. 139
3.17 Maull and Polyblank, James Scott Bowerbank, c.1854, albumen silver photograph in *R. C. A. Prior Album* (assembled from 1854). Collection: Linnean Society, London. 140
3.18 Maull and Polyblank, Samuel Stevens, c.1854, albumen silver photograph in *R. C. A. Prior Album* (assembled from 1854). Collection: Linnean Society, London. 141
3.19 Maull and Polyblank, J. O. Westwood, c.1854, albumen silver photograph in *R.C.A. Prior Album* (assembled from 1854). Collection: Linnean Society, London. 142
3.20 Triptree and Sons, *Lovell Reeve*, c.1860, albumen silver *carte-de-visite* portrait photograph. Collection: Linnean Society, London. 143
3.21 John McGregor, *James Hector*, c.1863, albumen silver *carte-de-visite* portrait photograph. Collection: Alexander Turnbull Library, Wellington (The Webster Collection). 144

4.1 'Encounter with a Moa' in P. H. Gosse, *The Romance of Natural History*, second series, James Nisbet & Co, London, 1861. 155
4.2 Rev. Richard Taylor, del. W. Dickes, *New Zealand Fauna and Flora of Past and Present Times,* 1870, chromolithograph. Collection: Alexander Turnbull Library, Wellington, New Zealand (A-247-037). 156
4.3 Monkhouse and Co., *Bones of a New Zealand Moa (Dinornis robustus* Owen*). Described and exhibited by Thomas Allis F.L.S. at a Meeting of the Linnean Society, June 16th 1864*, albumen silver photographs, Richard Owen Collection of Palaeontological and Zoological Drawings, f. 487. Collection: Natural History Museum, London. 157
4.4 Photographer unknown, *Dinornis robustus in York Museum*, c.1876, albumen silver photographs, Richard Owen Collection of Palaeontological and Zoological Drawings, f.488. Collection: Natural History Museum, London. 158
4.5 Julius Haast, *From Central Terminal Moraine of Great Godley Glacier, 5 March 1862*, pencil, ink and watercolour. Collection: Alexander Turnbull Library, Wellington, New Zealand (C-097-060). 162
4.6 John Gully, *From Central Terminal Moraine of Great Godley Glacier, 5 March 1862*, 1863, watercolour. Collection: Alexander Turnbull Library, Wellington, New Zealand (D-037-001). 163
4.7 Walter Mantell, *The Unfortunate Moa, What Was Mired in the Slough of Despond at Waikouaiti in New Zealand. Anno Mundi???*, 1849, pen, ink and wash. Collection: Alexander Turnbull Library, Wellington, New Zealand (E-281-q-037). 165
4.8 James Brown, *Refers to Mr. W. B. Mantell's Moa Bone Discoveries. The Gravel and Sand Bag Bespeak Him a Geologist*, c.1851, pencil on paper. Hocken Collections, Uare Taoka o Hakena, University of Otago (11.594). 166
4.9 Daniel Mundy, *Dinornis robustus* (frontal view), 25 October 1867, albumen silver photograph. Collection: Sir George Grey Special Collections, Auckland Libraries (589–77). 169
4.10 Daniel Mundy, *Dinornis robustus* (side view), 25 October 1867, albumen silver photograph. Collection: Sir George Grey Special Collections, Auckland Libraries (589–78). 170
4.11 Daniel Mundy, *Julius Haast Standing beside a Moa Skeleton on Display at Canterbury Museum*, 1867, stereo pair of albumen silver photographs in John Buchanan scrapbook, 1856-c1900. Collection: Alexander Turnbull Library, Wellington, New Zealand (E-207-q-043-1). 171
4.12 'The Canterbury Museum, New Zealand', *Illustrated London News*, 6 February 1868, p. 144. 177
4.13 'Leg of the Dinornis, or Gigantic Bird of New Zealand', *Illustrated London News*, 10 June 1865, p. 54. 178

4.14 Plates 79 and 80 (including Major Michael's specimen) in Richard Owen, *Memoirs of the Extinct Wingless Birds of New Zealand*, John Van Voorst, London, 1879. 181
4.15 Dr. A. C. Barker, *Johann Franz Julius von Haast at the Canterbury Museum Skeleton Hall*, c.1870. Collection: Canterbury Museum, Christchurch (1944.78.66). 182
4.16 Dr A.C. Barker, *Alfred Charles Barker at the Canterbury Museum Skeleton Hall*, c.1870. Collection: Canterbury Museum, Christchurch (1944.78.33). 183
4.17 A. C. Barker, *Julius von Haast and Taxidermist Frederick Fuller*, c.1866, albumen silver photograph. Collection: Alexander Turnbull Library, Wellington, New Zealand (PAColl-5381-01). 184
4.18 Photographer unknown, *Portrait of Julius Haast, 26 July 1859*. Collection: Sir George Grey Special Collections, Auckland Libraries (881-1). 186

Colour plates

Plate 1 James Glen Wilson, *Skull of Zygomaturus trilobus*, stereoscopic pair of salted paper photographs, 1857. Collection: Geological Society, London (LDGSL 203-3).

Plate 2 Photographer unknown, Construction of the new reading room cupola, British Museum, p.32 in *A Collection of Plans and Views of Portions of the Buildings and Contents of Old Montague House and the Present British Museum*, album assembled by John Edward and Maria Emma Gray, 1780-1858. Collection: British Library, London (Shelf mark Maps C.26.f.7)

Plate 3 Roger Fenton, *Dinornis elephantopus and leg bones of Dinornis giganteus in the Gallery of Fossils, British Museum*, c.1857, salted paper print from glass negative. Courtesy of the Museum of Modern Art, New York / Scala, Florence. Gift of Paul F. Walter (83.1988).

Plate 4 Roger Fenton, *Group of Shells from The Stereoscopic Magazine*, vol.3, no.38, 1861. Rare Books Collection, State Library Victoria.

Plate 5 Henry Barnes, *Gerard Krefft posed with holotype of* Ceratoptera alfredi *anonymous (Krefft)*, (manta ray), 1868, albumen silver photographs in *Photograph Album, no. 14, Fishes no.1*, AMS421-14, p.71. Collection: Australian Museum, Sydney.

Plate 6 Henry Barnes, *Echidna skeleton and Natural history specimens sent from the British Museum to the Australian Museum*, c.1864, albumen silver photographs, enclosed in a letter from Krefft to Owen, 21 October 1864: Owen correspondence, vol.27, f.3-5. Collection: Natural History Museum, London.

Plate 7 H.W. Pickersgill, *Portrait of Richard Owen* (holding the Dinornis bone), 1844, oil on canvas. Collection: Natural History Museum, London.

Plate 8 Maull and Polyblank, *John Edward Gray*, albumen silver photograph, 1855. Collection: Wellcome Library, London.

Plate 9 Daniel Mundy, *Skeletons of New Zealand Moa in Canterbury Museum, N.Z.*, 25 October 1867, albumen silver photograph. Collection: Sir George Grey Special Collections, Auckland Libraries (589–76).

Plate 10 Photographer unknown, *Major J. Michael and bones of Dinornis giganteus*, c.1865, albumen silver photograph, Richard Owen Collection of Palaeontological and Zoological Drawings, f.480. Collection: Natural History Museum, London.

Acknowledgements

I am indebted to many people who provided me with invaluable advice and assistance during the preparation of this book. Above all, I owe an enormous debt of gratitude to Mary Roberts, who supervised the doctoral dissertation from which this book is drawn. I would like to express my deepest thanks for her unfailing enthusiasm, wisdom and peerless guidance throughout the years of my research, for her careful reading and meticulous responses to the various drafts of my manuscript and for her enduring inspiration and generosity.

I am furthermore indebted to Tim Barringer, Julie Codell and Douglas Fordham who, in their reports on my dissertation, made many useful comments and detailed recommendations for working it into a book. I am also grateful to the two anonymous reviewers and to Barbara Larson, the series editor, all of whom offered constructive suggestions for improvements to my book manuscript.

An Australian Postgraduate Award enabled me to travel to Britain, New Zealand and the United States to undertake much of my primary research. I am very grateful for the award and the opportunities it presented. Also, post-doctoral visiting fellowships at the Harry Ransom Center, University of Texas at Austin, and the Yale Center for British Art, New Haven, inspired a fresh outlook and offered avenues for further investigation as I developed my book manuscript. The hospitality provided by friends and colleagues, especially in Austin, New Haven, Wellington and London, made those explorations immensely pleasurable and productive.

Individual staff members at numerous libraries, museums and archives in Australia, England, New Zealand, India and the United States have been extremely generous with their time and expertise. They include Stephanie Clarke and Francesca Hillier, the British Museum; Paul Cooper and Polly Tucker, Natural History Museum, London; Lynda Brooks and Elaine Charwat, Linnean Society, London; Caroline Lam and Wendy Cawthorne, Geological Society, London; Peter Barber and John O'Brien, British Library; Cecilia Cassingham, Science Museum, London; Colin Harris, Bodleian Library, University of Oxford; Michele Losse, Royal Botanic Gardens, Kew; Kerry McCarthy, Sally Burrage and Rachael Fone, Canterbury Museum, Christchurch; Pat Sargison, Christchurch City Libraries; Elspeth Orwin, Auckland Libraries; Jennifer Twist and Lissa Mitchell, Museum of New Zealand Te Papa Tongarew, Wellington; Chris Adam, Archives New Zealand; Vanessa Finney and Rose Docker, Australian Museum, Sydney; Jan Brazier and Jude Philp, Macleay Museum, Sydney; Robert Young, Museum of Comparative Zoology, Harvard University; Roy Flukinger and Bridget Gayle Ground, Harry Ransom Center, and Rhonda L. Evans, Centre for Australian and New Zealand Studies, University of Texas at Austin; and Amy Meyers, Gillian Forrester, Kraig Binkowski and Lisa Ford, Yale Center for British Art. I would also like to thank the Alexander Turnbull Library, Wellington; Hocken Library, University of Otago; National Portrait Gallery, London; Victoria and Albert Museum, London; Royal Society, London; National Archives, Kew; Zoological Society of London; Wellcome Library, London; Royal College of Surgeons, London; Mitchell Library, Sydney; State Library of Victoria, Museum Victoria, Melbourne; State Library of Queensland; National Library of Australia; National Gallery of Australia; Rare Books and Special Collections Library, University of Sydney; Royal Australasian College of Physicians, Sydney; Government Museum, Chennai; Getty Research Institute, Los Angeles; Beinecke Rare Books and Manuscript Library and the Yale University Gallery, New Haven, Connecticut.

Colleagues in the Department of Art History and the Power Institute at the University of Sydney have been extremely supportive throughout this endeavour. I would particularly like to thank Mark Ledbury, Roger Benjamin, Keith Broadfoot, Anita Callaway and Anthony Green.

I am fortunate to have had such excellent support from Ashgate and Routledge. Thanks to the commissioning editor, Margaret Michniewicz, and the Art History and Visual Studies editor, Isabella Vitti, for shepherding the book to completion.

On a personal note, special thanks to Ashley Mason for his indefatigable encouragement, patience and support; this book is dedicated to him.

Introduction

In 1859, Richard Owen, superintendent of natural history at the British Museum, published two papers that he had read before the Geological Society of London the previous year.[1] Both were framed around the subject of Australian marsupial fossils. This was a topic of immense importance to the new field of palaeontology, which had only recently begun to seize the imagination of a growing popular audience for natural history. Even for such a celebrated public figure at the pinnacle of imperial science, as Owen was, there was a great deal of prestige that could be gained from being involved in the discovery or classification of impressive new species. Yet, instead of identifying the subject at hand and immediately launching into his own authoritative interpretation of the specimen, as was the usual approach, Owen devoted his entire introduction to recounting the circuitous route taken not by the object itself – a skull of the giant extinct *Zygomaturus* – but by a number of photographs, stereographs and outline drawings that had travelled from the very edge of the empire into his possession in London and finally to the Geological Society (Plate 1). The production and transfer of these images was anything but straightforward. It involved colonial naturalists and museum trustees, a colonial governor and Royal Navy personnel, as well as the president of the Royal Geographical Society.

Owen's paper continued with two reports about the original fossil specimen: the first was written by the distinguished naturalist and Australian Museum trustee William Sharp Macleay, which had been published in the *Sydney Morning Herald* in September 1857; and the second was supplied by Dr. John Macdonald, assistant surgeon on the British survey ship *Herald*, undertaken during the ship's official survey of Sydney Harbour in 1857.[2] In his conclusion, Owen presented his detailed observations of the outline drawings and photographs of the skull. For the former, he was indebted to another colonial naturalist, his long-time associate George Bennett, who had produced 'four carefully-made, and apparently most accurate, outline-drawings of this unique fossil'.[3] The photographs, on the other hand, had an altogether more public provenance, having been executed by the official artist on HMS *Herald*, James Glen Wilson, and then conveyed, under the auspices of Governor Denison of New South Wales, to Sir Roderick Murchison, who, like Owen, was one of the foremost scientists of the empire. This exchange took place at the critical historical moment when Britain was consolidating its political control in the Pacific and also seeking to expand its influence in all directions across the globe as part of its informal economic and cultural imperium.[4] By drawing attention to these connections in his papers on Australia's fossil fauna, Owen highlighted the ways in which art and the sciences not only accompanied, but also influenced and were themselves shaped by, Britain's interests and activities at home, across its empire and globally.

Owen's presentation to the Geological Society pointed to the complex global network that facilitated such undertakings. As with many intellectual and cultural advances, the progress and increasing popularity of natural history during the nineteenth century owed less to individual genius than to the collective enterprise of an immense network of practitioners, enthusiasts, expositors and consumers. Within that labyrinth of connections, however, an awareness of who specifically had introduced or verified new knowledge and where it had originated was almost as important as its cumulative effect. For instance, colonial information was frequently devalued unless it came with the imprimatur of appropriate social or professional connections. Well-connected resident naturalists – such as Macleay and Bennett – acted as gatekeepers in the dissemination of local knowledge, often rendering other colonial figures invisible. As Owen's roll-call in his paper indicates, those who were acknowledged comprised an array of individuals, museums and other institutions located not only at the centre but also around the periphery of the British Empire as part of an elaborate, asymmetric web of knowledge and power. While the intersection of these political, economic and professional interests may not be readily apparent in natural history images, this network of influence – which Owen so purposefully disclosed – and the circumstances leading to the production of the images and their subsequent passage around the globe were explicit, even celebrated, in the very presence of these photographs and drawings at the Geological Society and certainly would have affected their reception.

The dissemination of images through different types of visual media and display was crucial to how they were interpreted. Some scientists were still uncertain about whether new visual technologies would be accepted by others in the field of natural history. For his own part, Macdonald had in fact submitted his written description of the *Zygomaturus* fossil for the precise purpose of supplementing the photographs taken by Wilson, the ship's artist. This was in anticipation, Macdonald suggested, 'of the many queries which will naturally arise in the mind of the anatomist inspecting them without the means of consulting the original'.[5] Owen's paper drew attention not only to the anatomist's practice of close, comparative observation but also to a widening interest in the very process of looking, which included the allure of new optical devices and other such innovations. Stereoscopes, for instance, even more than photographs, had a mass appeal and were far more pervasive in the commercial realm than they were in specialist domains. By replicating the solidity, depth and materiality of a scene depicted in close proximity, the stereoscope offered a seemingly tangible surrogate for the actual object or view.[6] Owen was acutely aware of how distinctive image technologies offered different kinds of viewing experiences and acknowledged that the three types of media – drawings, photographs and stereographs – were manifestly different in their conception, production and presentation. Outline drawings provided an effective visual summation of a skull, having already undergone a process of analysis, expurgation and synthesis by an experienced naturalist. Photographs and stereographs, on the other hand, presented a specimen as if it were being encountered for the very first time by each beholder; they entailed an ongoing and individualised process of exploration and discovery and represented a kind of intellectual and imaginative journey. Photographs achieved this in the guise of naturalism and in a way that was superficially independent of the observer, whereas stereographs functioned entirely through an illusion conjured optically by each viewer.

Using these media, an exotic or extinct specimen from nature could be experienced as uniquely as one's first encounter with the actual object – for instance, an unfamiliar animal or fossil – thereby signifying an absent presence for audiences separated by distance or time from each other and the object represented. Correspondingly, the combination of such discrete intellectual and popular pursuits opened the possibility of relating different aspects of the practice and promotion of natural history to other genres and modes of photography and an assortment of new visual and display technologies as rational entertainment.

One might question the extent to which Owen's address and the images he displayed could connect with a wider audience, given that London's Geological Society comprised a relatively small and privileged forum. Then again, the Geological Society was similar to other peak scientific bodies, all of which were situated in the metropole and wielded immense influence that stretched beyond that central location out and across Britain, into its colonies and to other parts of the globe, thereby directing the course of virtually all scientific activity. When photography was used in this context – by naturalists and other scientists for purposes ranging from the loftiest to the most mundane – it functioned as a cultural and political medium that was inextricably linked to the institutions of empire, with individual photographs operating 'as complex discursive objects of [imperial] power and culture'.[7] Photographs, even more so than other media, were also becoming available to a wide range of viewers. Taking on a variety of distinct but often complementary roles, they could be construed quite differently or could assume a new status as they moved between the centre and periphery, separate interest groups or social strata. Besides the documentation of specimens, photography had the capability to promote, transform or supplant other aspects of naturalists' practice. Due to its accessibility to a variety of audiences, photography also had the potential to draw together popular or commercial realms and natural history, the principal science of the empire. In this book, I examine how the production and circulation of photographs contributed not only to imperial and global science but also to the construction of local intellectual culture within colonial or provincial contexts through the more popular manifestation of the media.

The occasion of Owen's *Zygomaturus* lecture before the Geological Society remains a relatively overlooked episode in the globalisation of British science and culture. Yet, its emphasis on interconnections between imperial and colonial natural science – the professional and personal associations that led to many of Owen's scientific accomplishments – and the range of visual media used are emblematic of the concerns of this book. As such, Owen's lecture exemplifies several factors identified here as crucial to the evolving relationship between photography and natural history during the Victorian era: the significance of colonial and transnational participation for the empire and in the global economy of science and culture; the complexity and changing face of specialised networks and their role in the dissemination and advancement of scientific knowledge and expertise; and the emerging interest in new visual media, not merely as static repositories of cultural content and specialist knowledge but also in their capacity to function over space and time and to therefore engage different communities of interest.

By analysing this nexus between empire, museums, scientific method and new visual technologies, my aim in this book is to fundamentally rethink the ways in which photography and natural history intersected during the second half of the nineteenth century. This study analyses the emergence of the modern observing subject in relation to natural history photography according to two themes. The first examines the Victorian enthusiasm for natural history and considers the observer of natural history photographs with regard to the broad audiences – both domestically and across the empire – that were attracted to various types of rational amusement. These informative entertainments encompassed a multiplicity of physical settings and modes of presentation ranging from museums, international exhibitions, scientific soirées and lantern-slide lectures to assembled albums, illustrated publications, stereo views and studio portraits. From this perspective, photography is shown to epitomise the nineteenth-century acquisitive impulse and the attendant desire to be in immediate contact with all aspects of the world. The second theme considers the role of imagination in the production and reception of photographs and associated visual technologies. Rather than presuming photography to be an accurate, unmediated description of nature, this study explores the ways in which photographers and viewers involved themselves in the theatre of photography. In their sometimes substantial bodies of work, naturalists observed, collected and ordered the world of nature using a variety of visual and display regimes. In documenting the world around them, they could not help but also record their own presence and activities within the social and institutional sphere of natural history. Hence, naturalists' processes were neither wholly mechanistic nor entirely dispassionate but were subject to interpretation and extrapolation. In this context, nineteenth-century photographic processes are investigated, along with the creative interventions of photographers and naturalists to address the medium's possibilities and limitations. By situating these concerns within the context of Britain's imperial trajectory, this book aims to uncover and examine some of the complexity of photography's multifarious relationship with Victorian natural history, particularly with rise of the public museum.

Rethinking the role of photography in Victorian natural history

In considering photography in the context of natural history and the nineteenth-century museum, some clarification about the status of natural history in the Victorian era is useful here, given the immense changes that were occurring. While Darwin's theory of evolution by natural selection represented both a major epistemological break and the main point of departure between the practices of natural history and those of modern biology, the reality was more complex and far less abrupt. The professionalisation of the natural sciences in Britain and, concomitantly, their fragmentation into numerous sub-disciplines, with a greater emphasis on laboratory science including highly technical analysis and experimentation, were significant related factors. Despite these structural changes, the professionalisation of the natural sciences was a very gradual process. There was more continuity between natural history's emphasis on collecting and taxonomy and its successors in the experimental sciences than is immediately evident.[8] Public museums maintained this connection – due primarily to their extensive collections, which continued to be an unparalleled resource for many research areas in the natural sciences. Museums were also the public face of the natural sciences; museum keepers had begun to embark on dual arrangements of their collections, one arrangement devised for scientific study and intended to be representative of the main taxonomic groups, the other layout emphasising popular displays and comprising singular, spectacular specimens as an attraction for ordinary visitors.[9] As Peter Bowler observes, 'popular lectures and exhibitions were important vehicles by which science was promoted, and here the element of display was paramount'.[10] What this overlooks, however, are the myriad uses of popular and specialised images, together with the increasing application of new visual technologies, as part of the practice and diffusion of nineteenth-century science by museums and naturalists and more widely.[11]

In this book, I specifically examine the introduction and use of photography and related media in museums and by naturalists in Britain, Australia and New Zealand from the 1850s to the 1880s. Museums in imperial territories, of necessity, had different priorities to museums in London and other metropolitan centres.[12] The emergence of colonial museums and their diverse roles in various parts of the British Empire – in Africa, Australasia, Canada and Asia – has become a flourishing field of scholarship in recent decades. However, analysis of the role of visual culture and, more especially, the introduction and function of photography in early colonial museums in such studies is generally fairly cursory. My decision to focus on imperial–colonial parameters has been prompted by a dissatisfaction with a commonly held assumption perpetuated in histories of photography that, despite an ever-expanding market for commercial applications, the use of photography in the natural sciences was under-exploited in Australia and other settler colonies during the second half of the nineteenth century. The conventional reasoning behind this supposition is that photography was used by scientists essentially 'to discover, record and categorise the natural world'; while some individuals used photography for this purpose in the colonies, the needs of colonial science in this regard were sporadic, and consequently their endeavours were relatively solitary and piecemeal.[13] Some accounts suggest that changing perceptions of the medium due to the considerable popular success of early commercial photographers in colonial and provincial settings is a key factor in understanding why naturalists in those locations seemed to disregard photography as a credible means of pursuing and disseminating their work.[14] This goes some way towards explaining the hesitancy or ambivalence of some early amateur practitioners, particularly those for whom the pursuit of photography was predominantly an intellectual and creative pastime, practiced more for its novelty rather than as an applied technology with potential applications for a range of professional purposes. Nevertheless, a variety of naturalists who had no obvious connections with photography were motivated to consider the broader opportunities presented by the medium and developed different ideas on how to utilise it in their work to communicate information and to promote themselves and their achievements. In order to examine these issues in depth, I have been necessarily selective in my geographical scope by focussing mainly on Britain and its Australasian colonies. Within that context, a major focus of this study is the scientific and popular interest in palaeontology and geology during the Victorian era. Consequently, there are significant opportunities for further research into the role of photography in colonial natural history museums in other parts of the British Empire, and also as comparative empire studies.

Some of the ways that photography was taken up and used in Victorian science owed much to the groundwork established by commercial operators and their increasing effectiveness in a global context. The relationship between these different interests and distinct realms of practice are seldom acknowledged, let alone addressed in detail, and this aspect of nineteenth-century photography has remained either a source of perplexity or a subject to be skirted around in a way that it rarely has been for art and commerce. For instance, some historians note that, in contrast to imperial institutions, in the colonial philosophical and scientific societies – the main forum for local and visiting savants to share their interests and communicate their discoveries – there is little evidence of any concerted or organised effort made to advance the use of photography for natural history purposes during the medium's early years.[15]

Hence, in the Australian colonies, it is generally accepted that the greatest advance in the progress of photography came from commercial studio portraiture rather than from science.[16] Nevertheless, it has been demonstrated that some colonial scientists were not only quick to avail themselves of photography – with a number of them receiving instruction on photographic processes from local studios – but also that the importance of the medium for science was in fact advocated by prominent commercial photographers such as James Freeman, who realised that the greatest potential of photography in the age of empire was in the way it could be systematically applied to information of all kinds.[17] In most studies of photography during that period, the basis of such observations and claims has remained insufficiently understood.

Emerging sites of production: a comparative approach

One obvious factor impeding a better understanding of this area is the limitations imposed by the focus on ostensibly separate national histories. In reality, the cultural traffic between centre and periphery – and throughout the empire – was so fluid that the notion of an 'Australian' or a 'New Zealand' identity – or, for that matter, a 'British' identity – is far less obvious than many national histories indicate. Yet, by and large, accounts of the photographic interventions of colonial scientists, or of other practitioners on their behalf, have thus far emerged solely within the context of nationally focussed histories of photography, and even there relatively cursorily.[18] Typically, there is very little mention of the various ways in which contemporary science and related social circumstances influenced the production and reception of photographs. In biographical and other monographic studies, key colonial figures have been appraised in terms of their local significance – particularly in relation to emerging national cultures – and, only more recently, recognised for their contributions to the empire or as part of a dynamic intercolonial and global culture.[19] The many cultural and scientific endeavours that occurred in Britain's colonies did not take place in a vacuum but instead reverberated across the empire and played a vital constitutive role in the intellectual progress of Britain itself. Despite this, in the majority of metropolitan histories of visual culture, colonial influences are considered to be tangential or are simply disregarded.[20] As a result, the current literature on the intersection of photography and natural history in the nineteenth century is conspicuously short on case studies that adequately take into account both centre and periphery.

When scholars' attention has moved beyond Britain's shores, it has tended to focus on ethnography and the photographic incursions of travelling photographers and expeditionary scientists during the latter decades of the nineteenth century.[21] In fact, this has been a most prolific area of scholarship and in recent years has taken in the perspective of indigenous subjects as part of – or even productively involved in – these anthropological projects. Within these many studies, however, there has been minimal analysis of the interventions and contributory roles of metropolitan scientists' colonial counterparts or the substantial part played by individuals and museums at a local level.[22] The recording and classification of non-European peoples was merely one of a host of scientific explorations undertaken as part of imperial expeditions and colonial encounters. Landscapes were surveyed not only to integrate them geographically into the expanding space of empire and to dominate their native inhabitants but also to acquire their mineral resources and the many economic or unusual species of fauna and flora they offered for the purposes of scholarly research and to satisfy popular interest. As the nineteenth century progressed, the initial emphasis of imperial and colonial science on botany and zoology gradually shifted so as to encompass geology, palaeontology and anthropology. With the rise of these historical sciences, the dimension of time intersected with the mapping of space and became an increasingly important feature of natural history.

Formative visions of empire

As an investigation of the representation and dissemination of natural history in the British imperial and colonial context, this book draws on the approach developed in Bernard Smith's *European Vision and the South Pacific*. Smith's study addresses the relativity of European perception in relation to the exigencies of Enlightenment science and art, giving much consideration to the promotion of the biological sciences by the Royal Society through voyages of exploration to the Pacific, as well as to British reactions to Antipodean nature and the confluence of art and science as complementary pursuits in the Australian colonies. Smith describes the South Pacific as presenting many anomalies to existing paradigms of nature – evidenced in the images and specimens brought back from European voyages – which finally created the impetus for Charles Darwin's theory of evolution by natural selection.[23] In seeking to reconcile the imperatives of science with the art of the period, Smith contends that the material evidence from South Pacific expeditions was mediated in various ways: primarily in relation to the way artists' and scientists' perceptions were informed by their European sensibilities, but also through new visual practices and technologies.[24]

Smith's interdisciplinary study provides a model for examining the complex role of visual media in the work of empire and thus also for considering the different ways photography would be taken up for imperial science as part of that process. In *European Vision*, he describes a dual approach, which was directed, as needed, towards fulfilling an official agenda or satisfying the appetite of a European public eager for curiosities suitably composed to current tastes.[25] Hence, images of natural history subjects were produced not only as scientific documents or as artistic conceits but also to validate and promote Britain's expanding empire. That is, they did not exist solely as original works in the Royal Society or as engravings in the drawing rooms of armchair travellers, but as more capricious images that moved unpredictably between different spheres, sometimes fulfilling several functions simultaneously. In this book, I regard photography as an increasingly integral aspect of Victorian natural history practice; that is, I do not view it as merely a new form of illustration or authentication but as part of a broader cultural process. Indeed, natural history photography in the age of empire rapidly surpassed the limitations of its genre to become a viable means of exchange. Consequently, I not only examine the specific subjects depicted in photographs but also address the transmission and uses of photographs, including the way particular photographs can migrate across the boundaries of popular and professional discourse.[26]

Exchanging views of empire

The movement of people, ideas, objects and images that constitute the trajectory and expression of empire was not confined to Britain and her colonies but extended to other European cultures. This was nowhere more apparent than in the field of natural history, which developed through negotiations between far-flung individuals and museums across the globe and contributed to a culture of interdependency between centre and periphery. While most naturalists were part of an expanding web of first-hand information and expertise, they were also contributing to a body of knowledge and fields of endeavour that were being transformed in a variety of ways. A significant feature of this network was the array of transnational influences from cosmopolitan figures such as German-born Gerard Krefft and Julius Haast, curator and director at the Australian and Canterbury museums, respectively. Originating from different backgrounds than the British-born colonial naturalists, they nevertheless moved around the empire in similar ways, initially as expeditionary scientists and later as curators in the newly established colonial museums. Their exchanges very often involved informal communications with individuals rather than official dealings through government agencies, with the ensuing connections giving rise to further interactions with savants and museums in other centres of knowledge and power, including Germany, Austria, Italy, France, Sweden, Argentina, Canada, India and the United States, as well as Britain.[27] In this, as in other ways, the practice of natural history during the nineteenth century did not represent a simple dichotomy between the advance of British interests and their impact on the colonial subject. The presence of naturalists from various backgrounds who travelled extensively and took up positions of authority in significant numbers represents a more complex cultural terrain and, at the very least, introduced another set of interests in the form of traditions and loyalties that existed in parallel with those of the British Empire. However, the direct influence of German scientific research and practices in the context of this study is pertinent only to a certain point. The German colonial figures discussed in the following chapters were often wholly or partly self-taught naturalists and therefore were engaging with German scientific practices predominantly through their association with other men of science in the German expatriate network, such as Albert Gunther at the British Museum. While Haast studied geology at the Rhenish Friedrich Wilhelm University of Bonn, for instance, his practical knowledge of zoology and palaeontology was learned on the job while working with naturalist and collector Andreas Reischek and by corresponding with his network of colleagues. The exchanges that flowed from these interactions were essential to both the research of naturalists and the appreciation of natural history by the public at large. To regard the nature and extent of these transnational relationships solely in terms of the transfer of actual specimens, however, is to neglect other significant, although perhaps less prominent, factors that defined the practice of colonial natural history. Indeed, Germany's promotion of popular science motivated by 'a philosophical conception of science; professional interests; [and] political ambitions involving social criticism' encouraged the exposition of natural history through writing, illustration and display, and it fundamentally influenced the ways German expatriate naturalists disseminated their discoveries and to whom.[28]

As a number of surveys have shown, during the earliest days of settlement, the two activities of collecting specimens and scientific illustration were complementary endeavours.[29] In recent years, a number of studies of individual artists and episodes have sought to situate this genre within its social and historical context.[30] Collectively, these studies present a cogent picture of natural history illustration, which has become familiar and valued within the history of colonial art. While the major purpose of such images was to be as useful and informative to science as the specimens they represented, they were, nonetheless, also valued objects to be admired for their aesthetic effects and exceptional skill. The role of photographs, as another enterprise, was more ambiguous. Often they were considered to be marginal and unworthy of sustained consideration. Yet, the etiquette governing the way photographs were referred to and circulated between naturalists from the 1850s onwards indicates a considerable modification in the way the observations and speculations of correspondents were conveyed, shared or authenticated. As Martin Rudwick has noted, what explicitly connected a network of individuals can offer many insights, but 'what they held *tacitly* in common' can be particularly revealing.[31] Certainly, the distinct scientific, professional, social, economic and political factors in different countries and colonies had an influence on the adoption and reception of photography for that purpose.

The Australian Museum in Sydney in New South Wales is a case in point. At this time, the colony was embarking on an uneasy transition from English imperial rule towards colonial parliamentary self-government, a development that had many detractors in both locations and complicated the loyalties of those seeking to build or consolidate their authority in the colony and beyond. Governor William Denison, for instance, who had been instrumental in appointing Krefft to the Australian Museum, had experienced a decline of his own influence due to the winding back of imperial power.[32] Some of the more powerful museum trustees had similarly conservative views, which created a conflict for those trustees, officers and supporters who were ambitious for the museum's progress but also dependent on the local legislature. The museum trustees, like Denison and other British imperial administrators, attempted to not only replicate in the colony the social hierarchy established in Britain but to surpass it on account of their increased status within it.[33] Within this displaced and increasingly fragile hierarchy, the self-presentation of colonists from other European backgrounds, such as Krefft's German background, became all the more complex. Concurrently, in the case of New Zealand – which had taken on responsible self-government in 1856 and simultaneously maintained a number of powerful provincial governments and associated settler societies – Haast was establishing himself and the Canterbury Museum in Christchurch as the pivot for both local and – in particular – German-speaking and other European scientists, 'with whom he maintained a truly massive correspondence' and circulated numerous photographs and spectacular, sought-after specimens.[34]

Natural history practitioners and consumers were becoming increasingly diverse as a community during this period. With a burgeoning middle class, industrialisation and the concomitant expansion of communications, transport, trade and immigration throughout the nineteenth century, naturalists found themselves to be less constrained by geographical, national or social boundaries and, as a result, they were becoming less identifiable and cohesive as a group. However, the role of portrait photography for particular sectors – including knowledge or skill-based disciplines and their associated occupations – has not been given sufficient consideration. From the 1850s, portrait studios proliferated. In the cities, they displaced painters of miniatures who served the well-off and, for the first time, produced portraits for the new middle class. In provincial and new world settings also, entrepreneurial photographers provided opportunities for portraits where they had never before existed. The portrait photograph became one of the most accessible, efficient and effective forms through which an individual or group could signify their status and their interests. The prominent role portraiture played in the social construction of identity provides a salient starting point for observing fundamental changes within certain occupations and avocations during the Victorian era. As natural history gained popularity to become the most widely practiced nineteenth-century science, naturalists represented one such occupation, comprising as they did an increasingly heterogeneous and geographically dispersed community of enthusiasts and operatives ranging from the working-class to the elite. The production and circulation of photographic portraits by both formal and informal means was – along with the exchange of specimens – a way of establishing and consolidating professional affiliations. Moreover, it became a relatively expeditious way to build reputations and gain publicity for particular causes or projects within the broader public arena. When considering photography in the context of the wide-ranging practices of nineteenth-century natural history and its extensive community, a national or any other artificially delimited perspective is inadequate. Indeed, these activities and exchanges occurred across a range of uncharted locations, not just within the geographical boundaries of colony or nation-state.

From 'immutable mobiles' to 'boundary objects'

To fully appreciate the specifics of different locations and the ways they were connected, it is important to understand the kinds of spaces that separated them and the types of movement entailed in their traversal. Histories of photography have contributed considerably to our understanding of nineteenth-century networks and the circulation of images by identifying connecting threads that made up the complex cultural and social environment in which early photography was enmeshed.[35] Meanwhile, in histories of science, visual literacy and an appreciation of the diverse roles of visual culture, as part of the constitutive and rhetorical work of science, is an expanding area of investigation.[36] This includes valuable work that recognises the importance of spatial considerations in science and 'the situated nature of knowledge for understanding the history of intellectual and social change'.[37] In these studies, the context of scientific knowledge is considered of equal consequence to its substance and outcome, and accordingly, the sites of the production and consumption of scientific knowledge and its inherent mobility have become significant foci of study. Radically challenging the conventional stance of previous studies in the sociology of scientific knowledge, the anthropological approach formulated by Bruno Latour in his 'actor-network theory' has been fundamental to this paradigmatic shift.[38] Latour describes how scientists' work practices not only result in new, specialist knowledge but also influence the social world in which the scientists operate. A key aspect of Latour's theory is the notion of stable, portable data – 'immutable and combinable mobiles' – which provide the links between numerous different places and people. In a nuanced interpretation of this concept, Susan Star and James Griesemer have developed the notion of 'boundary objects', which are 'both adaptable to different viewpoints and robust enough to maintain identity across them'.[39] Applying these concepts to visual images in the practice and reception of natural history, Latour's term 'immutable mobiles' invokes the commonly held assumption that photographs provided ocular proof and thus incontrovertible photographic evidence. The notion of 'boundary objects', on the other hand, refers to visual representations and other material things that provide connections between different individuals and groups who nevertheless might view them, interpret them and use them in distinct ways or for different aims. Thus, boundary objects are not prescriptive in their influence, nor can they be enlisted to regulate behaviour between groups. Rather, they facilitate an ecological approach to knowledge making and sharing that 'does not presuppose an epistemological primacy for any one viewpoint' but instead involves translation, negotiation, debate, triangulation and simplification.[40]

Using this concept, studies in the history of science have analysed networks of activity and reception that extend across the boundaries of different social and political realms rather than simply focussing on the outward movement from a single source, such as in the way Victorian London – as the centre of empire – and other comparable sites have been considered to be exclusive sites for the accumulation and dissemination of scientific knowledge.[41] In their application to case studies, these interactions have very often been framed in terms of the exchange of natural history specimens or written observations and discoveries.[42] Similar activities around the production, circulation and reception of an expanded range of visual images in the realm of natural history in this period have, however, received considerably less attention. Yet, a new technology such as photography, as Jan Golinski argues, 'may make possible a new relationship between the source of images and those who view them'.[43] It is this potential for new or differently negotiated relationships that photography offered in the context of colonial–imperial natural history, particularly in public museums, that is central to my investigation and which distinguishes this book from previous studies of the intersection of photography and science in the nineteenth century.

Making and moving images

The vast and complex network operating in nineteenth century natural science would be inconceivable without the benefit of a range of visual and reproductive technologies, new modes of image processing and transmission and elaborate information systems.[44] What we regard as modern science would not have been possible without the types of information made viable by the earlier printing revolution and the cosmopolitanism this promoted.[45] However, this was not due solely to the printing press's facility to reproduce records precisely and in great detail – many earlier scientific traditions had that same capacity.[46] Rather, it was the unprecedented access to a breadth of scientific knowledge accumulated from different locations over time that made printing so important. In this sense, the mass of textual and visual information in libraries, study collections and museums became just as much an instrument of science as an anatomist's scalpel or a compound microscope.[47]

The volume of raw data in circulation expanded exponentially during the nineteenth century. For the Victorians, the place where scientific knowledge originated or was interpreted and deployed, and by whom, became at least as important – and sometimes even more so – as the methods by which it was obtained. As a consequence, issues around the trustworthiness, reliability and credibility of scientific information were closely related to what was known about the character and circumstances of those individuals or groups who were generating the new knowledge: their moral standing, their social or cultural background and their affiliation with museums, professional associations or eminent individuals. For those working in distant and unfamiliar local environments, information about their person and surroundings was a key aspect that contributed to their profile. Thus, the burgeoning trade in studio portraits and commercially produced views, far from being inconsequential to the core practice of nineteenth-century natural history – as has been commonly presumed – was in fact integral to ascertaining the respectability and authority of naturalists and thus a foundation on which the merit of their scientific observations could be considered.[48] This is abundantly evident in correspondence between naturalists, in which exchanges of portraits and views of local museums and civic sites were as numerous as other types of information. One conclusion that has been drawn from these genteel exchanges and social courtesies is that informal connections among members of the scientific community were of greater importance than formal institutional ties.[49] Accordingly, portraits and views were circulated as part of a complementary web of information that established and promoted networks of trust and facilitated the transmission of more conventional scientific data.

Commercial photography was not only expedient but became almost obligatory in fulfilling this role as mass popularisation of the medium coincided with the move towards professionalisation in the sciences and also with increased public access to the realm of natural history. Taking up this issue, I examine some of the correspondence protocols of naturalists operating at the centre and periphery of the empire. I consider the ways in which visual technologies, particularly photography, were employed to initiate or complement exchanges, to progress relationships and, ultimately, to establish or enhance one's credentials or those of other naturalists, especially in the emerging colonial museums. As I show, the particular medium used influenced the manner and extent of their subsequent circulation. Some visual and textual forms fed directly into the public domain, either as specialist papers delivered at metropolitan and colonial societies or as part of the burgeoning market for natural history publications.[50] Other media, including that circulated through official networks and between museums, was more ephemeral in the workings of Victorian natural history. By the same token, the rapidly changing terrain, the heterogeneity of early photography and the eclecticism of nineteenth-century visual culture meant that what has come to be commonly understood as the visual production of naturalists was much broader than has been recognised. The impetus for the various forms cannot be fully appreciated in terms of visual culture alone but must be considered in relation to changes in the practice and presentation of natural history itself – the rise of the modern museum, the professionalisation of the natural sciences and, simultaneously, the popularisation of natural history and the increased dissemination of its imagery to an expanding middle class. The ways in which natural history was represented became as diverse as its producers and consumers, leading to the incorporation and combination of new media and also different visual genres. The resulting categories of images might seem surprising if the visual work of naturalists is considered only in terms of showing or interpreting natural history specimens. Viewed in relation to the broader dictates of nineteenth-century natural history, however, it is in fact indicative of the expanding scope and changing purpose of a field that – more than any other science – was increasingly becoming part of the public domain and thus a reflection of other aspects of social existence at the time.

Peripatetic objects of empire and collections at the periphery

A group of albums in the archives of the Australian Museum records the introduction of photography into the practice of natural history and its application to an expanding range of programs and initiatives over a number of decades from the 1860s.[51] Encountering these albums early on in my research for this book, it became apparent that the photographs produced and used during the latter half of the nineteenth century were given a wider purview and acquired a different sort of authority compared to those of the medium's earlier years. It seemed clear that these images were viewed not only as documentary images or aesthetic objects but also as agents and emissaries. That is, the photographs were valued not solely for their evidentiary or descriptive capacity but increasingly in terms of the activities they signified and the networks they facilitated. This is not to say that aesthetic considerations or the ostensibly mechanical objectivity of the medium did not continue to add meaning or a certain frisson of authority to these images for those involved in the natural sciences, as it had for photography's pioneers. The albums and the photographs they contain reveal an assemblage of images and artefacts as well as evidence of a web of processes and relationships, and they provide a tangible example of the complex and progressively more dynamic role of photography in natural history and museums in this period compared to other visual media. Crucial to this analysis, however, is the way in which the photographs, and the albums themselves, were initially envisaged and how they have been interpreted and reinterpreted over time.

The museum albums consist of two broad types: those that comprise specialist subject areas related to the collecting priorities of the institution – including marine invertebrates, birds, mammals, fishes, molluscs, whales and ethnology – and those that represent more extemporised aspects of the museum's activities, including its relationships with external stakeholders and professional networks and its involvement in various expeditions. The series of albums bring together individual images that have been presented to the museum at different times and from any number of geographic, institutional and private sources. Some of the Australian Museum's early experiments with photography have also been incorporated throughout the albums, resulting in an eclectic and seemingly spontaneous combination of photographs produced in different circumstances and for a range of purposes. These include one-off images documenting very specific subjects, photographs taken to reproduce other visual media and views taken by commercial photographers for mass production and distribution. The albums are indicative of the many ways in which the subject categories were interpreted, with the albums becoming a focus not only for the designated category of scientific knowledge but also for a much wider realm of museum interests and networks of exchange. Fundamental to the production and reception of the albums and the various narratives that occur within their pages are the circumstances of their formation, which relied on the manual assemblage and imaginative arrangement by one or more individuals. A further crucial element is each viewer's unique mode of interaction with these albums. Created, viewed and reviewed in a variety of contexts and from various perspectives, the images clearly have multiple identities and, while providing an insight into the museum's activities and changing priorities in different periods, they also resist the grand narratives of 'science, progress and politics' that institutional documents usually convey.[52]

Navigating the archive

The circulation and reception of photographs and other visual materials, the specific ways they were accumulated and displayed as visual resources and their eventual formation as archives have significantly influenced the way I have approached this study. Various observations and questions arising from a close investigation of the Australian Museum albums have influenced my overall approach to the present study and generated some of the themes developed in individual chapters. I have treated this group of albums – along with their embellishments, omissions and idiosyncrasies – as characteristic of the different, and occasionally unorthodox, ways in which photography was viewed and utilised in the museum context and how this was mirrored elsewhere in the global community of nineteenth-century natural history. As I have indicated above, this approach to the archive involved not only a consideration of the objects themselves but also associated practices and conventions of photography. It also entailed examining the professional relationships that ensured the medium's ongoing presence in the larger realm of natural history.

Photography was pre-eminent among the nineteenth century's new visual technologies due to its social and cultural impact, including the way it markedly increased the quantities of reproduced images in circulation and progressively enabled individuals to produce and reproduce their own visual imagery.[53] Individuals could interpolate and seamlessly combine images from radically different sources and genres and for a wide range of purposes. In charting the emergence, scope and diversity of photographic applications in relation to natural history, this book brings to light a set of cultural practices that have remained largely invisible in previous histories of photography. My study examines some of the more pragmatic uses of photography – including commercial portraiture – which connected naturalists' practice with self-promotion and the popular dissemination of their achievements. My purpose is to provide a wider historical context for naturalists' uses of early photography in accumulating and shaping collections, museum building or undertaking collecting expeditions. In exploring and attempting to explain different – and often overlooked – practices as part of these endeavours, this study goes beyond an analysis of attributed works by well-known photographers or the recognised achievements of prominent naturalists.

This broader agenda has necessitated research across a diverse collection of ostensibly unrelated archives. Many apparent distinctions between archives of nineteenth-century materials relate more to organisational expediency and to the recent compartmentalisation of streams of knowledge than to the Victorians' own approach to science and art. In the Victorian period, it was possible, for instance, for the educated layperson to comprehend the 'wider range' of knowledge.[54] In a two-way traffic of ideas and imagery, Gillian Beer notes, 'scientists still shared a common language with other educated readers and writers of their time … [I]t was possible for a reader to turn to the primary works of scientists as they appeared, and to respond directly to arguments advanced. Moreover, scientists themselves drew openly upon literary, historical and philosophical material as part of their arguments'.[55] By bringing into view a range of sources that have seldom been considered in relation to each other, I have attempted to find new points of convergence between the histories of Victorian visual culture, museums and natural history. The broad-based approach I have taken in navigating the archives goes to the underlying purpose of this study, that is, to challenge the perception that despite their initial rhetoric about the medium, most naturalists were reluctant or unable to embrace photography as part of their practice. By going beyond their core scientific output, I assess the scope of their engagement with the medium at a time when photography had become a relatively accessible and practical technology and pervasive as a commercial mode of image production.

In their pursuit and dissemination of natural history, public museums were key sites for pioneering such uses of photography. Naturalists working in metropolitan, provincial and colonial museums were actively involved with the medium from early on; yet, the diverse ways in which they used photographs to promote their intellectual work and careers have tended to be underestimated. The following chapters alternate between metropolitan and colonial case studies. The first two chapters trace the emerging role of photography in museums as part of the making and circulating of new knowledge. The last two chapters examine the way photographs were envisaged and deployed to build reputations and leverage naturalists' careers and promote the interests of museums. In Chapter 1, I examine the introduction of photography into the British Museum in the 1850s and its diffusion beyond that institutional centre of imperial knowledge and power to a wider colonial and global public. The British Museum viewed the advent of photography as an opportunity to augment knowledge of their collections. Because photographs appeared to give privileged access to nature, for naturalists and a growing general audience for natural history, they were particularly intriguing and sought after as both visual records and collectable, aesthetic objects. One of photography's more enthusiastic supporters in the British Museum was Richard Owen, superintendent of the natural history collections, who, along with his colleagues in the zoology department, John Edward Gray and Albert Gunther, used it to supplement their collections – through their imperial networks – and to circulate information about their observations of specimens and natural phenomena. The museum's trustees initiated an elaborate venture in 1854, authorising the construction of a purpose-built, sunlit studio on the museum's roof and commissioning Roger Fenton, one of England's most talented early photographers, to photograph parts of the museum's collections. The implementation of this ambitious plan was problematic, however, and the medium was not systematically embraced at the museum. The most concerted attempts to use photography to promote the British Museum during this early period, albeit with some resistance from the museum trustees, were by independent entrepreneurs. The natural history publisher Lovell Reeve, especially, reproduced iconic objects from the museum collections and created stereoscopic views of the galleries in commercial publications for the growing ranks of middle-class consumers.

By the 1860s, the camera had become an indispensable apparatus in a number of colonial museums. In Chapter 2, I contrast the first ill-fated attempts to establish a photographic program at the British Museum with some of the ways in which the medium was taken up in the Australian Museum, Sydney; and focus on the museum career and scientific work of German-born naturalist Gerard Krefft. During this period, natural history represented the major part of most colonial museums' collections and was the principal activity that connected the British Museum to these modern museums being established across the empire. While photography was usually introduced into these museums with official approval, the ongoing application of the medium was typically due to the resourcefulness and professional networks of individual officers who foresaw its benefits for their careers as well as for their museums. Krefft's career and his substantial reforms at the Australian museum as secretary and one of its most accomplished curators exemplifies this professional trajectory and the diverse cultural traditions and loyalties underlying the practice of natural history in the British Empire. Due to substantial government support for the sciences and science education in the German states during this period, many German émigrés took up positions in Britain's metropolitan and colonial museums and joined British scientific expeditions. Some Germans arriving in Australia already had formal training and substantial expertise, while others, such as Krefft, used the expatriate network to obtain work as collectors and artists and to develop their scientific knowledge and skills on expeditions and, subsequently, in colonial museums. As one of the first colonial scientists to assert his independence from London's imperial overlords and to demand recognition for his own scientific discoveries, Krefft turned increasingly to photography to demonstrate and validate his expertise. The introduction of photography into the main functions of the Australian Museum presents an important case study of the complex imperial–colonial politics of natural history at that time. Whereas the most valuable specimens discovered on colonial expeditions were usually sent to London for identification and absorbed into imperial collections to ultimately become the triumphs of metropolitan savants, Krefft began to set aside the most rare and spectacular specimens for the Australian Museum's own collection and steadily replaced much of the earlier traffic in specimens with hundreds of meticulous photographs of the specimens – produced on site at the museum and in the field on expeditions – as a way of substantiating his scientific observations. Through this strategic use of photography, Krefft leveraged the Australian Museum and his own scientific standing away from a position of almost complete subservience to the dictates of the centre to claim a greater level of intellectual autonomy and, ultimately, international acknowledgement for the Australian Museum's significant role in the advance of imperial science.

The use of photography in the field of natural history had, from the late 1850s, begun to extend far beyond the documentation of specimens, with portrait photography assuming a key role in relation to the professional practices and reputations of naturalists. As was the vogue during that period, many communications between naturalists involved the exchange of *cartes de visite* and other types of photographic portraits, which were also reproduced in published portrait galleries of contemporary Victorian 'worthies'. In Chapter 3, I examine the role of photographic portraiture in Victorian natural history through the lens of contemporary notions of exemplarity. Focussing mainly, though not exclusively, on the leading figures of Charles Darwin and Richard Owen, I explore the production and reception of portraits of these two most prominent contemporaries of Victorian natural history. Although Darwin was England's most renowned amateur naturalist, he was nonetheless a reticent sitter for portraits and a reluctant public celebrity. Darwin's hesitancy in sitting for portraits and his subdued appearance when he did acquiesce, however, were not only due to his apprehension about the process of visiting a studio or his distaste for public attention. Rather, they epitomised his general approach to self-fashioning, whereby he purposely downplayed his subjectivity in order to assert his authenticity as a poised and dispassionate scientific investigator and theorist.

An alternative approach to self-promotion is evident in the extensive cycle of portraits of Richard Owen produced over the course of his career. Ever conscious of the imaginative and popular impact of certain visually impressive museum displays and specimens – particularly the immense extinct vertebrates that had distinguished his career – Owen actively exploited these attributes. His portraits often commemorate his role as the discoverer of many intriguing and spectacular species, collected on his behalf from locations all across the empire and brought together by him at the British Museum. Yet, they also convey other information of a very practical nature about the specimens, including their size and structure and how the anatomist has imaginatively and literally reconstructed them so as to present them to the public. The representation of specific, identifiable objects and practices of natural history as part of the portraiture genre performed two separate tasks: to explicitly identify sitters as part of the community of naturalists – increasingly diverse as that was – and to familiarise specialist and non-specialist viewers alike with some of the intricacies and minutiae involved in the practice of natural history. This use of portraits performed a similar rhetorical function to other natural history illustrations and thus represented a sophisticated amalgamation of genres that presented viewers with a more complex pictorial scheme than suggested at first glance.

In the same period that Owen's imperial portraits re-enacted and celebrated his scientific and imperial achievements – including his discovery of the extinct New Zealand moa – a series of photographs produced at the Canterbury Museum in Christchurch, New Zealand, portrayed its director, Julius von Haast, in a strikingly similar manner. As a counterpoint to Owen's discovery of the moa, which caused a scientific sensation in London, in Chapter 4, I analyse the origins of various representations of the moa and their significance in promoting colonial natural history and its custodians to metropolitan and local audiences, with a view to building the reputations of individuals and museums within the global scientific community. A spectacular addition to any museum display, the gigantic extinct moa was a coveted commodity and became a most valuable currency for Haast, enabling him to appear upon the world stage of Victorian natural history. His capacity to garner attention was due in large part to the fascination at that time with not only far-away and unexplored locations but also the distant past. Presenting the Canterbury Museum as the primary focus of this case study, I address the key role played by photographs and other images of extinct fauna as part of the institutionalisation of natural history: in promoting the desire for natural history as public spectacle and as a potent symbol of colonial aspirations that were emerging during the 1860s and which would become a counterpoint to the traditional imperial view of natural history.

A wide range of cultural forms conveyed the Victorians' fascination with unearthing and exposing – and also reanimating – the unseen and the unknown. It was no mere coincidence that the period in which the moa came to light and attained its greatest prominence comprised those same decades that witnessed the rise of modern museums, where the emergent historical sciences of geology, palaeontology and archaeology were coming under the public gaze in an entirely unprecedented way. The representation of extinct nature – such as the moa – held considerable significance for the prospects of colonial museums. Due to the extensive circulation of images and fossil specimens and the manner in which they were presented in public museums and popular publications, they also had extraordinary immediacy for contemporary audiences across the globe. The display of the New Zealand moa at India's Madras Museum, for example, was publicised through photographs reproduced in the London illustrated press in 1865 and revealed a quite perceptible gap in that local museum's collection, since Indian fossils were usually sent off to the British institutions as a matter of course. By tracking the trajectories of individual specimens and images, I chart the cultural biography and popular reception of the moa as a form of commoditisation and examine the ways the circulation of photographs generated a range of relationships across the empire between individuals and museums.

Exemplifying much of the material addressed in this book, these images are characteristic of the diverse ways photography was deployed and its growing currency in natural history. From an interrogation of photography in the museum context in the 1850s to field work in the 1860s and 1880s, this study not only recasts the parameters of what we actually identify as natural history photography in the Victorian era but also reconfigures how we might understand the very structure of empire in relation to this genre at that time. With the popularisation of the photography, the medium's role in natural history followed new directions that reflected the ways in which imperial and colonial science was being transformed. Ranging from scientific studies to naturalists' portraits, photography became a focus for various intellectual and cross-cultural encounters and thus presented not only an extraordinary new visual technology but also introduced a new means of discourse between individuals, museums and audiences in different parts of the British Empire and across the globe. By alternating between imperial and colonial structures, this book examines the intricacies of these relationships and offers a new perspective on photography's indispensable but complex role in Victorian natural history and the nineteenth-century museum. Finally, the comparative analysis that frames this study highlights the way naturalists in the Victorian era used the emerging medium of photography – alongside, but gradually also in place of, other forms of visual culture – to enhance imperial and global networks, support their investigations, advance their careers and promote natural history and their museums in novel ways for a new generation of consumers.

Notes

1 Professor Owen, 'On some Outline-Drawings and Photographs of the Skull of the *Zygomaturus trilobus, Macleay* (Nototherium, *Owen?*)', *Quarterly Journal of the Geological Society of London*, vol. 15, no. 1–2, February 1859, pp. 168–76 (Read 10 March 1858); Professor Owen, 'On a Collection of Australian Fossils in the Museum of the Natural History Society of Worcester; with Descriptions of the Lower Jaw and Teeth of the Nototherium Inferme and Nototherium Mitchelli, Owen; Demonstrating the Identity of the Latter Species with the Zygomaturus of Macleay', *Quarterly Journal of the Geological Society of London*, vol. 15, no. 1–2, 1859, pp. 176–86 (Read 23 June 1858).

2 'Donations to the Australian Museum during August, 1857', *Sydney Morning Herald*, 9 September 1857, p. 2; see also Andrew David, *The Voyage of HMS Herald to Australia and the South-West Pacific 1852–1861 under the Command of Captain Henry Mangles Denham*, Melbourne University Press, Melbourne, 1995, pp. 434–40.

3 Owen, 'On some Outline-Drawings and Photographs', p. 171.

4 See John Gallagher and Ronald Robinson, 'The Imperialism of Free Trade', *The Economic History Review*, series 2, vol. 6. no. 1, 1953, pp. 1–15; John Darwin, 'Imperialism and the Victorians: The Dynamics of Territorial Expansion', *The English Historical Review*, vol. 112, no. 447, 1987, pp. 614–42; Niall Ferguson, *Empire: How Britain Made the Modern World*, Allen Lane, London, 2003, pp. xiv and 314.

5 Owen, 'On some Outline-Drawings and Photographs', p. 169.

6 See Charles Wheatstone, 'Contributions to the Physiology of Vision – Part the First: On Some Remarkable, and Hitherto Unobserved, Phenomena of Binocular Vision', *Philosophical Transactions of the Royal Society of London*, vol. 128, 1838, pp. 371–94; also Jonathan Crary, *Techniques of the Observer: On Vision and Modernity in the Nineteenth Century*, MIT Press, Cambridge (Mass.), 1990, pp. 122–4.

7 Eleanor Hight and Gary Sampson (eds.), *Colonialist Photography: Imag(in)ing Race and Place*, Routledge, London, 2002, p. 1; see also Edwina Taborsky, 'The Discursive Object' in Susan Pearce (ed.), *Objects of Knowledge*, Athlone Press, London and Atlantic Highlands, 1990, pp. 50–77.

8 David E. Allen, 'Amateurs and Professionals' in Peter J. Bowler and John V. Pickstone (eds.),*The Cambridge History of Science Volume 6: The Modern Biological and Earth Sciences*, Cambridge University Press, Cambridge, 2009, pp. 15–33; Lynn K. Nyhart, 'Natural History and the "New" Biology' in N. Jardine, J. A. Secord, and E. C. Spary (eds.), *Cultures of Natural History*, Cambridge University Press, Cambridge, 1996, pp. 426–43.

9 J. E. Gray, 'On Museums, Their Use and Improvement', *Annals and Magazine of Natural History*, vol. 14. 1864, pp. 283–97.

10 Peter J. Bowler, 'Popular Science' in Bowler and Pickstone, *The Cambridge History of Science Voume. 6*, p. 623.

11 On photography's documentary and analytical role in science – see Ann Thomas, (ed.), *Beauty of Another Order: Photography in Science*, Yale University Press, New Haven & London, 1997; Kelley Wilder, *Photography and Science*, Reaktion, London, 2009; Lorraine Daston and Peter Galison, *Objectivity*, Zone Books, Cambridge (Mass.), 2010; also Carol

Armstrong, *Scenes in a Library: Reading the Photograph in the Book, 1843–1875*, MIT Press, Cambridge (Mass.), 1998.

12 John MacKenzie, *Museums and Empire: Natural History, Human Cultures and Colonial Identity*, Manchester University Press, Manchester, 2009, p. 5.

13 Anne-Marie Willis, *Picturing Australia: A History of Photography*, Angus & Robertson, Sydney, 1988, p. 21; Gael Newton, *Shades of Light: Photography and Australia 1839–1988*, Australian National Gallery and Collins, Canberra and Sydney, 1988, p. 4.

14 Newton, *Shades of Light*, p. 4; Alan Davies and Peter Stanbury, *The Mechanical Eye in Australia: Photography 1841–1900*, Oxford University Press, Melbourne, 1985, p. 1.

15 Newton, *Shades of Light*, p. 4.

16 Idem.

17 Willis, *Picturing Australia*, p. 13; James Freeman, 'On the Progress of Photography and its Application to the Arts and Sciences', *Sydney Magazine of Science and Art*, vol. II, January–February 1859, pp. 137–4.

18 Influential historical surveys of photography in Australia and New Zealand include Jack Cato, *The Story of the Camera in Australia*, 1955; Newton, *Shades of Light*; Willis, *Picturing Australia*; Hardwicke Knight, *Photography in New Zealand: A Social and Technical History*, 1971; William Main and John B. Turner, *New Zealand Photography from the 1840s to the Present*, Photoforum Inc, Auckland, 1993; also Helen Ennis, *Photography and Australia*, Reaktion, London, 2007; Angela Wanhalla and Erika Wolf (eds.), *Early New Zealand Photography: Images and Essays*, Otago University Press, Dunedin, 2011; David Eggleton, *Into the Light*, Craig Potton, Nelson, 2006.

19 See Ann Moyal, *The Web of Science: The Scientific Correspondence of Rev. W. B. Clarke, Australia's Pioneer Geologist*, Australian Scholarly Publishing, Kew, 2003.

20 Tim Barringer, Geoff Quilley and Douglas Fordham (eds.), *Art and the British Empire*, Manchester University Press, Manchester and New York, 2007, p. 3.

21 See James Ryan, *Picturing Empire: Photography and the Visualization of the British Empire*, Reaktion Books, London, 1997; Hight and Sampson *Colonialist Photography*; also Judy Annear (ed.), *Portraits of Oceania*, Art Gallery of New South Wales, Sydney, 1997.

22 A notable exception is Jane Lydon, *Eye Contact: Photographing Indigenous Australians*, Duke University Press, Durham & London, 2005 – see esp. pp. 93–8.

23 Bernard Smith, *European Vision and the South Pacific*, Yale University Press, New Haven and London, 1988, p. viii.

24 Ibid., pp. 335–7.

25 Ibid., pp. 13–14 and 27–29.

26 W. J. T. Mitchell, *Picture Theory*, University of Chicago Press, Chicago and London, 1994; W. J. T. Mitchell, *What Do Pictures Want?*, University of Chicago Press, Chicago and London, 2005; see also Georgina Born, 'Public Museums, Museum Photography, and the Limits of Reflexivity', *Journal of Material Culture*, vol. 3, no. 2, 1998, pp. 223–254.

27 Canterbury Museum Records: Donations & Exchanges (Duplicates) 1872–1887 (1869–1883); 'Exchanges', *Australian Museum Annual Report* – for the years 1861–1874.

28 Kurt Bayertz, 'Spreading the Spirit of Science: Social Determinants of the Popularization of Science in Nineteenth-Century Germany' in Terry Shinn, and Richard Whitley (eds., *Expository Science: Forms and Functions of Popularisation*, D. Reidel, Dordrecht, Netherlands, 1985, p. 222; Peter J. Bowler, 'Popular Science' in Peter J. Bowler and John V. Pickstone (eds.), *The Cambridge History of Science Volume 6*, p. 626.

29 See Smith, *European Vision*; Richard Neville, *A Rage for Curiosity*, State Library of New South Wales, Sydney, 1997; John Kean, *The Art of Science*, Museum of Victoria, Melbourne, 2013.

30 See Caroline Jordan, *Picturesque Pursuits: Colonial Women Artists and the Amateur Tradition*, Melbourne University Press, Melbourne, 2005; Tim Bonyhady, *Colonial Earth*, Miegunyah Press, Carlton, 2000.

31 Martin Rudwick, *The Great Devonian Controversy*, University of Chicago Press, Chicago, 1985, p. 10 (emphasis in original).

32 Kingston, *The Oxford History of Australia, vol. 3,* pp. 237–8; C. H. Currey, 'Denison, Sir William Thomas (1804–1871)' in *Australian Dictionary of Biography*, Online Edition, Australian National University, 2006, http://www.adb.online.anu.edu.au/biogs/A04008b.htm (accessed 11 December 2010).

33 See David Cannadine, *Ornamentalism: How the British Saw Their Empire*, Oxford University Press, 2002, p. 130.
34 James Braund, 'German-Speaking Scientists in New Zealand, 1773–1951: Research Past, Present and Future', *European Studies* 21, 2005, pp. 179–81.
35 See Grace Seiberling, *Amateurs, Photography and the Mid-Victorian Imagination*, University of Chicago Press, Chicago, 1986; Jennifer Tucker, *Nature Exposed: Photography as Eyewitness to Science*, John Hopkins University Press, Baltimore, 2005; Elizabeth Edwards, *Anthropology and Photography 1860–1920*, Yale University Press, New Haven and London, 1992; Elizabeth Edwards, *Raw Histories, Anthropology and Museums*, Berg, Oxford, 2001.
36 See Jennifer Tucker, 'The Historian, the Picture, and the Archive', *Isis*, vol. 97, 2006, pp. 111–20; also Martin Rudwick's histories of British geology and palaeontology; Barbara Maria Stafford's explorations into art, science and intellectual history; Lorraine Daston and Peter Galison's studies of scientific objectivity; and Renato Mazzolini's edited volume *Non-verbal Communication in Science Prior to 1900*, 1993.
37 *Geographies of Nineteenth-Century Science: An Interdisciplinary Conference*, held at the University of Edinburgh, 2007 – http//www.eeo.ed.ac.uk/geography/geog19c (accessed 29 November 2016); see also David N. Livingstone, *Putting Science in its Place: Geographies of Scientific Knowledge*, University of Chicago, Chicago, 2003; David N. Livingstone and Charles W. J. Withers (eds.), *Geographies of Nineteenth-Century Science*, University of Chicago, Chicago and London, 2011.
38 Bruno Latour, *Science in Action: How to Follow Scientists and Engineers through Society*, Harvard University Press, Cambridge (Mass.), 1987; Bruno Latour and Steve Woolgar, *Laboratory Life: The Construction of Scientific Facts*, Princeton University Press, Princeton, 1986.
39 Susan L. Star and James R. Griesemer, 'Institutional Ecology, "Translations" and Boundary Objects: Amateurs and Professionals in Berkeley's Museum of Vertebrate Zoology, 1907–39', *Social Studies of Science*, vol. 19, 1989, pp. 387–420.
40 Star and Griesemer, 'Institutional Ecology', p. 389.
41 See Ian Inkster, 'Scientific Enterprise and the Colonial "Model": Observations on Australian Experience in Historical Context', *Social Studies of Science*, vol. 15, 1985, pp. 677–704; also Roy McLeod, 'On Visiting the "Moving Metropolis": Reflections on the Architecture of Imperial Science', *Historical Records of Australian Science*, vol. 5, no. 3, 1982, pp. 1–16.
42 See Jim Endersby, *Imperial Nature: Joseph Hooker and the Practices of Victorian Science*, University of Chicago Press, Chicago, 2008; Anne Secord, 'Corresponding Interests: Artisans and Gentlemen in Nineteenth-century Natural History', *British Journal for the History of Science*, no. 27, December 1994, pp. 383–408; Susan Sheets-Pyenson, *Cathedrals of Science: The Development of Colonial Natural History Museum during the Late Nineteenth Century*, McGill-Queens's University Press, Kingston and Montreal, 1988.
43 Golinski, *Making Natural Knowledge*, pp. 146–7.
44 Doreen Massey, *Space, Place, and Gender*, Polity, Cambridge, 1994, p. 159.
45 Elizabeth Eisenstein, *The Printing Revolution in Early Modern Europe*, Cambridge University Press, Cambridge, 2005.
46 Ibid., p. 470.
47 Ibid., p. 519.
48 Martin Kemp and Craig Ashley Hanson, 'Science and Art', *Grove Art Online – Oxford Art Online*, Oxford University Press, http://www.oxfordartonline.com/subscriber/article/grove/art/T076962 (accessed 28 November 2016).
49 See Golinski, *Making Natural Knowledge*, pp. 29–30.
50 See Barbara T. Gates, 'Why Victorian Natural History?', *Victorian Literature and Culture*, vol. 35, 2007, pp. 539–49.
51 Photograph albums, AMS 421/1–19: Archives and Records, Australian Museum, Sydney.
52 Janet Hoskins, *Biographical Objects: How Things Tell Stories about People's Lives*, Routledge, London, 1998, p. 5.
53 Ruth Pelzer, 'Technical Reproduction and its Significance' in Matthew Rampley (ed.), *Exploring Visual Culture*, Edinburgh University Press, Edinburgh, 2005, p. 197.
54 Samuel Johnson, *The Works of Samuel Johnson, with an Essay on his Life and Genius by Arthur Murphy*, vol. 1, George Dearborn, New York, 1837, p. 464.
55 Gillian Beer, *Darwin's Plots: Evolution Narrative in Darwin, George Eliot and Nineteenth-Century Fiction*, Cambridge University Press, Cambridge, 2000, pp. 4–5.

1 Paper museums

Photography and natural history at the British Museum

Introduction

Flourishing in the Victorian era, the natural history museum is incontestably linked to the expansion of industrial nations across the globe and the great wealth and power that accrued from that process. The resulting spoils, coupled with the Enlightenment quest for encyclopaedic knowledge, promoted a fervour for collecting that was unprecedented and remains virtually unsurpassed even today.[1] Even as museums bulged with innumerable assorted specimens, there was a proliferation of manuals offering guidance to individuals on how to collect and preserve ever more specimens, and scientific societies continued to offer museums parts of their collections accrued from expeditions and colonial correspondents.[2] Yet, the resulting museum collections were far from comprehensive. In his classic work *Museum without walls*, André Malraux points to the inherently accidental nature of all museums and the contingent status of their collections: 'Even when the greatest zeal and enormous resources have gone into its making, a museum owes much to opportunities that chance has thrown its way'.[3] Destined always to be incomplete, Victorian-era museums acquired and exhibited an extraordinary array of natural and cultural objects. These artefacts were highly prone to a variety of meanings in different contexts, for the museum served not only as a site for display but also became a focus for re-presenting and reinterpreting rare and prized objects through other visual media, including photography.[4] The implementation of photography programs in museums from the mid nineteenth century enabled museum staff and audiences, plus other stakeholders, to organise, supplement, publicise and circulate museum collections by means of virtual specimens documented on site or sent from correspondents across the globe.

This chapter examines the introduction of photography at the British Museum with particular reference to its role in re-presenting and promoting the museum's natural history collections. As an early participant in the endeavour to formalise the documentary functions of the medium, the British Museum viewed the advent of photography as an opportunity to augment knowledge of the collections.[5] Photographs appeared to give privileged access to nature. For naturalists, therefore, they were particularly intriguing as records of specimens and views, and they were also desirable objects in their own right.[6] The practice of collecting as much as recording pervaded photography, with naturalists adding photographs to 'paper museums' (which included drawings and engravings) that complemented – but sometimes also reinterpreted and complicated – official collections of specimens.[7] Some naturalists also embraced photography as a popular medium to promote their collections to a general audience and participated in openly commercial ventures.

Structured in three sections, this chapter describes the emergence and role of photography at the British Museum, as well as its diffusion beyond that institution to an imperial and global public. The first section examines the British Museum's attempt to introduce a formal photographic program during the 1850s and compares the approaches taken by staff in the departments of natural history and antiquities. The natural history collections and the antiquities department's Assyrian collections each occupied an uncertain position in the hierarchy of the British Museum at that time. Despite institutional ambivalence and outright resistance by some museum officers, the keepers of these particular collections generally recognised the potential benefits of photography for their respective fields. Section two analyses some of the specific efforts made by museum officers to utilise photography in their professional and imperial networks and to use it to further their own research agendas, cultivate new knowledge and address deficiencies within their collections. The need to access and methodically organise a range of specimens so as to interpret and compare their characteristics contributed to the rise of a new mode of production whereby photographs of fossils and other exceptional specimens were exchanged rather than the objects themselves. The circulation of these images, although not usually publicly acknowledged, was widespread across various networks. Most prominently, Richard Owen, superintendent of the museum's natural history collections at the time, was actively involved in negotiating with his many contacts across the empire for photographs in lieu of original specimens or other types of illustrations. The concluding section of this chapter describes commercial initiatives that were aimed at promoting not only the museum's official program but also an appreciation of natural history to an emerging consumer class. While the British Museum failed to develop any clear policy related to the photographic reproduction of its collections, it did open up avenues for individual staff members, private entrepreneurs and colonial institutions and individuals, who were more enterprising in their use of photographs as both documents and objects of aesthetic value and popular appeal.

From institutional priorities to imperial interests

While the British Museum became a public institution in 1753, it was not until the 1830s that the parliament, followed by the trustees and staff, began to address its public face and the access it provided for general visitors.[8] Prior to the 1830s, the museum had remained very much in the mould of an Enlightenment cabinet of curiosities, devoted to augmenting its collections and duchessing a coterie of specialists rather than promoting itself as a place of public education or rational entertainment. Official histories note that 'early illustrations and descriptions give a rather secretive approach to the objects [and] visitors were expected to come in awe, to wonder and to not ask too many questions'.[9] With collections based on Sir Hans Sloane's extensive holdings and maintaining his universal approach to collecting, a substantial part of the British Museum's activities centred on the accumulation and study of natural history specimens, in addition to the antiquities, works of art and the other kinds of cultural artefacts that predominate today. George Scharf's watercolours portray the situation at the British Museum in 1845. Depicting visitors in the circulation areas between the galleries, one drawing reveals stuffed antelopes and a reconstructed tapir or small hippopotamus cheek by jowl with grand neo-classical landscapes and a celestial-themed trompe l'oeil in one staircase (Figure 1.1), while

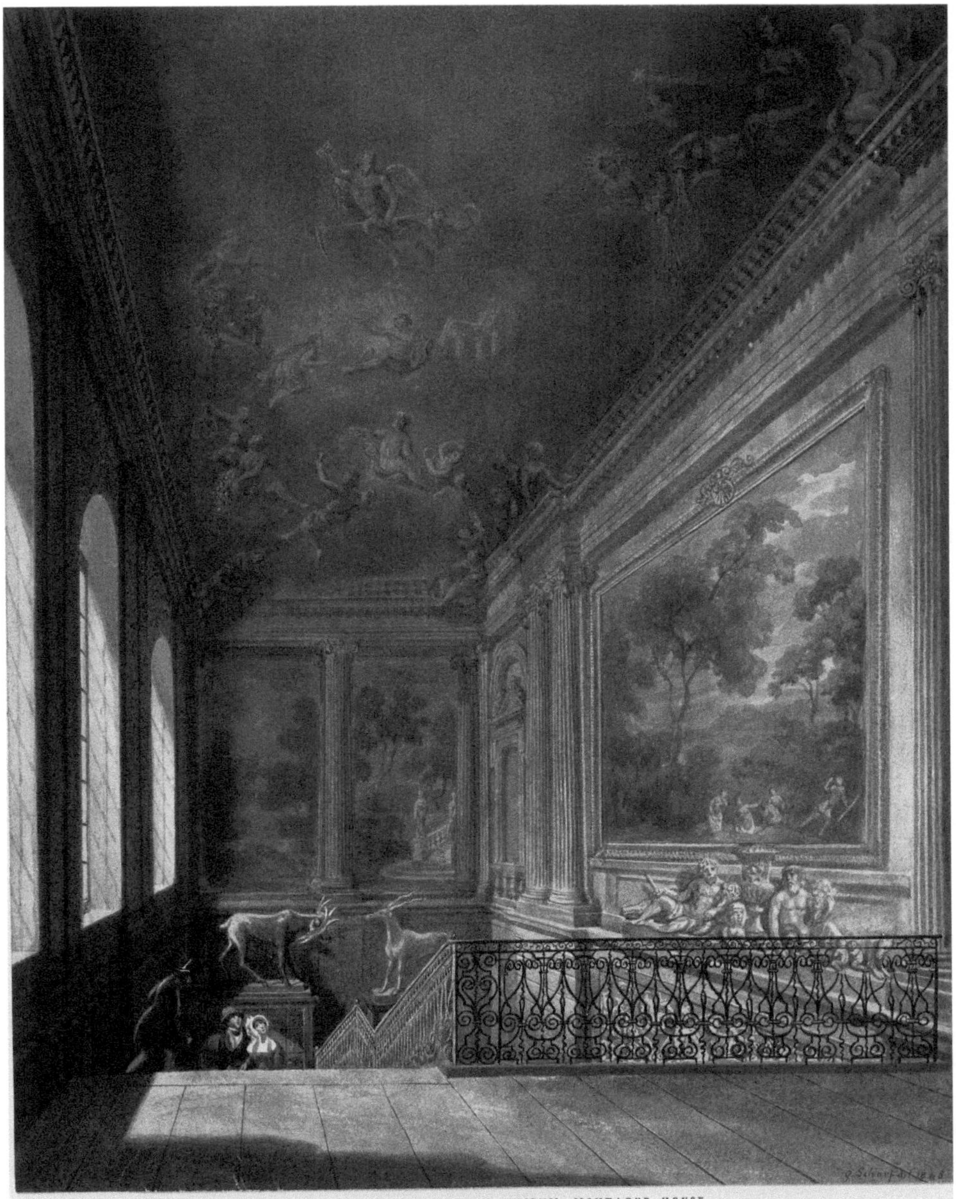

Figure 1.1 George Scharf, *Staircase of the Old British Museum, Montague House*, 1845, watercolour. Collection: British Museum, London (1862, 0614.631).

another image showed a display of giraffes crammed onto a landing. Rather than being exhibited in discrete or sequential galleries, these disparate collections were displayed in whichever spaces could accommodate them, and incongruent objects were often placed together.

An album assembled by the keeper of zoology, John Edward Gray, during the mid-nineteenth century codifies in visual form the range of issues that beset the museum during this period and is a revealing record of official and popular attitudes to the presentation and re-presentation of the museum and its collections.[10] The natural history departments were acknowledged to be in a lamentable state, as there was simply not enough room to display or store all the specimens that the museum had managed to accumulate, let alone fulfil its aspiration of creating a physical index of the entire natural world. The British Museum, like other nineteenth-century museums, was also a locus of scholarship, and, although the general public was largely unaware of these activities, it was regarded by British scholars as the national, imperial and global centre for the practice and advancement of natural history. The move from Montague House to the new museum building, which was completed in 1852, did little to accommodate the overflow of the natural history collections and behind-the-scenes activities. In a report to the trustees in 1851, Gray advised that, while the space for zoological specimens had almost trebled, increasing from 13,745 to 36,600 square feet, the collection had grown tenfold since 1836, such that 'about half the increase was boxed up in the basement cellars'.[11] By 1857, even the objects that were on display were so unsuitable for study or general viewing compared to those in the principal museums of Europe that it had become an issue of national pride:

> The whole of these collections are more strictly systematically arranged, more uniformly named, and in accordance with the modern state of science than any of the public collections in Europe.... At the same time all parts of the collection are so cramped for space, and the specimens are so crowded that those which are even exhibited in the public rooms are not seen to advantage.[12]

Along with Robert Smirke's architectural plans for the new building and several cuttings from the illustrated press portraying some of the museum's new galleries, the album compiled by Gray includes nine calotype photographs of the exterior of the museum. The earliest calotype is little more than a faded silhouette of old Montague House, while the others images present more detailed views of the reading room cupola under construction, the portico and the new building's main façade dating from the 1840s to the mid1850s (Plate 2). Representing probably the earliest photographs of the British Museum, they demonstrate Gray's interest in the physical transformation of the building. Essentially depicting a colossal building site, the scenes portray a great national monument in the making. The very scale of the buildings in these exterior views suggests the magnitude of the collections and reflects the imperial power and prestige of the British Museum. Equally significant is the manner in which Gray has carefully juxtaposed the photographs not only with official documentation of the museum building but also with popular images from the *Illustrated London News* of the Easter Monday crowd in the Great Zoological Gallery, 1845 (Figure 1.2); and the museum's new Coral Room, 1847.[13] The artist's impressions of the new gallery spaces – appearing considerably more congested than the photographs of the museum's colossal exterior might have promised – portray the general public promenading, conversing and viewing the massed displays of specimens and sundry objects in the zoological galleries. The two perspectives seem contradictory: the photographs depict the British Museum as a monumental edifice dedicated to the privileged realms of knowledge and research, which exclude the general visitor, while

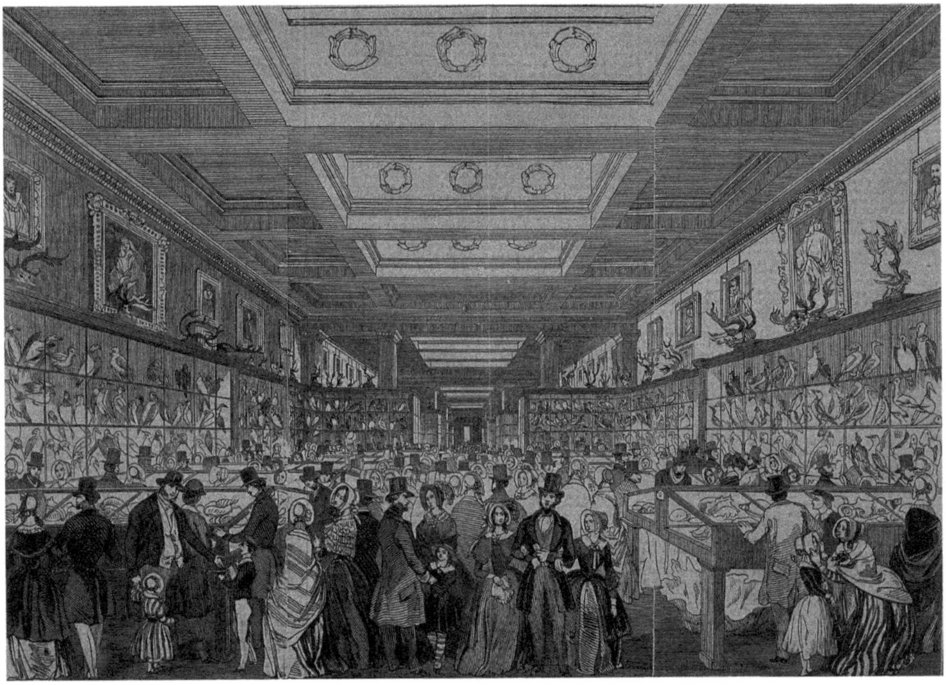

Figure 1.2 'Easter Monday – The Great Zoological Gallery, British Museum', *Illustrated London News*, 29 March 1845, p. 201. Collection: Wellcome Library, London.

the illustrated press images reflect rational amusement and social reforms of public service and unrestricted access.[14]

It is no coincidence that the introduction of photography at the British Museum occurred at that historical juncture when issues of access, promotion and collection development were being aired in public as well as discussed behind the scenes at the museum. The public mood for reform had the potential to reduce the museum's authority as the ultimate arbiter of its priorities and public face. The advent of photography presented an opportunity to enhance but still control public dissemination of the collections. By using photography to document selected objects and displays, it was presumed that the museum trustees and officers would maintain control over the way the collections – and thus the museum itself – were represented beyond their immediate domain.

The studies of the Assyrian tablets produced by Roger Fenton in the spring of 1854 have been identified as the earliest extant photographs of the British Museum's collections.[15] Other objects chosen to be photographed during the six-year period of Fenton's association with the museum included other antiquities, such as the Elgin friezes; and art treasures by the Old Masters. His photographic style was not only effective in documenting each object but also emphasised their value as cultural treasures, reflecting aspirations of the time.[16] Fenton's photographs of selected objects from the natural history collections and their display in the museum galleries are less numerous and have received scant attention.

The choice of the Assyrian collections for the museum's first foray into photography is significant in terms of the institution's organisation and governance, and in many respects it mirrors official attitudes to the natural history collections and their wider reception. Just as the status and official support of the natural history departments depended on the museum's ongoing connection with its own heritage in Sloane's cabinet of curiosities, the Assyrian collections occupied an indeterminate position in the hierarchy of collections due to their distinctiveness and extreme novelty at the time. Excavated by the English archaeologist Austen Henry Layard during the 1840s and 1850s, Assyria was the great archaeological sensation of the nineteenth century and, for the Victorians, it held an irresistible fascination. Indeed, there are certain parallels between the way the Assyrian and the natural history collections were regarded, both officially and in terms of their visual appeal and interest to general audiences. With the rise and growing popularity of the historical sciences, naturalists and geologists were keen to draw an analogy between the 'antiquities of nature' and the 'monuments' of past civilisations. Likening themselves to antiquarians or archaeologists, these scientists pointed to the way in which their work contributed to an historical account of the earth comparable to the way in which human history was being recast with the unearthing of previously unknown civilisations. As the French founder of comparative anatomy, Georges Cuvier, had envisaged, it was necessary for systematists, or taxonomists, 'to do for the history of nature what antiquarians [were doing] for the history of the techniques and customs of peoples':

> The former will have to go and search among the ruins of the globe for the remains of organisms that lived at its surface, just as the latter dig in the ruins of cities in order to unearth the monuments of taste, the genius, and the customs of the men who lived there. These antiquities of nature … will provide the physical history of the globe with monuments as useful and as reliable as ordinary antiquities provide for the political and moral history of nations.[17]

In both fields of study, a great many of the objects being discovered had a visual immediacy, and they were alluring to general audiences as symbols of national achievement while also embodying the historical exoticism of revealing past eras for modern-day viewers. Whereas Cuvier had chosen to show fossils excavated from the gypsum quarries around Paris to make his comparison and to render his subject appealing to a local, general audience, the British Museum during the Victorian era increasingly had access to impressive discoveries of long-extinct species from across the British Empire.[18] Critically, it was the Assyrian and natural history collections that figured most prominently in the early negotiations for the British Museum's photographic commissions. While the Assyrian discoveries were visually spectacular and a prize in terms of Britain's recent excavations in the Middle East, some influential trustees were dismissive of their importance compared to the classical antiquities. As a consequence, the museum's presentation of these artefacts was haphazard and sporadic.[19] On the other hand, from their first appearance in public in 1847, the artefacts were prominently featured by the *Illustrated London News*, conveying their significance.[20] Although the Assyrian artefacts were viewed by the public with fascination, therefore, the institution's official attitude to the collections remained ambivalent, and it was not until 1853, due to the initiative of the museum's managing architect and following considerable public interest, that the collections were installed in purpose-designed galleries and interpreted in relation to the other antiquities.[21] This decision gave the Assyrian collections considerable status as well as a new prominence in the institutional framework. Having finally determined the way these objects were to be presented and interpreted within the museum, it became imperative to reclaim and regulate the way they were disseminated in the broader public arena.

The museum's attitude towards the collection of Assyrian objects can be characterised, Frederick Bohrer suggests, as 'something between a proper object of study, a trophy, and a curiosity', and this echoed the range of views about the status of the natural history collections as well.[22] The main issue turned on the prestige of the museum collections as a focus for elite scholarship and research. For a number of trustees and prominent museum officers, the broad appeal of natural history to the general populace – compared with more rarefied branches of learning – seemed to diminish not only the importance of those collections but also to taint the decorum and reputation of the museum itself.[23] For this reason, it is perhaps revealing of the trustee's intentions that, at the British Museum's Standing Committee meeting of 11 February 1854, it was Edward Hawkins, keeper of antiquities, who 'reported that the Photographic Room was completed, although some of the Apparatus had not yet been delivered, and that Mr. Fenton had taken several pictures, specimens of which were laid before the Board'.[24] The minutes do not record the subject matter of those initial photographs, nor are there any comments on how they were received by the committee. However, on receiving Hawkins's advice, the trustees' immediate response was a request to the museum's principal librarian, Sir Henry Ellis, 'to acquaint Dr. Gray, that it is their wish that as soon as the Photographic House and apparatus have been completed, that he should lend some Shells for the purpose of taking them photographically, as also some Corals, and some Insects'.[25] The instruction implied that they were cautiously impressed with the photographs submitted to them and had decided to extend the application of photography to other museum collections. As it happened, however, upon receiving Fenton's estimate of photographic costs the following month, along with a recommendation from Hawkins 'that twenty new impressions should be taken of each of the Assyrian Tablets and five of ordinary new acquisitions', the trustees advised that the enterprise was 'merely an experiment for the present'.[26] Agreeing to pay Fenton for the next two months, they went on to request that he 'devote himself chiefly during that time to the Nineveh Monuments, with the understanding that all his works shall be the property of the Trustees, and that at the end of the two months, the subject shall be reconsidered'.[27]

Whether the anticipated photographic studies of shells, corals and insects from the natural history collections were ever undertaken is unclear. From these brief instructions, it is apparent that, despite their official directive, the trustees were uncertain about the type of subject matter that would be most suitable for photography, and thus which collections would benefit most from the enterprise. Consequently, while initially decisive in their request that natural history specimens be photographed, the trustees then appeared to equivocate and seemed more inclined to be influenced by specific requests from individual departments and keepers. Following this initial, and somewhat cautious, response to the increasingly frequent requests received from historians, naturalists and other scholars for photographs of the museum's treasures to assist in their studies, it is remarkable that no rationale was ever clearly articulated for photographing the different collections.[28] Regardless of this, the museum continued to receive applications from naturalists, antiquarians and historians – including William

Henry Fox Talbot, the British inventor and promoter of photography, who was also well known for his studies in etymology and hieroglyphics.[29] Consequently, neither Gray's demonstrated interest in the medium nor the directive from the trustees to arrange for some shells, corals and insects to be photographed signalled the beginning of any purposeful program by the museum to utilise photography in documenting and disseminating its natural history collections.[30]

Instead, it was Hawkins's application for photographs to be taken of the Assyrian objects that would provide a model for documenting the natural history collections. Submitted to the Standing Committee in March 1854, his proposal called for the Assyrian inscriptions to be transcribed and printed in the same manner as others had been previously produced; and the photographic impressions sent, for verification of the printed copies, to those scholars who were known to be engaged in the translation and interpretation of the inscriptions.[31] While the study of Assyria was still a fledgling field, the volume of photographs of Assyrian artefacts produced was quite sizeable compared to those of 'ordinary antiquities'. Hawkins considered that five copies would generally be sufficient for the latter: three to be retained for the use of the museum, and two to be sent to any museum 'which may be peculiarly interested in the object represented'.[32] Perceiving these photographs to be near perfect facsimiles of the artefacts, Hawkins anticipated that photographs of 'ordinary antiquities' could be used to cultivate knowledge, expertise and connections with a wide range of museums. A report from fellow antiquarian, Augustus Franks, to the museum's trustees in November of that year reveals the magnitude of Hawkins's plans for the Assyrian artefacts:

> The photographic operations in our department during the past year may be divided into two branches – Assyrian Tablets and miscellaneous antiquities acquired during the year. Of Assyrian Tablets we have received from Mr. Fenton 181 photographic plates forming 3,620 pictures. On these are represented – where they are inscribed on both sides – 120 tablets. A considerable number of negatives have been made of other tablets which have not yet been printed off and which make the total number of tablets which have been transferred to glass about 200.[33]

This left approximately 250 tablets, which Fenton was preparing to complete at a reduced rate of payment owing to the volume and consistency of the work.[34] While the scale of the initial commission was substantial, its scope remained nonetheless limited, as the photographs were never intended for dissemination within the public realm. As Hawkins had made clear, the photographs were to be distributed exclusively to scholars for the purpose of promoting specialist knowledge of the Assyrian collections. The specific choice of objects to be 'transferred to glass' demonstrated a further, more tacit, constraint on their representation and circulation. While the colossal Nineveh sculptures provided spectacular subject matter for the popular illustrated press and could be easily adapted to different media for the purposes of visual commodification, the numerous tablets of cuneiform script were not only far more esoteric but also virtually devoid of visual interest for all but the most dedicated and learned antiquarians.[35]

The photographs of the Assyrian tablets were most likely conceived to replace the use of squeezes – papier-mâché impressions – which had long provided archaeologists and historians with copies of incised or relief designs but which proved to be exceedingly time consuming. The improvements attained by Fenton's photographic reproductions included their infinite reproducibility; the facility to circulate accurate images of entire tablets in a reduced, portable form; and the ability to convey detailed views of the cuneiform text in an enlarged form. Photographs facilitated examination, translation and verification by the entire fraternity of antiquarians, not just those members of the museum coterie who had privileged access to the original objects or a very limited number of facsimiles. The implications for the museum and its burgeoning collections were significant. Nevertheless, the trustees were still seeking both purpose and direction for the photographic program some three years later, as evidenced by their request to the heads of departments for suggestions of objects to be photographed from their respective collections.[36] In response, a variety of submissions were presented. For the most part, officers in the various natural history departments gave the impression of having had little acquaintance with the medium and were quite ambivalent about its potential use; the majority of their recommendations, therefore, ranged somewhere between caution and indifference. For instance, among the first to reply was Robert Brown, keeper of the Banksian Department, who begged 'to state that he [had] not sufficient knowledge of the capability of that process to enable him to judge with confidence that it will succeed on the specimens he has selected so as to exhibit structure as they do at present, and which it was his object to shew [sic] in preparing them for Exhibition'.[37] Interestingly, Dr. Gray expressed similar sentiments, reporting that: 'Photography is only applicable to a very small part of the specimens in the Zoological Collection, and [I am] so much occupied by [my] routine duties and the over-looking of the scientific works ... that [I am] not desirous of undertaking any fresh works, until those have been completed.'[38] He recommended that no more than 50 positives be printed from each plate, in the first instance, as he reasoned that 'more copies could easily be printed from the negative, if there was found to be any sale for the work'.[39] As such, the extent of the recommendations was a total of seven predominantly Australian specimens from the herbarium under Brown's charge, along with some 10 or 12 large specimens of corals and a series of five or six skulls of the manatee (sea cow) from Gray's zoological collections. There are no extant copies of these photographs, and it is very probable that the submissions were never acted upon.

The departments in charge of the art collections were somewhat more optimistic, with William Carpenter, keeper of the Department of Prints and Drawings, suggesting that Durer's 'fine carving' of the Birth of St John 'may be photographed, it being one of the most interesting objects in his custody, as well as being peculiarly well-suited for such a purpose, & there can be no doubt of its meeting with a most extensive sale, if a faithful representation be given of it'.[40] Moreover, he took the opportunity to recommend that:

> [A] further portion of the Drawings of the Old Masters should be proceeded with: one third of the expenditure on such as have already been executed, having been returned by such as have been sold. If good faithful copies can be produced by Mr. Fenton from some of the finer specimens of the Engravings of Marc Antonio, it will be highly desirable to have them done, as rendering such fine objects attainable at a comparatively small cost will tend greatly to elevate the taste & feeling of the public.[41]

Carpenter's stated aspiration of cultivating the general public into an educated and discerning audience was overshadowed, however, by the initial motivation to use photography as an exclusive and privileged means of exchange between connoisseurs and other acknowledged experts.

This convoluted and protracted planning – the conflicting instructions, half-hearted submissions and problematic negotiations – that continued to hinder the production of Fenton's natural history studies for the British Museum is significant for understanding the ambitions of individual figures and the aims of the institution with respect to the status and future of photography. Of all the scientific officers, it was Richard Owen, newly appointed to the museum as superintendent of the natural history departments and already a prominent public figure due to his discoveries in comparative anatomy, who appeared to immediately comprehend the general public's immense interest in the medium and thus its potential to advance both his own interests and those of the museum. In contrast to Brown and Gray, who made it clear that their submissions were presented simply 'in compliance' with the trustees' request, Professor Owen considered it an 'honour' to submit his '"First List" of objects' from the natural history departments 'as being suitable for photography and desirable to be photographed'.[42]

The appointment of Owen to the British Museum the previous year had been an unmistakably political move designed to elevate the status of the natural sciences in Britain's foremost public museum. The impetus for this came not only from the scientific fraternity but also from the highest echelons of British society in the form of Prince Albert who, as well as instigating the Crystal Palace exhibition, had promoted a range of other initiatives and reforms to advance the cause of science with the hope of supporting the national economy. Accordingly, Owen's submission in response to the trustees' request was extensive; stretching over ten pages of closely spaced lists, it nominated numerous species and entailed at least three photographic viewpoints for whole or partial specimens for each of those species (Figure 1.3 *a* and *b*). In all, this single, initial submission called for well over 300 photographic views. The immense scale of such an undertaking for Fenton – let alone any of the less accomplished photographers of the time – suggests that, while Owen may have been quite astute about the usefulness of photography for his research and the museum's broader interests, his knowledge of the practicalities and constraints of the medium using the wet-plate process was rather more limited. For example, in a letter submitted in April 1857 to Antonio Panizzi, who had replaced Ellis as principal librarian, Fenton sought to point out some of the practical constraints of the still nascent medium by advising that if any photography was to be undertaken that year it should 'be done as soon as possible, the light at this time [mid-Spring] being better than at any other period of the year'.[43] Certain objects, Fenton suggested, were not only desirable subjects for photography but were well-situated or could easily be re-positioned for the purpose: 'The Assyrian sculptures which are in a good light, on the east side of the galleries; some of the [slabs] which are waiting for the completion of the new room, might when they are in course of removal, be halted for a short time in a convenient light for the purpose'.[44] Such considerations were particularly important in the museum context, where the objects to be photographed were either unique or rare and often extremely fragile or too unwieldy to relocate. Fenton's practice of photographing in a 'glass room on the leads' of the roof – where all available light could be harnessed to minimise the camera's exposure time – meant that the procedure of photographing elaborately articulated skeletons and other large or intricate specimens was anything

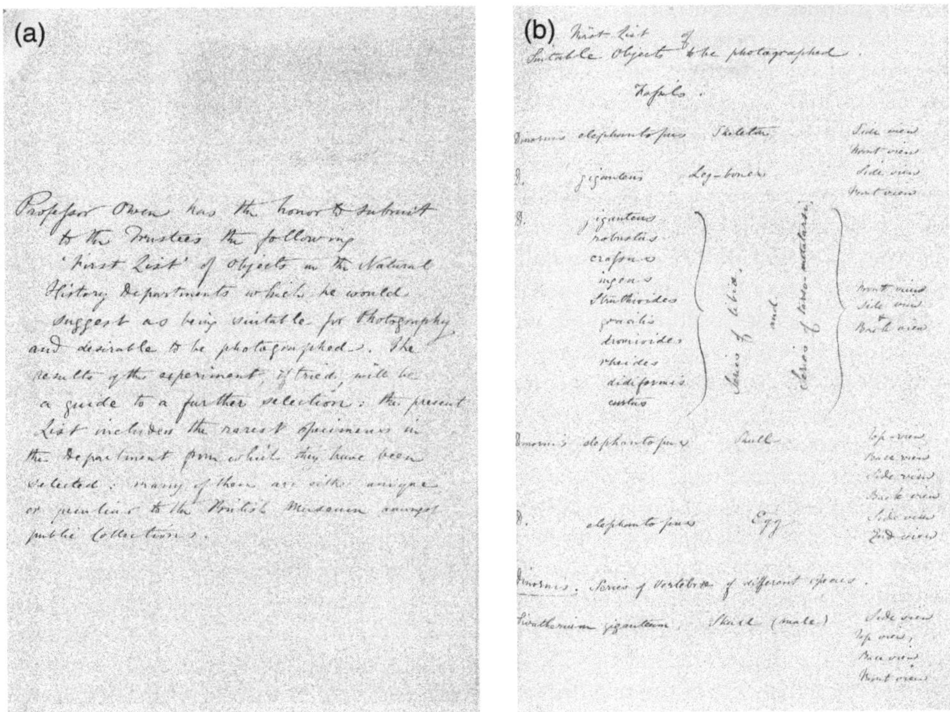
(a)

Professor Owen has the honor to submit to the Trustees the following 'First List' of Objects in the Natural History Departments which he would suggest as being suitable for Photography and desirable to be photographed. The results of the experiment, if tried, will be a guide to a further selection: the present List includes the rarest specimens in the Department from which they have been selected: many of them are either unique or peculiar to the British Museum amongst public collections.

(b) First List of Suitable Objects to be photographed.

Fossils.

Figure 1.3 Richard Owen, 'List of Suitable Objects to Be Photographed', British Museum: Officers Reports, 23 April 1857, f.367, pp. 1–2 of ten pages. Courtesy of the Trustees of the British Museum.

but straightforward. Certainly this process did not always offer the efficiencies or convenience that the medium might have suggested to anyone unfamiliar with the finer points of the photographer's practice.[45] This is not to say, however, that skeletons or specimens were necessarily any more fragile or cumbersome than antiquities or works of art. Rather, trustees and officers may have been more attuned to the value of these cultural artefacts and treasures, which they probably considered more worthy than natural history objects of the effort involved. Yet, in neither of these realms would the process be a quick or casual enterprise; instead, it would be one that necessitated considerable deliberation in view of the relative importance of the objects selected to be photographed, the possibility of any detrimental impact on them and, ultimately, the production and circulation of the photographs.

Owen's initial recommendation for over 300 photographs of the natural history collections would certainly have been challenging. Yet, in no way could his choice have been construed as an arbitrary assortment of specimens. At that time, other keepers considered photography only in terms of their current projects – in the case of Brown, his imminent botanical exhibition; and for Gray, a forthcoming publication.[46] In contrast to his colleagues, however, Owen was already contemplating the ways in which photography could generate greater patronage for the natural history collections and promote his own career. As his memorandum noted, 'the present list includes the rarest specimens in the Department from which they have been selected: many of them are either unique, or peculiar to the British Museum amongst public collections.'[47] His desiderata, while necessarily selective, were nevertheless quite thorough in respect to those parts of the palaeontological and geological collections that corresponded with the professor's own research areas and the specialist interests of influential figures such as the museum's two scientific trustees: Sir Roderick Murchison, geologist and president of the Royal Geographical Society; and Sir Philip Egerton, parliamentarian and palaeontologist. For instance, the list included the fossilised fish *Holoptychius nobiliformus*, a particular research interest of Egerton's (Figure 1.4).[48] Similarly, a group of Selenite crystals that had been presented to the museum by Prince Albert, but which otherwise may have been thought unremarkable, was also considered a priority for photography.[49] Most notable, though, was the prominence given to certain specimens pertaining to Owen's own professional triumphs in natural history, especially those relating to the identification and reconstruction of extinct and visually spectacular animals such as the giant *Dinornis* and *Megatherium*.

The fossil remains of the *Dinornis*, or moa, were to be the subject of more than 74 separate photographic negative plates, with Owen stipulating that each part was to be photographed from multiple perspectives. No other submissions came anywhere near Owen's level of detail, with other officers leaving issues of viewpoint, composition and other formal elements to the expertise of the photographer. The number of prints solicited was also unparalleled and somewhat audacious given the fact that the museum's photography enterprise was still at a very early stage. Indeed, a month after his first recommendation in April 1857, Owen appended his report to include a further 500 prints from just one of the negatives, a single side view of the *Dinornis elephantopus*. For this additional work, he had already received a costing from Fenton of £2.2.0 for photographing the negative plate of each view and 1/6 for each positive copy, with the expectation of those copies being sold for three shillings each in anticipation of a profit of 100 per cent for each print produced. In spite of this, however, there is no

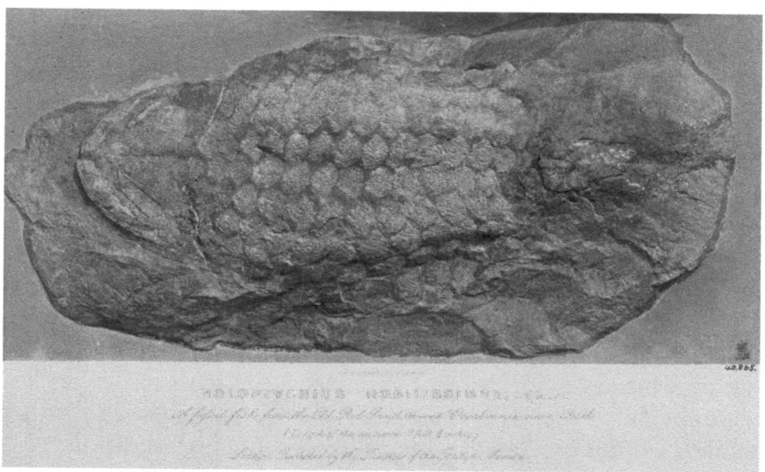

Figure 1.4 Roger Fenton, *Fossil fish, Holoptychius nobiliformus*, 1858, salted paper print. Collection: Victoria and Albert Museum, London.

evidence that any plan for promoting or commercially distributing the photographs was ever considered.[50]

The sheer ambition and audaciousness of Owen's request was questioned almost immediately by Panizzi, to whom the trustees had assigned overall control of the photography program. Consequently, Owen was compelled to justify his selection of specimens and extensive list of photographic views and, in particular, his request for 500 prints of just a single side view of an extinct wingless bird, the giant New Zealand moa. He explained his reasons for selecting the skeleton of the *Dinornis elephantopus*: first, it was unique; secondly, it was of extraordinary proportions, 'indicating habits of life as peculiar in the extinct bird'; and, thirdly, it was structurally – and thus visually – complex.[51] Owen initially resisted providing any further explanation for the large number of prints requested, except to add a rejoinder the following day that, to some degree, he had been influenced by the scale of prices indicated in the estimate of costs provided by Fenton, and the result of his calculations was that 'the lowest price of the "Positives", with a due consideration of the chance of sale, is that attached to the number 500'.[52] Thus, he anticipated that the considerable expense involved in producing numerous photographic plates of individual fossil bones – which would hold little appeal for a general audience – could be recouped by applying economies of scale to more marketable images. He conceded, however, that the museum's photographic enterprise, and thus his own proposal as a part of that, was very much 'the beginning of an experiment, and that the conditions of success [could] not be clearly discerned. Some risk must attach to first steps, and the result of those [would] be the best guide for future proceedings'.[53]

A virtual inventory of specimens: circulating new knowledge and filling in the gaps

While the British Museum's first venture into photography was undoubtedly experimental, the types of uses for the photographs proposed by individual officers, including Richard Owen, were not without precedence. The British Museum's initiative coincided with the formation of early amateur photographic associations during the 1850s. The interests of these associations often complemented activities in other areas and groups, such as the Photographic Exchange Club, which evolved out of the Antiquarian Photographic Club.[54] The *Journal of the Photographic Society* enthusiastically promoted the medium for the purpose of exchange, suggesting that the large numbers of prints that could be obtained from negatives and circulated among members would help enlarge or even complete individual collections:

> Most valuable results would … be reaped from a system of exchange of positives, by applying the art to Architectural and Archaeological subjects; they would derive from it advantages similar to those which naturalists now experience from the interchange of local natural productions, and it is scarcely too much to say, that when once adopted *exchange of photographs will become as indispensable to antiquaries as the exchange of plants is to botanists.*[55]

The subjects of these circulated photographs reflected a wide range of member interests. Moreover, the shared fascination for both natural history and antiquarian subjects provided a common link between many of the members in the Photographic

Exchange Club.[56] Those members who were occupied in the natural sciences, such as John Dillwyn Llewelyn and George Shadbolt, envisaged the club as a forum through which they could disseminate their scientific observations and complement their specimen collections, or even merely demonstrate their dedication to the field by portraying themselves engaged in their scientific investigations.[57]

For others, the subjects depicted in the exchanged photographs were of lesser consequence than their interest in experimenting with the medium itself and collecting examples illustrating new or enhanced processes and applications from other photographers. 'Like the naturalists', Grace Seiberling observes, 'photographers referred to the objects they collected as "specimens", and for both groups the amassing of collections was an activity of interest in itself.'[58] The guidelines for exchanges were reasonably clearly set out and generally adhered to by members. While the photographic clubs were informally structured and included both amateur and professional photographers, there was little ambiguity about the members' individual interests, their commitment to photography, and the purposes and expectations of their exchanges. The impetus for these associations came out of the simple desire of likeminded individuals to acquire, or at least have access to, new knowledge and techniques as well as examples of photographs. The clubs were characterised by an exchange economy with which naturalists were already very familiar. Indeed, the basis for naturalists' and photographers' transactions overlapped in many ways since both pursuits had an aesthetic side but also involved conveying one's values and sophistication in addition to creating and disseminating new knowledge.[59] As a founding member of the Photographic Society, Fenton probably envisaged that at least some of his commissions for the British Museum would be used for similar exchange purposes and conducted under comparable terms, which if not openly stated were implicitly agreed between gentleman scholars. For the antiquarians, at least, this was almost certainly the case. While discoveries and displays of works from past civilisations, such as the spectacular Assyrian collections, attracted considerable popular interest, the field of archaeology remained an insistently elite and specialised area of endeavour.[60] In contrast, the natural sciences were becoming increasingly pluralistic and dynamic throughout the Victorian age, and those who participated in and contributed to the study of natural history represented a much more socially heterogeneous community compared to their antiquarian colleagues.[61] The use of photography to cultivate and communicate new knowledge about natural history was therefore a relatively complex and speculative proposition. This was not only due to naturalists' hesitancy or lack of know-how about the emergent technology but also to the many different individuals across the empire and the globe who contributed or interpreted images and other information. The ways in which such a range of stakeholders approached the photographic medium was bound to be correspondingly diverse, and thus less consistent and more contentious than in more exclusive knowledge sectors, including archaeology.

The degree of detail in Owen's request to the British Museum trustees reflected his painstaking approach to producing and commissioning natural history illustrations. It was also indicative of his skill in verbally and visually dissecting specimens in his communications with other naturalists and his ability to rhetorically construct previously unknown animals in his lectures for the benefit of both scholarly and general audiences. The latter, in particular, demonstrated his virtuosity and showmanship as a functionalist restorer of extinct fauna.[62] As Adrian Desmond establishes, Owen did not merely recognise earlier species, 'giving taxonomic expression to the "objective" fact; more properly, he designed them – invented them, in a sense'.[63] Owen's approach underlined a growing tension in modern museums between portraying natural history as a series of ongoing processes as opposed to an accumulation of objects to be comprehended only in their existing form.[64] It also highlighted distinctions between different kinds of museum audiences. Consequently, while it is apparent that the items to be photographed were selected with deliberate objectives in mind, the prospective audience for these images was not predetermined. Owen's appended request indicates that he had a popular and commercial purpose in mind for at least some of the photographs. For example, while at least one large extant print presents a comparative view of two *Dinornis* species – *Dinornis elephantopus* and *Dinornis giganteus* – another spectacular image of the visually impressive *Dinornis elephantopus* apparently was to be disseminated to a general audience of natural history consumers (Plate 3, Figure 1.5). The main purpose of this latter, an elegantly composed and richly toned salted-paper print produced by Fenton, was evidently not to document the skeleton simply as part of an inventory but rather to portray the *Dinornis* primarily as an object of cultural value and aesthetic interest and to present it for the visual pleasure of the ordinary viewer.[65] Nonetheless, a rational framework underpins the actual image. The clarity of the photograph and the orchestration of the scene draw the viewer's eye to the studio trappings, which have almost as much visual power and imaginative possibilities as the moa itself. Further, while the impressive skeleton appears to be monumentally self-contained, the photographer has ensured a precise alignment between the skeleton's different anatomical sections and the folds and fall of the drapery, which has the effect of visually dividing the skeleton in a way that reminds the viewer that the various segments of fossil bone were previously separated. The schema also highlights the incongruity of the different parts of this strange bird, with its tiny skull, substantial hips and long solid thighs providing the jigsaw pieces in this palaeontological puzzle. The discoveries of the different fossilised parts, along with Owen's original reconstruction and naming of the giant bird, had been widely publicised and would have been familiar to many in the general reading public. The recipients of photographs on Owen's initial list, however, would almost certainly have been confined to a coterie of specialists and amateur savants with a particular knowledge of fossil osteology.

This more systematic use of photography in the museum reflects the uses proposed three years earlier, when Fenton began photographing at the British Museum; and, in particular, when Hawkins had requested that multiple photographic impressions be taken of each of the Assyrian tablets and of any new acquisitions.[66] By regarding those earlier photographs as near perfect facsimiles of the original Assyrian tablets, Hawkins had anticipated that antiquarian scholars' access to the museum via these photographs would be greatly enhanced. The British Museum would similarly use photographs of ordinary antiquities to cultivate knowledge, expertise and connections with a range of other museums. Thus, while photography had once been considered as simply another mode of visual notation – an aid for the complex skills of observing and drawing – from the mid-1850s onwards, objects and photographic images were starting to be considered equivalent, which contributed to a rhetoric of substitution that served to increase the authority of photographs in different knowledge areas.[67] In accordance with this, photographs of natural history specimens gradually began to stand in for the objects they represented and were likewise collected, scrutinised and identified not only in the context of scientific meetings but also, for some naturalists, increasingly as part of their day-to-day practice.[68] Yet, like any solitary specimen, a single photograph

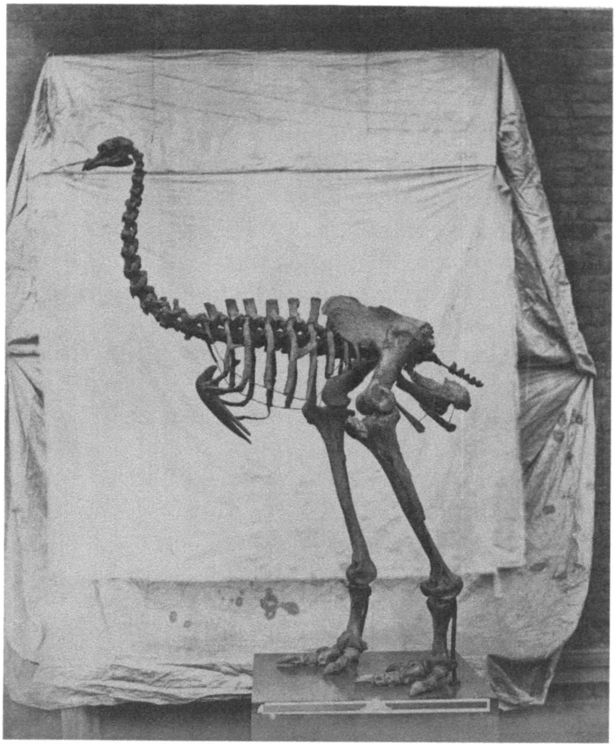

Figure 1.5 Roger Fenton, *Elephantine Moa (Dinornis elephantopus), an Extinct Wingless Bird, in the Gallery of Fossils, British Museum*, c.1857, salted paper print. The J. Paul Getty Museum, Los Angeles (84.XP.452.3). Digital image courtesy of the Getty's Open Content Program.

tended to act more like a cipher and could assume several meanings or serve a range of purposes for the various producers and audiences of Victorian natural history. As Stephen Asma notes, 'the significance of the collected object does not inhere in the specimen itself, but is socially and theoretically constructed'.[69] Consequently, even the rarest and most beguiling of objects from nature do not mean anything unless they can be viewed in the context of a system.

Of the officers at the British Museum, Owen and Gray were among the first to become actively engaged in negotiating different types of photographs in lieu of original specimens or other modes of illustration. A precursor for this use of photographs by naturalists can be found in the production and exchange of separately published prints of fossils, a sub-genre of natural history illustration that, as John Thackray observes, was at its height in Britain between 1815 and 1845.[70] Like the earliest photographic images that succeeded them and distinct from larger edition prints, separately published prints were multiple images intended for circulation within a small coterie of specialists or aficionados; they did not always strictly follow conventions of scientific illustration but often were a means of self-promotion within that network. The original purpose of these limited, tightly controlled antecedents to naturalists' photographs was to parade and celebrate an exceptional or recently discovered specimen, as evidenced by the inclusion of an owner's or discoverer's name as part of the title of a print. Their subject matter was not as popular, however, as illustrations depicting, for example, elaborately arranged collections of ornate shells.[71] Thus, these intermittently produced prints initially remained within the domain of discerning collectors. Nevertheless, they contributed to the nineteenth-century penchant for accumulation.[72] Other examples of this genre showed off the skill of the artist or the versatility of the printer or demonstrated innovations in visual technologies, with the intricate forms of rare fossils considered an ideal subject for these purposes. The most prescient were those using photographs as a visual source for the resultant printed images, thereby anticipating the eventual shift from separately published prints in the form of lithographs, etchings and engravings towards paper-based photographs.[73] Mostly, separately published prints of natural history subjects were produced to announce acquisitions of rare or exquisite objects and to educe information about the specimens portrayed.[74] In contrast to the published literature and public museum displays, which were the end products of accrued natural history knowledge, separately published prints played a significant role in the ongoing, behind-the-scenes process of data assimilation by naturalists which, together with letters, notebooks and drawings, formed part of their more informal exchanges and discussions with colleagues. Although there was an established protocol for showing specimens in certain orientations, the representation of individual fossils could differ between these various types of prints; for instance, some features might be either downplayed or emphasised depending on the print's purpose. Connoisseurs, for example, often minimised or omitted imperfections in prints of their specimens to convey idealised versions of their prized possessions, and sometimes they included ornamental elements or other visual conceits. On the other hand, those prints produced specifically to assist and promote the practice of natural history tended to be more faithful to the original object through careful attention to individual details and a more naturalistic presentation. The audiences for these prints were very often personally known to those producing, or at least circulating, them. Captions, or other textual information, therefore, tended to be concise or to take the form of incidental marginalia; alternatively, this information was absent so as to not influence or otherwise pre-empt the viewer's own observations about the specimen. Lengthy descriptions or other substantial information, if this was supplied, tended to be included in the accompanying correspondence between the sender and recipients.

The emergence of these informally circulated prints, and subsequently many of the photographs produced by and for naturalists, was due in part to the rise of geology as an influential field of study; the discovery of fossils and their relationship to different rock strata became a subject of vital importance.[75] The increased interest in and cachet of discovering or owning fossil remnants or entire fossilised life forms accounts for this being the predominant subject matter of this print genre. Yet, the imperfection of the geological record was seen to jeopardise the triumph of geology as a 'noble science', one that had been largely invented by the British but which raised as many questions as it answered in relation to contemporary theories of evolution. It was a puzzle that institutions and individual collectors keenly sought to address by filling in the gaps.[76] As Charles Darwin aptly described it, 'the crust of the earth with its embedded remains must not be looked at as a well-filled museum, but as a poor collection made at hazard and at rare intervals'.[77] No museum could ever hope to attain anything approaching comprehensiveness in its representation of the historical sciences. In nature also, as

Darwin pointed out, 'the blank intervals' between the successive stages of geological upheaval were of vast duration and thus reduced the prospect of geologists and palaeontologists unearthing enough fossil remains to satisfactorily represent organic evolution, especially from any single geographical region.[78] To compensate for these deficiencies in knowledge, as far as was possible, public and private collections came to be regarded as a single but distributed resource. Naturalists sought to not only acquire distinctive objects for public collections but also to replicate unusual specimens in the possession of others around the globe in order to establish and augment their own 'paper museums' as comprehensive visual and intellectual inventories.

Paper museums had a venerable history before the advent of photography, which Martin Rudwick describes with regard to Cuvier's vast collection of 'proxy specimens'.[79] In the Victorian era, however, naturalists' paper collections were augmented as photographic technology became progressively more accessible. The decline of separately published prints from the mid-1840s compared with the upsurge of photography was influenced by the expediency, informality and flexibility of the new technology. The production and controlled circulation of these limited prints –and then photographs – supplemented more conventional natural history drawings and prints and contributed to the construction of composite collections of rare and unique specimens that could never have been brought together under a single roof. Furthermore, it was a tacit acknowledgement of the distinctiveness of peripheral sites and the relational – rather than monolithic and programmatic – nature of the metropolitan 'centres of calculation'.[80]

The development of paper museums was due not only to the methodological imperatives of particular fields of inquiry, such as geology and comparative anatomy, but also to the rise of museum-based professional naturalists. In some respects, the relatively new discipline of comparative anatomy was only possible due to the institutionalisation of science and the advent of the modern museum. From the perspective of scientific inquiry, this period in general could be typified as 'museological' insofar as the analytical basis for a broad spectrum of scientific work, as John Pickstone suggests, was 'historically embedded in the emergence of specialist professional groups with command over large bodies of material'.[81] As the cynosure of a global network of correspondents, Owen, Gray and other officers at the British Museum were communicating continually with other naturalists and contributors with varying backgrounds, knowledge and expertise. Moreover, they were receiving and sending photographs – at first cautiously but then more confidently – as part of their exchanges. Gray's efforts in this area resulted in a relatively modest two-volume album of comparative illustrations. This album – which became an invaluable record of unusual and useful images of whales and dolphins dating back to the eighteenth century – comprised prints, drawings and, later, numerous photographs sent from individuals and institutions from across the world.[82] Clearly intending Gray to be the sole viewer for many of these photographs, correspondents, including Gerard Krefft, curator at the Australian Museum in New South Wales, would often add their observations and descriptions as fleeting, partial notations around or across the images, which were then cross-referenced and expanded in long accompanying letters.[83]

Likewise, during the course of his work at the museum from 1857 to 1895, Gray's colleague and successor in the zoology department, Albert Gunther, compiled volumes of prints, drawings, photographs and printed ephemera of terrestrial and marine fauna from all regions during the 1870s and 1880s (Figure 1.6).[84] Even more sophisticated, however, was Owen's paper museum of palaeontological and zoological illustrations – encompassing works dating back to 1681 and numbering more than 3,500 items by the end of his career – built up from an archive begun by William Clift, his former mentor at the Hunterian Museum.[85] Owen's collection was indicative of his own, more extensive imperial network and the immense contribution made by naturalists and artists from all backgrounds.[86] Under Clift's guidance, Owen had learnt to preserve and dissect specimens and to record and compare his observations in notes and drawings; he had also become an accomplished and prolific draughtsman.[87] Owen's extensive collection – including his own productions – covered a variety of subjects from single-celled organisms to giant fossil reptiles, birds and mammals, the most spectacular of which would become icons of the museum and allow him to promote his own remarkable achievements in reconstructing extinct species.[88] Relatively few of Owen's studies and reconstructions were initiated, progressed or even settled without the input of his numerous contacts across the empire, and his paper museum was fundamentally eclectic, pragmatic and incomplete, despite the fact that comprehensive collections remained the aspiration of Victorian museums and the key to metropolitan authority.[89]

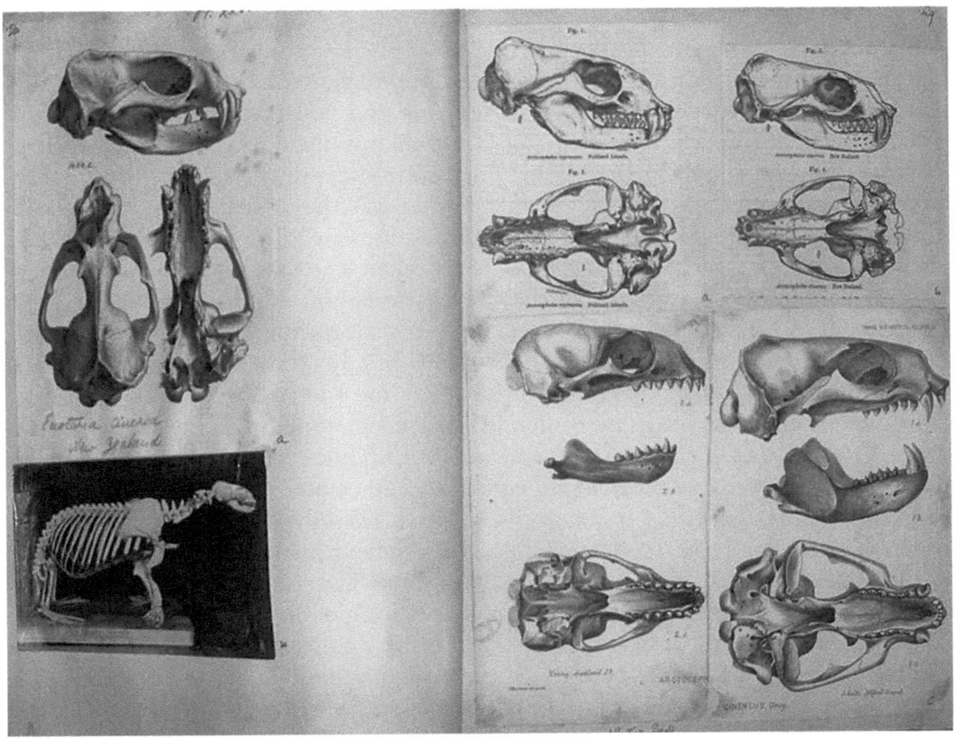

Figure 1.6 A. C. Gunther, *New Zealand*, vol. 17, from the A. C. Gunther series of albums, 1870s–80s, case 22: mss. 551–2, f.28–29 (album opening showing lithographs and drawings of various skulls and jawbones, and a photograph of skeleton). Collection: Linnean Society, London.

Many of Owen's contributors were regular correspondents who, as well as sending written observations and related information, intermittently dispatched specimens or included images in their letters in lieu of the actual objects. Tens of thousands of letters received over the course of a career was not unusual for a Victorian scientist of some standing.[90] These letters were received from other museum professionals but also from accomplished, or merely interested, amateurs who were sometimes seeking information for their own studies or who simply assumed that their particular specimens might be of interest to the British Museum.[91] One such correspondent was William Bedford, an assistant colonial surgeon in Hobart Town, Tasmania, who sent an echidna and a platypus along with other specimens that he hoped would provide Owen with useful information. A proficient naturalist himself, Bedford apologised to Owen for the delay in sending the specimens, bemoaning the duties of his hospital appointment and the lack of books that made progress in his own natural history studies difficult.[92] Although these very real constraints prevented some of the keener naturalists in the colonies from more rigorously pursuing their own work in natural history, they nonetheless managed to provide a rich and seemingly endless resource for imperial scientists such as Owen in the metropole.

As a result of these networks, countless original specimens, and casts from specimens, found their way to the British Museum. Yet, the comparative ease with which drawings and other images were exchanged meant that paper museums, including those at the British Museum, had the potential to be far more comprehensive than collections that relied on actual specimens. This applied not only to the totality of the collection but also to the separate representations of individual animals, especially those that were reconstructed from fossil remains and which, more often than not, were incomplete or imperfectly articulated. In some instances, it is likely that illustrations were produced when Owen or other museum officers only had temporary access to specimens, such as those occasions when Owen was the intermediary for colonial naturalists forwarding specimens to other destinations in Europe. For example, the German explorer, naturalist and artist Ludwig Becker sent Owen shipments of specimens from Melbourne requesting that they be forwarded on to Johann Kaup, Becker's former teacher in Darmstadt and, by then, a noted palaeontologist.[93] On other occasions, correspondents sent specimens to Owen to briefly study and document while they were negotiating with other institutions.[94] Some specimens were so treasured by their owners that they only ever intended to lend them temporarily to the British Museum. Owen also had cause to visit and view other collections by virtue of his position, most notably, the osteological collection and associated paper museum formed by his great mentor, Georges Cuvier, at the Musée d'Histoire Naturelle in Paris.[95] It is probable that, for his research purposes and to augment his proxy collection, Owen opportunistically created or commissioned images of materials identified during his travels. The paper museum also became a resource for a circle of gentlemen amateurs and other close associates of Owen, who relied on this collection to pursue their own research and, in turn, to contribute to Owen's ongoing individual and collaborative projects.[96] Generally, though, the movement of images, as with original specimens and other forms of information, continued to be in a one-way direction, from periphery to centre.

Owen's paper museum consisted mainly of drawings and prints, with comparatively few photographs. Clearly, his interest in using photography to document collections did not displace other media in any concerted way. Moreover, the British Museum trustees' early foray into photography did not seem to have much of an impact on the way collections were represented and disseminated and thus in regulating the public image of the institution itself. Yet, this overlooks the more complex and ambiguous status given to photographs – including those images that formed part of Owen's own collection – compared to other media as part of the museum's public and back-of-house activities. The influence of photographs in the museum context, Elizabeth Edwards notes, 'was not simply in the nature of the objects themselves but dependent on the rhetoric of their photographic performance', reflecting 'not only informational rhetorics', she argues, 'but also disciplinary and institutional rhetorics.'[97] Photographs showing the separate sections and articulated skeleton of *Dinornis elephantopus* – including those produced in 1858 as part of Fenton's commission for the British Museum – were especially significant for Owen's career, not only scientifically but also for his status as a museum reformer and as one of the most influential figures of the empire. A number of these photographs formed part of an extraordinary cycle of portraits over many years in which Owen posed with the moa. Presenting his 'discovery' and restoration of this giant extinct bird as the defining motif, he exemplified his comparative methodology, considerable professional achievements and unrivalled status in imperial science. His publications on the subject of the moa brought him early and wide recognition, which continued throughout his career. Starting with the cautiously titled 'Exhibition of a bone of an unknown struthious bird of large size from New Zealand', published in the *Proceedings of the Zoological Society of London* in 1839, he followed up with his detailed studies 'On Dinornis', published in the society's *Transactions* over 25 parts and concluding in 1886.[98] However, these studies seemed to progress in an arbitrary fashion, with each paper analysing one or two aspects of the osteology of the genus *Dinornis*. That is, the series did not evolve systematically or in a logical sequence; rather, the papers appeared according to the order in which the fossil remains were unearthed and found their way to the British Museum either as objects or images. For example, part 7 (published in 1858) addressed the leg and foot of *Dinornis elephantopus*; part 9 (published in 1866) described the skull, atlas and scapula of *Dinornis robustus*; and part 18 (published in 1873) interpreted the pelvis and leg bones of *Dinornis gravis*. Over time, Owen would gradually build up a composite account of a generic moa from its constituent parts. From these serialised descriptions of a complex behind-the-scenes process, Owen set out to accomplish the visual trope of depicting an animal that had hitherto never been seen.[99]

Owen's scientific monograph, *Palaeontology or, A Systematic Study of Extinct Animals and their Geographical Relations*, published in 1861, includes an illustration after one of Fenton's photographs of the *Dinornis*, giving the earliest indication of how Owen would employ photographs initially – such as combining them with other images from his paper museum – to illustrate his line of reasoning and make his conclusions more persuasive (Figures 1.7–1.8).[100] In the development of his extensive paper museum and its use in his research, Owen owed a great deal to Cuvier, whose method was essentially 'actualistic': what was known – and more importantly what could be shown – to be characteristic of living species in the present time became the basis for determining affinities in fossilised animals from the past.[101] The extensive fossil collection of Paris's Musée d'Histoire Naturelle – unrivalled in its scope when Owen visited in 1831 – used in combination with Cuvier's pioneering collection of paper-based specimens, were the basis for this methodology. Combining his own drawings with images sent to him from naturalists from all over Europe, Cuvier would create a collage of images on a single sheet, which was then reproduced as a single engraved plate that was not only visually harmonious but also highly coherent as a piece of visual rhetoric.[102] Similarly, while the historical depth of Owen's inherited collection was invaluable, the global scope of his contacts and the breadth of his paper museum were indispensable. To generate this breadth, he used photographs, drawings and printed illustrations received from a range of correspondents, as well as some of Fenton's studies that were produced during the British Museum's early foray into photography and which featured in his monograph *Palaeontology*. Later publications showed how photographs sent by correspondents could be directly translated into a uniform visual style in order to prosecute his argument. For example, Owen seamlessly incorporated into his series on the *Dinornis* two photographs of a fossilised femur of *Dromornis australis* sent by the New South Wales colonial geologist Rev. W. B. Clarke, which were probably photographed at the Australian Museum (Figures 1.9–1.10 *a* and *b*).[103] Modifying the different styles and media of the various original images used in each printed plate to make them visually coherent obscured the way Owen – and most other imperial figures – incorporated the work of others into their

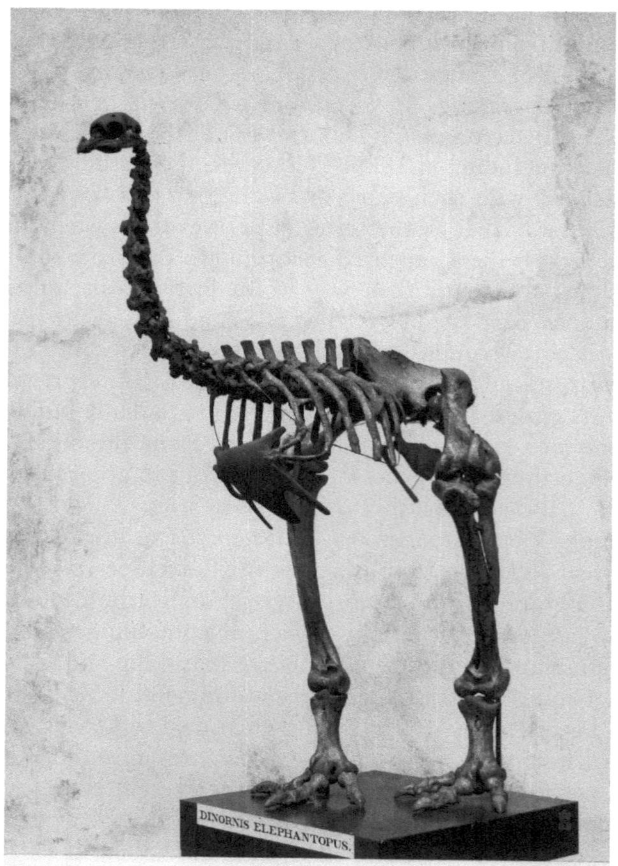

Figure 1.7 Roger Fenton, *Dinornis elephantopus*, c.1855, salted paper print. Collection: Natural History Museum, London.

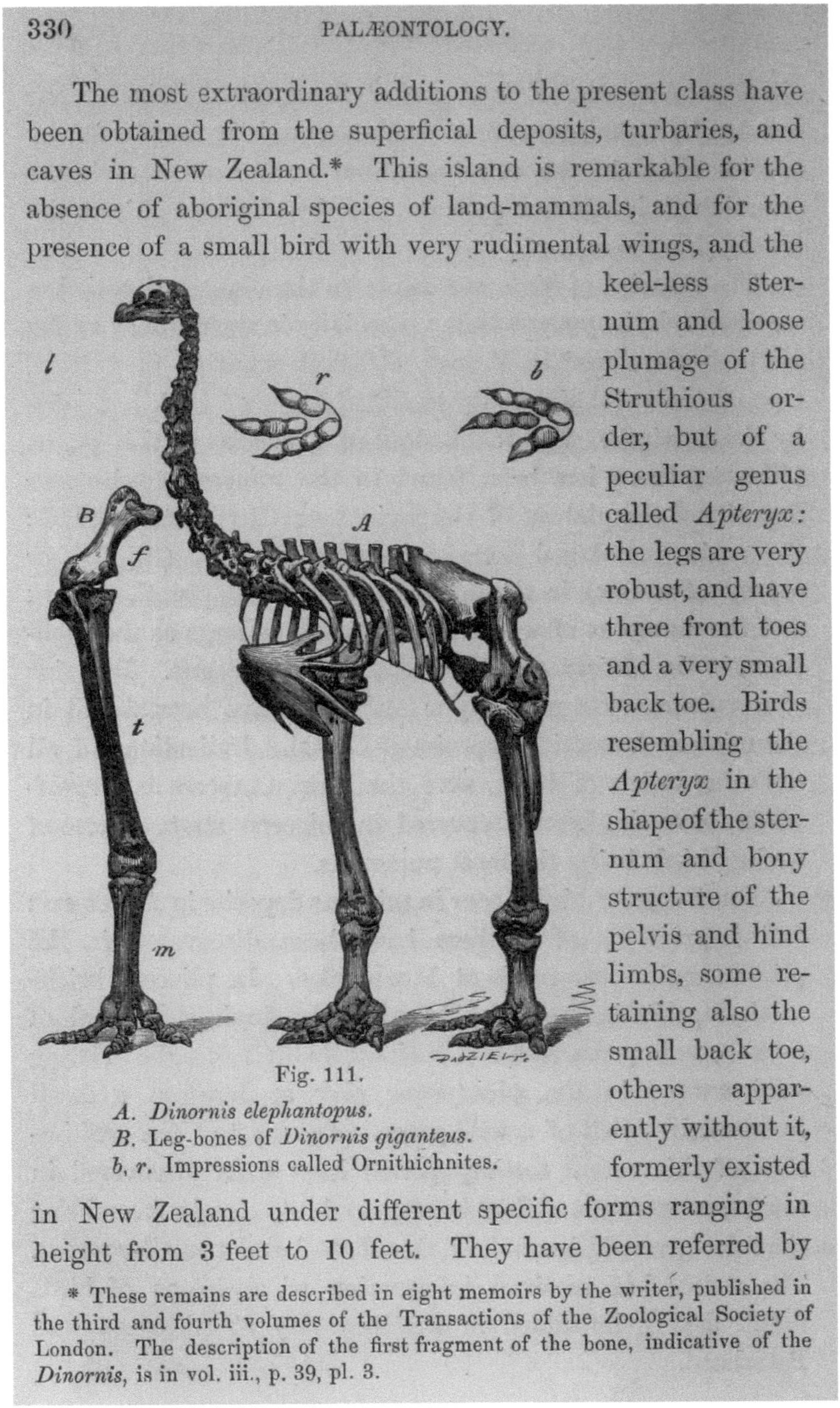

330 PALÆONTOLOGY.

The most extraordinary additions to the present class have been obtained from the superficial deposits, turbaries, and caves in New Zealand.* This island is remarkable for the absence of aboriginal species of land-mammals, and for the presence of a small bird with very rudimental wings, and the keel-less sternum and loose plumage of the Struthious order, but of a peculiar genus called *Apteryx:* the legs are very robust, and have three front toes and a very small back toe. Birds resembling the *Apteryx* in the shape of the sternum and bony structure of the pelvis and hind limbs, some retaining also the small back toe, others apparently without it, formerly existed in New Zealand under different specific forms ranging in height from 3 feet to 10 feet. They have been referred by

Fig. 111.

A. Dinornis elephantopus.
B. Leg-bones of *Dinornis giganteus.*
b, r. Impressions called Ornithichnites.

* These remains are described in eight memoirs by the writer, published in the third and fourth volumes of the Transactions of the Zoological Society of London. The description of the first fragment of the bone, indicative of the *Dinornis*, is in vol. iii., p. 39, pl. 3.

Figure 1.8 Richard Owen, *Palaeontology or, A Systematic Study of Extinct Animals and Their Geographical Relations*, A. & C. Black, Edinburgh, 1861, p. 330. Image from the Biodiversity Heritage Library. Digitized by Smithsonian Libraries.

own projects and also tacitly claimed them as their own observations.[104] Although individuals' contributions were recognised privately – for instance, in correspondence and other personal exchanges – in public demonstrations, natural history displays and the publication of scientific research in finished form, any trace of these 'invisible technicians' was usually eradicated.[105]

These elaborate paper museums, which became the mainstay of naturalists' comparative studies, comprised large numbers of images originally produced and circulated for a wider range of purposes both within and beyond the field of natural history. In their aspiration to form representative – if not comprehensive – collections, professional and amateur naturalists alike were relatively pragmatic about the visual material they acquired.[106] Comparatively few of the images in the paper museums entered into the public domain or were in fact ever intended for general dissemination as illustrations in scholarly publications or as part of lectures wherein naturalists presented their research findings. Yet, this eclectic collection of surrogate specimens was crucial to their work and, while largely unacknowledged, tended to be deployed with at least the same expediency with which they had been acquired. Occasionally, substantial paper museums were inherited by naturalists, for example, from their predecessors in museums and other institutions, as Owen had inherited from Clift. Usually, the beneficiaries continued to add to those collections, either along the same lines as their precursors or deviating so as to accommodate new research interests. As such, the range of sources and interpretations of individual images could be modified or completely altered over time and throughout the life of the collection. These variations can be attributed to increased knowledge about the specimens or to changes in the cultural, ideological or intellectual circumstances underpinning the overall collection. Furthermore, the images themselves contributed to these changes through the different

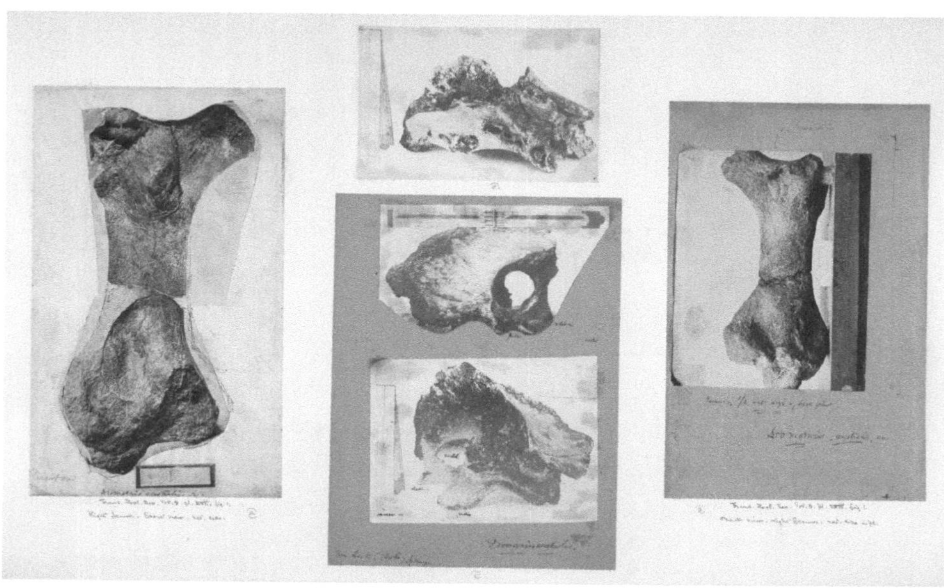

Figure 1.9 Henry Barnes (attrib.), *Dromornis australis*, (photographs received from Rev. W. B. Clarke, Sydney, 1872), Richard Owen Collection of Palaeontological and Zoological Drawings, f.506. Collection: Natural History Museum, London.

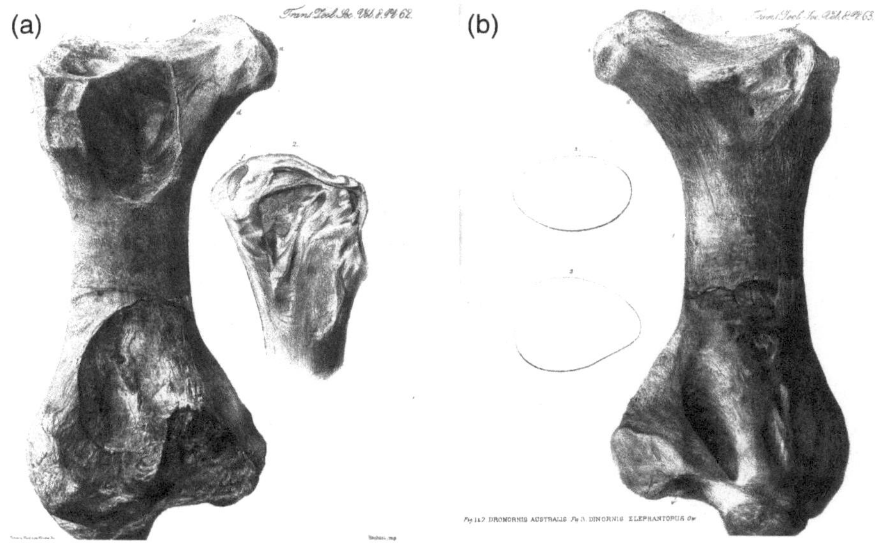

Figure 1.10 Plates LVIII and LXI in Richard Owen, 'On Dinornis (Part XIX): containing a Description of the Femur indicative of a Large Wingless Bird (*Dromornis australis* Owen) from a post-tertiary deposit in Queensland, Australia', *Transactions of the Zoological Society of London* vol. 8, no. 6, 1873. Courtesy of John Wiley & Sons.

modes of representation and new visual technologies used; thus, these images could be seen to constitute as well as demonstrate significant epistemological changes, particularly away from Enlightenment thinking and practice towards the science of the Darwinian era. For instance, during the Enlightenment period, visual representations played a crucial role in taking intellectual 'possession' of nature.[107] Victorian science, on the other hand, was increasingly concerned with the broad geographical sweep and sheer diversity of nature – and its transformations over time – and how best to represent these. Yet, in the nineteenth-century museum, the public displays were often based on obsolete classifications dating back to the time of the collection's establishment and had not been updated in line with new thinking and contemporary approaches.[108] For example, despite considerable advances in palaeontology and geology, fossil displays were still arranged according to a method that had been devised in the early part of the century. In contrast with the public galleries in the museum, where displays tended to be uniform and unchanging, the naturalists' private paper museums were characterised by their informality and flexibility.

Private operators and entrepreneurs

Any popular success enjoyed by the British Museum's early photography program was largely due to several discrete commercial initiatives, prompted initially by Roger Fenton and then pursued by several publishers and print sellers with the cooperation of individual museum keepers, including Richard Owen and William Carpenter.[109] Initially, the distribution of photographs was extremely ad hoc, consisting of the

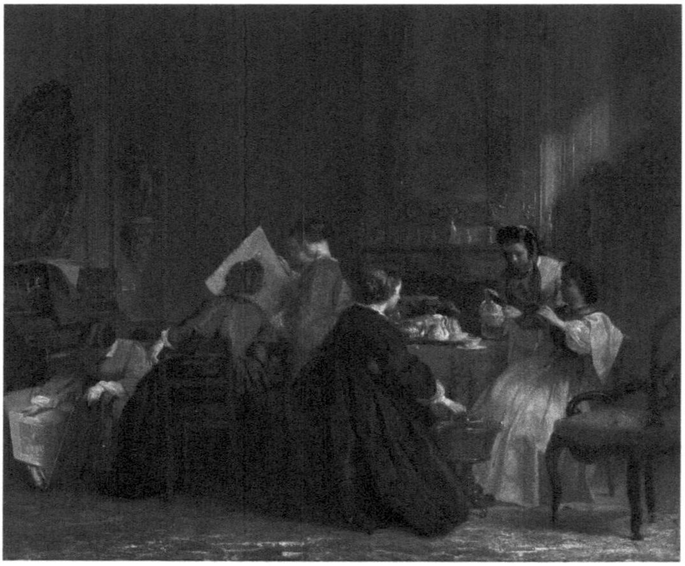

Figure 1.11 Jacob Spoel, *Group of women looking at stereoscope photographs*, c.1868, oil on panel. Courtesy of the Rijksmuseum, Amsterdam.

occasional sale of individual photographs at the museum's entrance. Fenton put a number of commercial proposals to the trustees over several years, mainly to cover losses due to the museum overextending its budget for the photography project. In August 1857, he offered to take photographic negatives of certain iconic objects at no charge and sought authorisation to produce prints from them for his own commercial purposes.[110] He envisaged that these extra copies might be marketed to the *Art Journal* since his 'photographs of the Greek marbles & of the drawings of the Old Masters would ... be considered very suitable for circulation among their subscribers' and that this was an opportunity 'to circulate as widely as possible the photographs which [had] been produced under the auspices of the Trustees of the Museum'.[111]

Another proposal was submitted in November 1857 by William Carpenter, keeper of Prints and Drawings, who enlisted the firm of Colnaghi & Co to distribute Fenton's photographs of Old Master drawings in the museum's collection at prices varying from five shillings and sixpence to four pence each.[112] As the *Art Journal* reported, these photographs were of the 'highest class and of the largest size and in every possible variety', and their marketing heralded 'a new era in art'.[113] By far the most radical decision – and galling for Fenton – was an agreement with Henry Cole to undertake large-scale production from all the British Museum glass-plate negatives, without the photographer's imprimatur, and to sell them through the South Kensington Museum. This was part of a proposal put by the Department of Science and Art in 1858 aimed at centralising all metropolitan museum photography.[114] Yet, with the recent expansion in mass-publishing – due to more efficient print technology, a burgeoning middle class and a population that had doubled in the first 50 years of the century and which was set to double again by the turn of the century – the potential market was far greater than these proposals anticipated, extending beyond England's shores to foreign and colonial audiences. The possibilities for wide distribution were only hampered by practical constraints, such as the limited numbers of photographs that could be included in publications as this was before the advent of photomechanical reproduction. Despite these technological limitations, however, the British Museum found a large general audience though one of the most pervasive and fascinating of visual technologies: the stereoscope.

From even the first issues of his *Stereoscopic Magazine*, Lovell Reeve, the prominent London publisher and naturalist, had planned to include representations of important objects and displays from the British Museum.[115] This publication was the earliest magazine to be illustrated entirely with stereo views, and it was first advertised in the *Athenaeum* in June 1858.[116] In this announcement, as well as in its subtitle, Reeve promised that the *Stereoscopic Magazine* would offer 'A Gallery of Landscape Scenery, Architecture, Antiquities and Natural History'.[117] The allusion to a gallery underlined the unmistakably visual nature of the publication as well as the sequential presentation of photographs in each issue as successive instalments. The fundamental purpose of the magazine was for subscribers to accumulate the stereo photographs and assemble their own private collections of otherwise unobtainable or inaccessible objects and views. In this respect, the *Stereoscopic Magazine* connected the experts' world of rarefied, abstruse knowledge – which public museums still represented for most Victorians – with a broader social and mercantile world such as the commercial galleries that were beginning to emerge in the mid-Victorian period, and it brought new and diverse audiences, including those people either disinclined or unable to visit a museum in person.

The publication was rather like a cabinet of curiosities in its mix of subjects from natural and human history and took its inspiration from the eclectic collections of the British Museum, which came to feature prominently across its pages. With each monthly edition containing three original stereo views, images of a variety of rare and exotic objects in museums and other public sites could be acquired for the cost of two shillings and sixpence per issue. Following the appearance in several early issues of interiors in key natural history institutions, including the Museum of Practical Geology, the Succulent House at Kew and the Osteological Museum of the Royal College of Surgeons, photographs of the British Museum collections began to appear from July 1859.[118] The museum's first foray into the magazine, published in July 1859, was dedicated to views of the museum's Temple collection of Egyptian antiquities. In subsequent issues, views of the museum's antiquities and galleries of natural history became a familiar feature in the magazine and almost entirely supplanted representations of other public collections. The year-long delay before the British Museum collections' appearance in the magazine was due to a series of unsuccessful applications by Reeve to the museum trustees requesting permission for Fenton to access the collections and use the photographic room to produce the negatives Reeve desired for his publication.[119] This was symptomatic of the trustees' ongoing uncertainty about their own photographic program and, at the same time, their disinclination to allow private operators and entrepreneurs to access the collections for commercial purposes, as the trustees would be unable to regulate how the collections and the institution were portrayed.

Tapping into the burgeoning market for stereographs, Reeve's publication was likely to do well. Robert Hunt observed at the time that the stereoscope could be a subject for discussion within philosophical societies, a tool of learning or rational amusement in any drawing room and a magical entertainment in children's nurseries (Figure 1.11).[120] Yet, stereoscopy was still controversial, and it seemed to have been its very ubiquitousness that led commentators to deride the medium and disregard the actual effect of its optical capabilities and, thus, its imaginative potential.[121] Reeve himself admitted that it was impossible to disguise the fact that, despite the progress and triumphs of photography, when 'considered in its relationship to the stereoscope, [the medium] had not been generally applied to the best uses'.

> Our drawing-room tables are strewed ... with stereographs, but they are generally expensive, with subjects being often vulgar, or, at least, unmeaning, whilst those of more real interest have not been accompanied with sufficiently intelligible descriptions.[122]

Two factors would be crucial to the success of the *Stereoscopic Magazine*: the identification of suitable subjects and their mode of presentation. Most importantly, the subjects were chosen to optimize the stereoscopic experience. Usually architecture was considered pre-eminent for this purpose, followed by three-dimensional works of art such as statuary. Because of his own interests, Reeve was also keen to have stereos of natural history displays and specimens, which he guaranteed would be both 'curious and beautiful as pictures, and valuable as scientific illustrations'.[123] He also advised that 'Eminent men in the several departments of Art, Science, and Literature' would be commissioned to write scholarly descriptions to accompany each stereo-pair and their professional judgment called on for the selection of subjects.[124] Reeve believed that critics' growing disdain for the medium, due to poor presentation and frivolous subject matter by other publishers, meant that the value of stereoscopic photographs for serious study had been entirely underestimated.

A portable stereoscope, adapted for viewing stereoscopic illustrations in books, was to be sold with the magazine.[125] Alternatively, they could be viewed with an ordinary instrument that would still provide an 'above the average' experience.[126] Both the stereoscopic apparatus and the photographic prints would be supplied by London's most reputable firms: stereoscopes manufactured by Negretti & Zambra, and photographs printed by the society photographer Arthur Melhuish under the supervision of astronomer and meteorologist James Glaisher.[127] Reeve's stereographs were thus designed to provide modern audiences with tangible experiences of sites and views across Britain, and, increasingly, its colonies.[128] To these nineteenth-century viewers, the union of photography and stereography brought a new form of realist representation that opened up a range of possibilities for combining the factual, the didactic and the experiential. Reeve advocated the use of the medium not merely as a passive reflection of the external world but also to stimulate a more profound engagement with the processes and products of exploration and science.

Reeve's natural history stereo views were perfectly suited to the materialism and acquisitive inclinations of the nineteenth century, due to their commercial appeal and affordability and the object-filled galleries and extravagant displays they frequently portrayed. Natural history studies, in particular, through their attention to intricate and contrasting natural forms, gave the viewer a heightened sense of the spatial character and tangibility of these often unfamiliar subjects. Formal pictorial conventions, which photographers had been previously keen to adopt in order to legitimise their work in artistic circles, became less imperative than achieving this sense of tangibility.

In accordance with this, photographers and publishers envisaged the effect of their stereographic images primarily in terms of the medium's capacity to evoke a sense of depth.[129] Yet the technical aspects were challenging, as the spatial illusion created by the stereoscope was inconsistent. It depended on the choice of subject and also the way the camera was positioned to create a point of view that would maximise the impression of three-dimensionality. This effect relied on prominent objects or forms appearing to push out into the foreground or middle area of the composition. The more of these visual cues in the image, the more the optical effect was amplified.[130] When there were insufficient objects, the simulation of three-dimensional space was less satisfactory. Correspondingly, the impression of tactility for the viewer relied on the physical characteristics and intricacy of the objects portrayed. The illusion created by the stereoscope was most compelling, therefore, when it incorporated both spatial and tactual effects. The most effective experience of the stereoscopic image for the Victorian viewer derived from chaotic spaces crammed with architectural features, furniture and elaborate objects, such as those that characterised contemporary interiors, from domestic drawing rooms to public museums.[131] For middle-class consumers, anything less engaging to the senses would have been akin to 'ocular deprivation'.[132]

The first view of the British Museum's natural history collections appeared in the *Stereoscopic Magazine* in August 1859.[133] *The Megatherium, British Museum* exemplified Reeve's approach to the medium and effectively demonstrated his stated rationale for the publication. The image showed the articulated skeleton of the Megatherium, not as a discrete, solitary object to be viewed in every detail, but within the cluttered Gallery of Fossils with which Victorian audiences would become increasingly familiar (Figure 1.12). When examined through the stereoscope, cases of bones surrounding the Megatherium not only protrude forward but also stretch back into the gallery. This leads the eye into the depth of the space where the smooth ivory tusks of a Mastodon push towards the viewer. Surrounded by cabinets crowded with countless

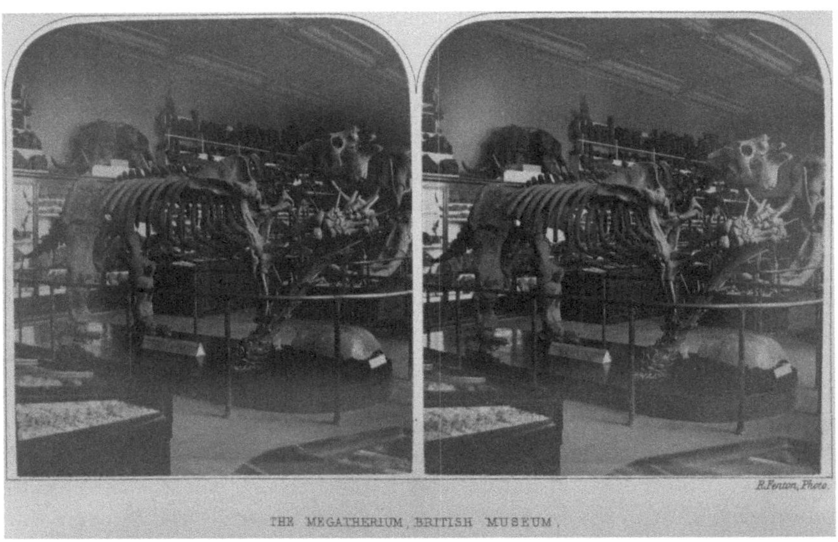

Figure 1.12 Roger Fenton, *The Megatherium, British Museum* from *The Stereoscopic Magazine*, vol. 1, no. 14, 1859. Rare Books Collection, State Library Victoria.

fossils, the viewer has the opportunity to experience the claustrophobic effect of the British Museum's natural history galleries as a virtual encounter. Taken by Fenton, the stereograph reprises one of the photographer's studies of the articulated *Megatherium americanum* originally requested by Richard Owen in 1857.[134]

Owen was, in fact, one of the experts enlisted to provide recommendations for subjects and accompanying text for the *Stereoscopic Magazine*, and he had almost certainly facilitated access to the collections for Reeve and Fenton for the purpose of producing stereographs. He also was more likely to be open to the possibilities of the medium than either the trustees or many of his colleagues at the museum.[135] His vision for the role of photography in representing and promoting the British Museum collections largely corresponded with Reeve's thinking; their differences on this issue would have been more a matter of degree than of kind. For his own part, Reeve was also well established in natural history circles – accomplished in the study of shells, fêted for the excellence of his own conchological collections and noted as a successful publisher and dealer of natural history.[136] His published work was typically intended for collectors and general audiences as well as discerning scientists.[137] His challenge, therefore, was to appeal to both of these audiences. By combining the optical technology of stereoscopy with descriptive, scholarly commentary, he could use his expertise to enhance the way natural history was presented in his new magazine so as to broaden knowledge and interest in the subject in a way that was neither soberly didactic nor merely entertaining but a satisfying merger of both. In other words, he adopted a heuristic approach to cognitive understanding by way of visual perception – a process for which, Reeve reasoned, the stereoscope was better suited than any other visual medium.

The desired effect of the stereoscope required viewers to conjure up the illusion of depth and three-dimensionality through their own optical perception of the two slightly different views. Thus, each individual had the autonomy to generate their own unique permutation of the combined images. Innovations in the stereoscopic apparatus introduced the facility for each viewer to adjust the focus to suit their own optical requirements, enabling an even more personalised viewing experience. Stereoscopes also began to incorporate blinkers that excluded all peripheral visual stimuli. This development emphasised the exclusiveness of the viewing experience since, for the time in which the viewer was engaged with the stereoscope, they were essentially cut off from any other external visual experiences, even those within their own domestic domain. This had the effect not so much of re-presenting the world within one's own domestic realm but of virtually transporting the viewer to another place and time. Some viewers, however, could not assimilate the stereo images to create the effect of depth and three-dimensionality; therefore, viewing the British Museum collections in the *Stereoscopic Magazine* was a highly variable and particularised experience.

A key component of Reeve's approach was the commentary supplied by the relevant experts, which formed the major part of the magazine's written content. Its purpose was to provide facts about the objects portrayed and also to guide the viewer as to how best to examine and fully appreciate their physical qualities in order to get the most out of the stereoscopic experience. This was not always the case, though. Owen's text for the *Megatherium* stereo, for instance, only makes passing references to the actual objects encountered in this view.[138] Yet Owen clearly understood that the performance of the stereoscope was reliant on the visual primacy of the objects closest to the viewer – as if almost within touch. Subsequently, he attempted to engage the viewer with the immediacy of the specimens portrayed and with the processes of viewing and picture making. In his description of the Fossil Gallery, published in December 1860, he draws attention to the location of Fenton's camera which, he points out, has the effect of bringing very conspicuously into the foreground a 'well articulated skeleton of the extinct Gigantic Irish Deer (*Megaceros hibernicus*)' (Figure 1.13).[139] The stereoscopic effect transforms the skeleton into an intricately articulated three-dimensional object; its substantial, imposing form dominates the space and must be negotiated in order for the viewer to move in any other direction through the gallery. The constraints of photographing in the long galleries of the British Museum, however, and the desire to show the magnitude of the natural history collections meant that this immersive effect was not always successfully realised. The extended view in this image, down onto and across rows of symmetrical display cabinets, produces a less than absorbing experience when seen through the stereoscope. Although not particularly alluring or useful for the viewer in terms of the tactual characteristics of the objects on display, this perspective serves to situate the viewer within the gallery and demonstrates how the stereoscope conveyed the orientational aspect of the tactile by creating a heightened awareness of just where and how the body of the viewer is positioned within the physical space.

Reeve's presentation methods and his choice of subjects were generally endorsed by the press. *The Leader* noted that 'amusement is not the only object of this serial; it aims at a loftier purpose – that of imparting scientific instruction; and, judging from the execution of the whole work, it is not unlikely to find a large audience willing to give it a hearty welcome monthly'.[140] The *Literary Gazette* reviewer was slightly more circumspect about the inclusion of scientific subjects and ventured a suggestion that 'if the editor wishes his work to find extended favour, let him take care that these are selected so as to be of general and not merely of special interest'.[141] Moreover, as the *Art-Journal* declared in 1858, while the 'general idea of this "Stereoscopic Magazine" may … be a sound one, the ultimate success of the Magazine itself, however, must mainly depend upon the subjects of the stereoscopic views which it will be

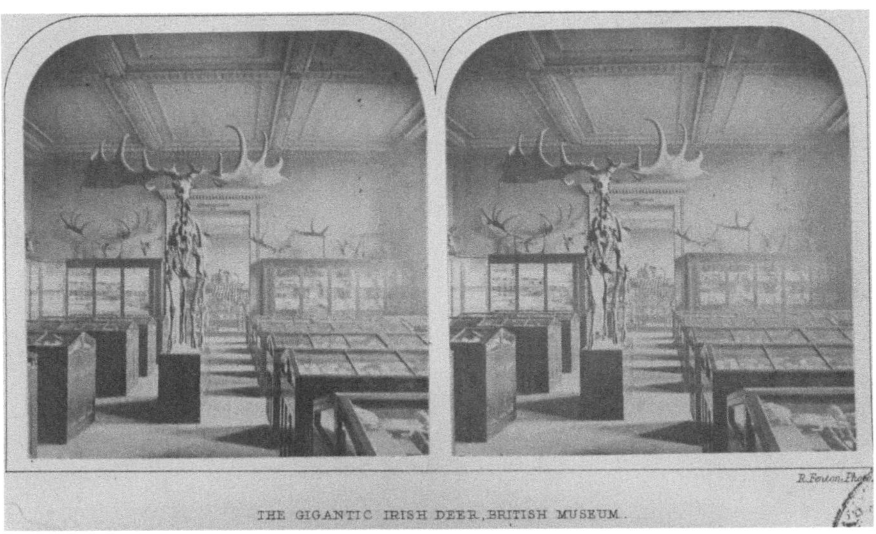

Figure 1.13 Roger Fenton, *The Gigantic Irish Deer, British Museum* from *The Stereoscopic Magazine*, vol. 2, no. 17, 1859. Rare Books Collection, State Library Victoria.

found to contain, upon their merit as photographs, and also upon the character of the letter-press that is to be associated with them'.[142] In particular, the *Art-Journal* was concerned that some of the written sections were incompatible with the magazine's format as they did not directly address the subjects represented in the stereographs. This would diminish the long-term usefulness of the publication as a printed museum of desirable objects to be collected and augmented over time:

> The descriptive notices are much too long and much too ponderous. If they are to be read, they must be brief and attractive; they must be 'descriptive notices', and not historical or biographical essays. Such articles as those that appear in this first part may be valuable, but they are out of place; and they will only ensure the views being cut out for the stereoscope, while the letterpress is thrown aside unnoticed. If the work is to be kept together, its literary department must be recast.[143]

More engaging for the viewer – and probably a more useful heuristic in Reeve's estimation – was the description accompanying Fenton's view of the Reptile Room, most likely contributed by Albert Gunther (Figure 1.14). The writer meets the viewer at the threshold of the gallery and recounts the path the photographer has taken 'in order to get a display of objects of … conspicuous … definition in the foreground'.[144] This accords with the stereo view, which locates the viewer at one end of the long room in preparation for inspecting the collection. Then, carefully guiding the viewer through the space, the writer describes the contents in sequence. Most striking is the way the text contrasts the tactility of different natural objects, such as the description of the rattlesnake's tail with its 'row of horny joints fitting loosely one into another'.[145] Likewise, the writer notes, that 'the table-cases … so well seen in the stereograph' will enable the viewer to appreciate 'the Echini … armed with long spines; the Sea-Pancakes, so depressed that there scarcely seems to be any room for internal organs;

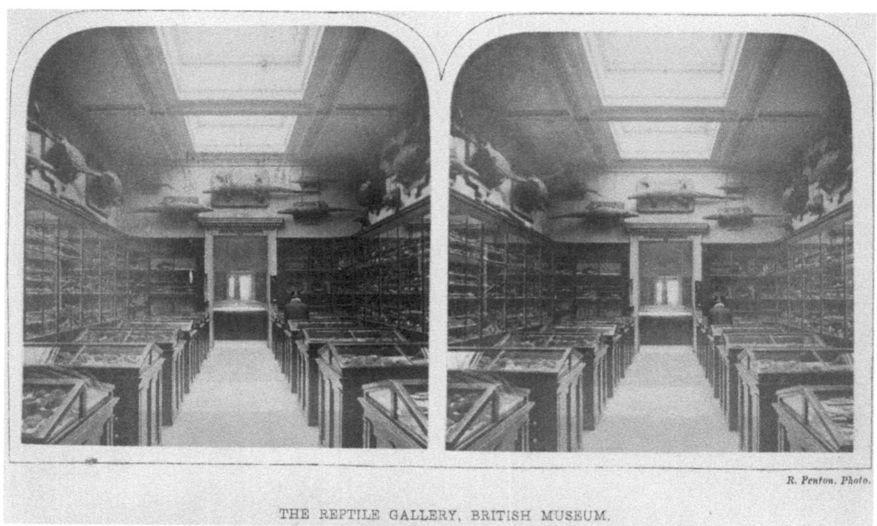

Figure 1.14 Roger Fenton, *The Reptile Room, British Museum* from *The Stereoscopic Magazine*, vol. 2, no. 20, 1860. Rare Books Collection, State Library Victoria.

and the curious Gorgon's Head, with its innumerable branches'.[146] Inspired by the sense of physical immersion created by the stereoscope, the writer attempts to direct the viewer's eyes – and body – through a space of rendered perspectives and to evoke the sensation of touch using the visual cues provided. Unlike the view of the Fossil Gallery, here the camera has been positioned lower so that the viewer can look right through the glass-sided display cases. While some objects inside are too small in the photograph to identify from Gunther's description, the cabinets are evidently filled with natural curiosities. When seen through the stereo viewer, their tactual characteristics become more pronounced, and they are clearly perceived as objects that one could pick up and feel their weight and surface textures. This evocation of volume and tactility is reinforced through the rotund forms and scaly hides of a collection of crocodiles, mounted around the top of the walls – with their harpoon-like tails and rows of sharp ridges along their backs that jab out into the gallery space.

The gratification derived from stereo views due to the illusion of touch was immensely attractive to Victorian audiences and helped define the *Stereoscopic Magazine* as a commodity contrived for both visual consumption and tactual delight. It was perhaps most effective, therefore, when it combined the study of natural history with the genre of still-life. *Group of Corals, British Museum*, which was published in October 1859, prefigured a series of still-life studies (Figure 1.15).[147] Of all the British Museum studies, this arrangement most successfully conveys the stereoscopic effect. Significantly, the gallery vista – with its sometimes odd perspective and resulting lack of detail – has been abandoned. The viewer's entire attention is directed to the elaborate form and convoluted structure of the specimens. The profusion of fine spicules pokes out towards the viewer and draws the eye back into the depths of each coral mass. When viewed through a stereoscope, close-up views of such intricate objects are far more successful at conveying volume than images that rely on groups of more austere objects. For example, *The Assyrian Gallery*, published in May 1860, resembles a

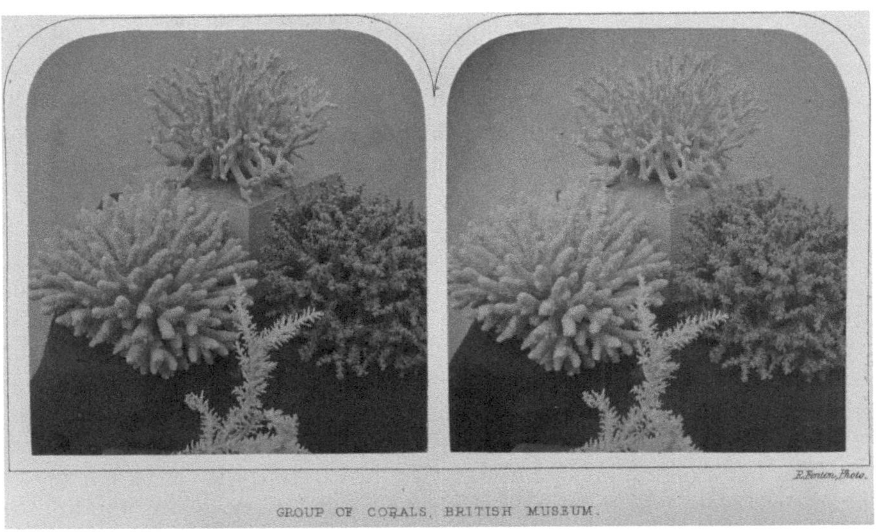

Figure 1.15 Roger Fenton, *Group of Corals, British Museum* from *The Stereoscopic Magazine*, vol. 1, no. 16, 1859. Rare Books Collection, State Library Victoria.

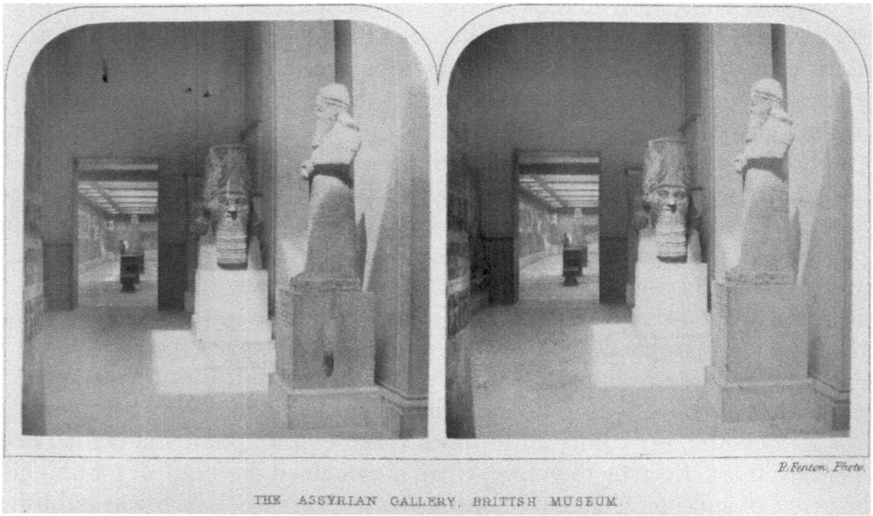

Figure 1.16 Roger Fenton, *The Assyrian Gallery, British Museum* from *The Stereoscopic Magazine*, vol. 2, no. 23, 1860. Rare Books Collection, State Library Victoria.

stage set of two-dimensional planes placed one behind the other in the receding space (Figure 1.16). The simple forms of the Assyrian sculptures provide very few points of visual convergence and, coupled with the sparseness of the gallery, offer a paucity of details for the stereoscopic effect to work satisfactorily.[148]

By including still-life studies in the *Stereoscopic Magazine*, Reeve ensured that the natural objects portrayed – very often shells – were enticingly displayed. These were drawn from Reeve's own collection and often were duplicates of specimens held in the British Museum. In this way, nature's curiosities procured from all corners of the globe were made not just intelligible to the general public but collectible as well. *Group of Shells by Fenton* appeared in the magazine in August 1861 (Plate 4). It provides an object lesson in attentive looking by emphasizing the desirability of the shells' multi-faceted forms and surfaces, with their knobbly, jagged exterior shapes and smooth spiral structures – displaying symmetry and asymmetry within a single form – and their coarse exterior armour shielding lustrous mother-of-pearl interiors.[149] Distinguished by the ornamental quality of their individual forms and delicate surface textures, this group of shells was gathered together for two apparent purposes: to highlight the diversity of these intriguing objects, and to contrast and thereby maximise the visual and tactual allure of individual specimens. Every shell in this ensemble is described in the accompanying text. Some were first named by Reeve himself, including the *Helix cambojiensis*, or fine reversed snail, which, having been carefully positioned in the centre of the composition – 'resting on the outer furbelowed surface of [a] *Tridacna* valve' – stares out disarmingly like a glass eye, inviting the viewer to visually caress the pleated surface on which it lies.[150] With the shells set also within the folds of an 'embroidered China cloth' and a sturdy British plaid, the image exemplifies Britain's global reach. Moreover, by placing several shells on the open pages of *Conchologia Iconica* – Reeve's magnum opus, which he both wrote and published and which is

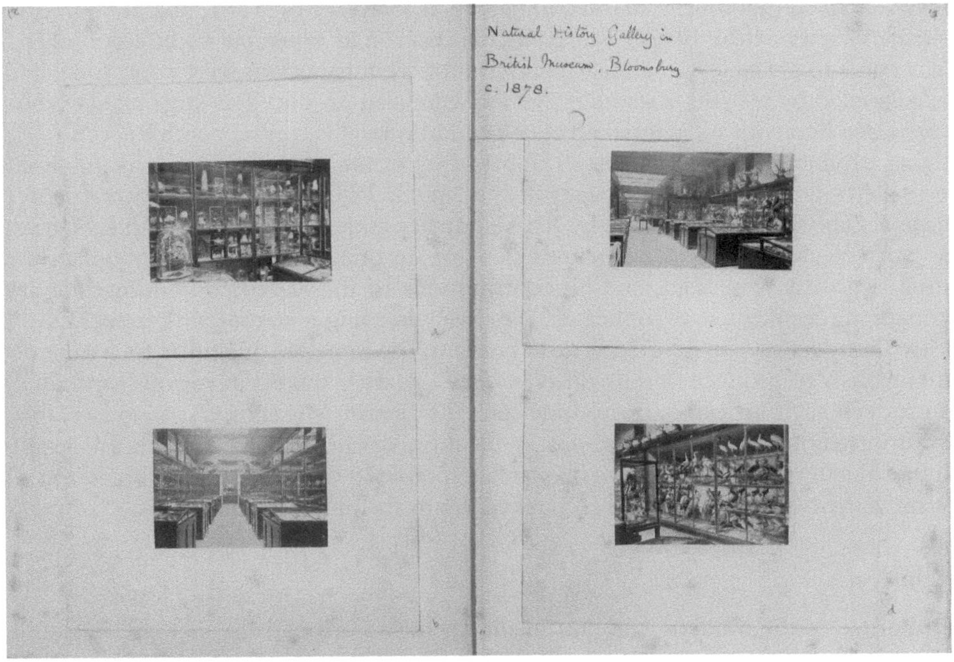

Figure 1.17 A. C. Gunther, *Europe Mammals,* vol. 8, from the A. C. Gunther series of albums, 1870s–80s, case 22: ms. 534, f.18–19, (album opening showing photographs of the British Museum natural history galleries, c.1878). Collection: Linnean Society, London.

identified in the accompanying text – Fenton emphasises Reeve's intellectual and commercial achievements in the field of natural history.[151] The nineteenth century was, in fact, something of a golden age for shell collecting. Eye-catching conchological collections – acquired from imperial expeditions and private collections – became a feature of natural history displays in the new public museums during that period, which in turn ignited a veritable shell craze. This popular pastime, which coincided with the increasing prominence of the natural sciences, encapsulated the pleasure and prosperity that the Victorians derived from nature. Fenton's studies of shells and corals in the British Museum and in private collections are representative of the way that, at the same time as aspects of the world were becoming more accessible for public collections and discriminating individuals, the whole world was being made visible and desirable through innovations in the field of the visual.[152]

Photographs of exotic specimens and artefacts collected and displayed in metropolitan museums and in other key collections or assembled as opulent arrangements in still-life studies were the only signs of Britain's imperial interests and global reach in the first four years of the *Stereoscopic Magazine*. In the initial issues at least, Britain's unparalleled influence around the world, due to its political and economic expansion, was not represented by images of colonial life or scenery as viewed or experienced by locals but by material possessions brought home for domestic consumption and edification in public museums. This preference was certainly not due to any paucity of suitable material;

indeed, at precisely this time, photographers were starting to flourish in the colonies and were actively building a market for colonial views.[153] By 1861, the *Stereoscopic Magazine's* views of the British Museum collections and other public collections had already come to a conclusion and, as far as Fenton's contributions were concerned, had been succeeded by his still-life studies, which were also produced as large single prints and provided him with substantial revenue in addition to his contract with Reeve.[154] The changing production and commercial imperatives of the photographer and publisher were spelled out in the opening pages of the March 1861 edition of the *Stereoscopic Magazine*: 'The subjects furnished by Mr Fenton are taken by a camera with six lenses, three pairs of views being taken simultaneously on one sheet of glass, by which the printing is greatly expedited, and all contributions to the *Stereoscopic Magazine* are now made in duplicate and triplicate'.[155] As well as using a stereoscopic camera with increased production capabilities, Fenton began to use three or sometimes four cameras simultaneously to produce slightly different views of each subject in various formats.[156] To introduce such an elaborate routine into the British Museum was inconceivable; such an intrusion in the galleries, not to mention the brazenness of such an openly commercial enterprise, would have been unacceptable to the trustees and threatened to undermine their fundamental role as gatekeepers to the museum collections.

Conclusion

With the rise of the modern museum in the Victorian era and the increase in public access to collections, one of the most challenging issues regarding the objects thus privileged was not what they were but rather what they became through the processes of assimilation and display in the museum environment and their re-presentation back out into the world through different media.[157] In the museum context, a fossil skeleton, shell or freshly killed whale was as much a cultural artefact as any painting, mask or manuscript. In the modern public museums, the course of the empire contributed materially to the advance of particular collections and thus played a significant role in the way these objects were socially and theoretically constructed. This complex socio-political and intellectual background perceptibly influenced the avenues and methods according to which collections were disseminated and the different ways they were encountered and understood by audiences over time.

This period in which photography became a viable and widespread technology coincided with a phase of substantial reforms in the British Museum. The medium was, therefore, a momentous choice for the trustees and officers who wished to influence the wider dissemination of their collections, although this was never officially articulated or entirely resolved. While the British Museum had initially contracted Roger Fenton – one of the most talented of Britain's early photographers – to embark on an important commission to document the collections, Fenton's relationship with the museum was fraught with practical and financial difficulties and, ultimately, with a degree of inaction. Whereas the British Museum was somewhat ineffective in its own use of the medium and failed to develop any clear policy for the photographic reproduction of its collections, it did in fact open up other avenues for production. Several unofficial programs of photography undertaken by individual staff members, private entrepreneurs and colonial institutions and individuals were more enterprising in their use of photographs as documents and also as objects of aesthetic value and popular appeal.

For the departments of natural history and antiquities at the British Museum, photography offered an excellent opportunity to reproduce select objects from the collections. As near-perfect facsimiles of the objects they represented, photographs contributed to a rhetoric of substitution that allowed the museum's collections to be augmented and their reach extended to a greater number and unprecedented range of viewers. Photographs quickly became an essential part of the paper museum collections that existed in parallel to the museums' holdings of actual objects, and they were intended to fill gaps in the collections and to circulate new knowledge. Through this process, objects in public and private collections at different sites around the globe were recorded, consolidated and re-circulated in different permutations, and they came to be regarded not as separate, or geographically discrete, entities but as part of a single, expanded resource. As a consequence, the identity and significance of objects would ultimately be determined by a much wider field of stakeholders than the coterie of specialists that had previously been given access to metropolitan collections. In the context of the British Museum, this was particularly significant given the imperial scope of Richard Owen's professional aspirations – including his work in comparative anatomy, which largely focussed on the fossil fauna of Britain's colonies – and his ambitions for a dedicated, comprehensive national museum of natural history.

Private operators and entrepreneurs also hoped to play a role in popularising the British Museum collections and sought to use photography as a new mode that had the capacity to manufacture an experience of the real through published images. Nineteenth-century natural history photography found a wide audience in one of the most fascinating of the new visual technologies, the stereoscope. From the earliest issues of his pioneering publication, the *Stereoscopic Magazine*, Lovell Reeve – publisher, conchologist and dealer in natural history – included images of objects and displays from the natural history collections of the British Museum and other collections, both public and private. The stereoscope occupied an indeterminate cultural position during the second half of the nineteenth century: its ubiquity led commentators to deride it and underestimate the impact of its visual capabilities and imaginative potentialities, while at the same time, Reeve's natural history stereo cards were perfectly suited to the acquisitive inclinations of the Victorians. Its attractions were evident in the way images of object-filled galleries and eclectic displays of specimens by Fenton and other photographers could be made to appear vivid and tangible through the stereo viewer. As in the naturalists' informal paper museums, many of Fenton's British Museum studies not only recorded singular, iconic objects but also re-presented them as part of the material world for new kinds of visual consumption.[158]

Following Fenton's foray into the British Museum, the only photographers to produce extensive studies of its collections until well into the twentieth century were Frederick York and Stephen Thompson. York had a photographic business in Cape Town during 1853–61 and understood the importance of photography for connecting British metropolitan centres with those working in the colonies and points overseas.[159] York wrote to the director of the British Museum in 1875, advising that he had been taking a series of photographs of London over the previous six years and had accomplished over 600 subjects in four different sizes: 'I now come to the British Museum [seeking your] permission to take a few general views of the different courts'.[160] York had recently been lauded in the press for his 'marvellous photographic achievement' in capturing the 'wild beasts' at the Regent's Park Zoological Gardens; his extensive studies of the natural history galleries, commenced in 1875, seemed comparatively straightforward at the outset, especially following his zoo work and his years photographing in the colonies.[161]

While his business, York & Son, was one of the largest English producers of lantern slides in the second half of the nineteenth century, York's British Museum venture became another casualty of the museum administration's antipathy to its natural history departments and the trustees' reluctance to promote the museum's collections. As a result, his British Museum images were not widely distributed.[162] His magnificent 'York Album', compiled by the photographer for presentation to the museum's trustees, contained 250 proofs of his gallery views and represented a unique visual record of the museum at that point in time – some 20 years after Fenton's commission. Yet, this impressive work resulted in only a small number of *cartes de visite* and cabinet and stereoscopic photographs being produced for sale to the public.[163] Like John Edward Gray's album, which contains perhaps the earliest extant photographs of the British Museum, some of York's views of the natural history galleries found their way into keepers' private collections, including Albert Gunther's research albums (Figure 1.17). These photographs provided museum officers with a point of comparison when viewing images of their own and other collections over time.[164] Soon after, Owen's vision for a national museum of natural history was realised at South Kensington, and the enormous and popular natural history collections of the British Museum began to be relocated and re-evaluated for the new museum's opening in 1881.

Stephen Thompson's photographs, in contrast, were predominantly formal studies of antiquities and coins in the British Museum. The resulting publication, *Catalogue of a Series of Photographs by Stephen Thompson, from the Collection in the British Museum*, 1872, comprised 929 photographs in seven parts spread over 12 volumes. It represented different sections of the antiquities collections and included introductory essays by the respective keepers. The entire project was initiated and underwritten by Charles Harrison, a London lawyer, politician and amateur archaeologist.[165] Notwithstanding the more immersive experience of Reeve's stereoscopic views of the British Museum, the inclusion of explanatory texts by expert antiquarians in Thompson's book was similarly intended to make the plates 'accessible to the public' and 'at a fixed moderate cost'.[166]

The main motivation for Thompson's commission, perhaps, was Harrison's desire to circulate photographs in the same manner as Owen and his colleagues in the natural history departments, which would result in a paper museum of antiquities collections from near and afar. As Harrison explained in his introduction, he hoped that each local museum would have its objects photographed and that a system for the mutual exchange of copies would be established between these museums and their 'great national collection':

> By thus collecting our materials, especially if the plan is extended to continental collections, we may acquire at a glance some idea of [earlier civilizations]. There is an abundance of materials at hand ... The present selection of subjects from the National Collection has now led the way. If the several local museums take up and develop the plan, it will be found that Photography, when so applied, will occupy the same place as regards circulating information of the past that newspapers of the day now fill in diffusing the daily records of the period we live in.[167]

Harrison's publishing venture challenged a common characterisation of the paper museum as involving only a centripetal movement in which objects and images were propelled towards the centre and consolidated into one great metropolitan archive, a view very much in line with the museum's conventional approach.[168] There were others who also visualised a new direction for their national museums that was more like a centrifugal force, with not just the political power of the centre but also its accumulated knowledge moving outwards to inspire and inform the general public and to benefit museums around the globe. The *Athenaeum* summarised the enormous potential of paper museums in a review of Thompson's *Catalogue* that same year: 'this publication is neither more nor less an attempt to make the British Museum portable'.[169] This binary model of imperial–colonial relations through the flow of images and objects was only part of the picture, however. Local institutions, including colonial museums, were at the centre of other types of networks.[170] As John Gascoigne notes, 'the British Empire could be viewed from other angles than London and was wide enough and fluid enough to encompass many networks'.[171] In a networked conception of empire, even so, Joseph Hodge cautions that '[c]areful attention must be paid ... both to the unevenness and inequalities of power that operated within specific networks of imperial science, and the complex and often hybridized nature of scientific knowledge making in particular geographical and historical contexts'.[172]

In the following chapter, I will consider these directional forces and various networks from the perspective of the periphery. By examining the traffic in natural history specimens and representations between individuals and institutions at the centre and periphery of the British Empire, I uncover the complex meanings and expectations that accrued as part of these exchanges through a case study of the introduction of photography in the Australian Museum. Photography was promoted by the museum's curator, Gerard Krefft, not only as a way of documenting the museum's objects and collections but also a means to reorient colonial natural history away from its traditionally one-way trajectory – based on the colonial–imperial binary of Sydney and London – towards a more cosmopolitan perspective that involved other European traditions and centres. My intention is to bring to light and evaluate some of the different circumstances and conventions surrounding specimen exchange, for instance, the various protocols relating to correspondence between naturalists at the centre and periphery of empire; and the way in which photography was employed to initiate or complement exchanges of objects and information, to foster relationships between naturalists and to establish or enhance their credentials.

Notes

1 See Thomas Richards, *The Imperial Archive*, Verso, London, 1991; also Tony Bennett, *The Birth of the Museum*, Routledge, London and New York, 1995.

2 Jim Endersby, *Imperial Nature: Joseph Hooker and the Practices of Victorian Science*, University of Chicago Press, Chicago and London, 2008, p. 54; Edward Miller, *That Noble Cabinet: A History of the British Museum*, Andre Deutsch, 1973, p. 232.

3 André Malraux, *Museum without Walls*, Secker & Warburg, London, 1965, p. 11.

4 Frederick Bohrer, 'The Times and Spaces of History: Representation, Assyria, and the British Museum' in Daniel Sherman and Irit Rogoff (eds.), *Museum Culture: Histories, Discourses, Spectacles*, Routledge, London, 1994, p. 198

5 David Wilson, *The British Museum: A History*, British Museum Press, London, 2002, p. 128.

6 Robert E. Kohler, 'Finders, Keepers: Collecting Sciences and Collecting Practice', *History of Science*, vol. 45, 2007, p. 432.

7 On 'paper museums' before the advent of photography – see Martin Rudwick, 'George Cuvier's Paper Museum of Fossil Bones', *Archives of Natural History*, vol. 27, no. 1, 2000, pp. 51–68.

8 David M. Wilson, *The British Museum: Purpose and Politics*, British Museum, London, 1989, pp. 13–14; Richard Altick, 'National Monuments' in David Boswell and Jessica Evans (eds.), *Representing the Nation*, Routledge, London, 1999, pp. 240–57.
9 Wilson, *The British Museum: Purpose and Politics*, p. 20.
10 *Collection of Plans and Views of Portions of the Buildings and Contents of Old Montague House and the Present British Museum*: Maps C.26.f.7, British Library.
11 Albert E. Gunther, *A Century of Zoology at the British Museum: Through the Lives of Two Keepers, 1815–1914*, Dawsons, London, 1975, p. 140.
12 *Annual Report*, Zoological Department, British Museum, 7 January 1857 – quoted in Gunther, *A Century of Zoology*, p. 141.
13 'Easter Monday – The Great Zoological Gallery, British Museum', *Illustrated London News*, 29 March 1845, p. 201; 'The New Coral Room, at the British Museum', *Illustrated London News*, 3 April 1847, p. 221.
14 Altick, 'National Monuments', p. 240.
15 See Anthony Hamber, *'A Higher Branch of Art': Photographing the Fine Arts in England, 1839–1880*, Gordon and Breach, Amsterdam, 1996, p. 380; Christopher Date and Anthony Hamber, 'The Origins of Photography at the British Museum, 1839–1860', *History of Photography*, vol. 14, no. 4, 1990, pp. 309–25; John Hannavy, *Roger Fenton of Crimble Hall*, Gordon Fraser Gallery, London, 1975, p. 34; Christopher Date, 'Photographer on the Roof', British Museum Society Bulletin, July, 1989, p. 10.
16 Elizabeth Edwards, *Raw Histories, Anthropology and Museums*, Berg, Oxford, 2001, p. 53.
17 Georges Cuvier, 'Extract from a Memoir on an Animal of which the Bones are Found in Plaster Stone around Paris, and which Appears No Longer to Exist Alive Today' in Martin Rudwick, *Georges Cuvier, Fossil Bones and Geological Catastrophes*, University of Chicago Press, Chicago, 1997, p. 35.
18 Rudwick, *Georges Cuvier, Fossil Bones and Geological Catastrophes*, p. 34.
19 Miller, *That Noble Cabinet*, p. 192; Wilson, *The British Museum: A History*, p. 109.
20 'The Nimroud Sculptures', *Illustrated London News*, 26 June 1847, p. 409; 'The Nimroud Sculptures, Just Received at the British Museum', *Illustrated London News*, 31 March 1849, p. 213.
21 Wilson, *The British Museum: A History*, p. 109; Bohrer, 'The Times and Spaces of History', p. 204.
22 Bohrer, 'The Times and Spaces of History', p. 203.
23 Miller, *That Noble Cabinet*, pp. 220 and 234–35.
24 British Museum: Standing Committee minutes, 11 February 1854, C.8652–3.
25 Idem.
26 British Museum: Standing Committee minutes, 11 March 1854, C.8659–60 and C.8660–1.
27 British Museum: Standing Committee minutes, 11 March 1854, C.8660–1.
28 John Hannavy, 'Roger Fenton and the British Museum', *History of Photography*, vol. 12, no. 3, 1988, p. 193.
29 See Mirjam Brusius, 'From Photographic Science to Scientific Photography: Talbot and Decipherment at the British Museum around 1850', in Mirjam Brusius, Katrina Dean and Chitra Ramalingam (eds.), *William Henry Fox Talbot: Beyond Photography*, Yale Center for British Art and Paul Mellon Centre for Studies in British Art, New Haven and London, 2013, pp. 219–44.
30 Gray exhibited two calotypes taken by an unidentified photographer and described as 'Specimens of the Application of Photography to Natural History' (cat. nos. 778 and 779) at the 1852 photographic exhibition of the Society of Arts, London.
31 British Museum: Standing Committee minutes, 11 March 1854, C.8660–1.
32 Idem.
33 Augustus Franks to British Museum Trustees, November 1854 – quoted in Hannavy, 'Roger Fenton', p. 194.
34 British Museum: Original Papers (OP 63), 9 June 1859 (received), f.317a.
35 Miller, *That Noble Cabinet*, p. 203.
36 British Museum: Index to Trustees' minutes, 28 March 1857, C.9178.
37 British Museum: Officers Reports, 27 April 1857, f.389.
38 British Museum: Officers Reports, 6 May 1857, f.472.

39 Idem.
40 British Museum: Officers Reports, 8 April 1857, f.394.
41 Idem.
42 British Museum: Officers Reports, 23 April 1857, f.367.
43 British Museum: Original Papers, 4 April 1857, f.185.
44 British Museum: Officers Reports, 4 June 1857, f.543.
45 British Museum: Officers Reports, 16 May 1857, f.588. The ease with which objects could be photographed influenced the prices charged for photographic negatives; also, Fenton explicitly spells out the price differential between photographs taken in the galleries (4 guineas) and those of objects removed to the 'glass room on the leads' (2 guineas) – British Museum: Officers Reports, 4 June 1857, f.544.
46 British Museum: Officers Reports, 8 April 1857, f.385.
47 British Museum: Officers Reports, 23 April 1857, f.367.
48 British Museum: Original Papers, 18 March 1858, f.243.
49 British Museum: Officers Reports, 23 April 1857, f.375.
50 British Museum: Officers Reports, 2 June 1857, f.587.
51 Idem.
52 British Museum: Officers Reports, 3 June 1857, f.590.
53 Idem.
54 Grace Seiberling, *Amateurs, Photography and the Mid-Victorian Imagination*, University of Chicago Press, Chicago, 1986, p. 10.
55 'Notice to Members: Exchange of Positive Pictures', *Journal of the Photographic Society*, no. 1, 1 April 1853, n.p. – quoted in Seiberling, *Amateurs, Photography and the Mid-Victorian Imagination*, p. 10 (emphasis in original).
56 Seiberling, *Amateurs, Photography and the Mid-Victorian Imagination*, p. 11.
57 John Morris, *John Dillwyn Llewelyn, 1810–1882: The First Photographer in Wales*, Welsh Arts Council 1980; George Shadbolt, 'On the Mode of Producing Extremely Minute Photographs for Microscopic Examination', *Journal of the Photographic Society*, no. 4, 1858, pp. 75–81.
58 Seiberling, *Amateurs, Photography and the Mid-Victorian Imagination*, p. 61.
59 Idem.
60 Frederick N. Bohrer, *Orientialism and Visual Culture*, Cambridge University Press, Cambridge, 2003, p. 74 passim.
61 Steven Shapin and Arnold Thackray, 'Prosopography as a Research Tool in History of Science: The British Scientific Community 1700–1900', *History of Science*, no. xii, 1974, pp. 4–5.
62 Nicolaas Rupke, *Richard Owen: Victorian Naturalist*, Yale University Press, New Haven and London, 1994, p. 130.
63 Adrian J. Desmond, 'Designing the Dinosaur: Richard Owen's Response to Robert Edmond Grant', *Isis*, vol. 70, no. 2, 1979, pp. 224.
64 Ken Arnold, 'Presenting Science as Product or as Process: Museums and the Making of Science' in Susan Pearce (ed.), *Exploring Science in Museums*, Athlone, London and Atlantic Highlands N.J., 1996, p. 57; Tony Bennett, *Pasts Beyond Memory: Evolution, Museums, Colonialism*, Routledge, London and New York, 2004, pp. 12–35.
65 See Elizabeth Edwards, 'Between Specimen and Imagination: Photography in the Age of Evolution' in Bernard Lightman and Bennett Zon (eds.), *Evolution and Victorian Culture*, Cambridge University Press, Cambridge, 2014, pp. 64–93.
66 British Museum: Standing Committee minutes, 11 March 1854, C.8659–60.
67 Joel Snyder, 'Nineteenth-century Photography of Sculpture and the Rhetoric of Substitution' in Geraldine A. Johnson (ed.), *Sculpture and Photography: Envisioning the Third Dimension*, Cambridge University Press, Cambridge, 1998, pp. 21–34.
68 Ann Thomas (ed.), *Beauty of Another Order: Photography in Science*, Yale University Press, New Haven and London, 1997, p. 81.
69 Stephen T. Asma, *Stuffed Animals and Pickled Heads: The Culture and Evolution of Natural History Museums*, Oxford University Press, Oxford, 2001, p. xiii; see also Ken Arnold, 'Presenting Science as Product or as Process: Museums and the Making of Science' in Susan Pearce (ed.), *Exploring Science in Museums*, Athlone, London and Atlantic Highlands N.J., 1996, p. 66.

70 John Thackray, 'Separately-published Prints of Fossils in Nineteenth-century Britain', *Archives of Natural History*, vol. 12, no. 2, September 1985, pp. 175–99.

71 J.M. Chalmers-Hunt (ed.), *Natural History Auctions 1700–1972: A Register of Sales in the British Isles*, Sotheby Parke Bernet, London, 1976, pp. 32 and 45; S. Peter Dance, *A History of Shell Collecting*, E. J. Brill, Leiden, 1986.

72 While not reaching anything near the print run of traditionally published images, they were also occasionally offered for sale and even advertised in a very limited way by natural history dealers and bookshops – Thackray, 'Separately-published Prints of Fossils', p. 181.

73 One of the earliest extant examples of this innovation is a work entitled *Fossils, Engraved on a Daguerreotype Plate*, printed in 1840 using a process developed by amateur geologist L. L. Boscawen Ibbetson – Thomas Spenser, 'An Account of some Experiments Made as to How Voltaic Electricity may be Usefully Applied for the Purpose of Working in Metal', *Westminster Review*, vol. 34, 1840, p. 459; William Buckland, 'President's Address', *Proceedings of the Geological Society*, iii, 1841, p. 476.

74 Thackray, 'Separately-published Prints of Fossils', pp. 177–9.

75 Ibid., p. 183.

76 Simon Knell, 'The Roller-Coaster of Museum Geology' in Susan Pearce (ed.), *Exploring Science in Museums*, Athlone, London and Atlantic Highlands N.J., 1996, p. 29.

77 Charles Darwin, *On the Origin of Species by Means of Natural Selection*, (ed.) Joseph Carroll, Broadview, Peterborough, 2003 (1859), p. 396.

78 Ibid., pp. 274–278; 396.

79 Rudwick, 'George Cuvier's Paper Museum of Fossil Bones', p. 59.

80 On 'centres of calculation'– see Latour, *Science in Action,* esp. Chapter 6.

81 John Pickstone, 'Museological Science? The Place of the Analytical/Comparative in Nineteenth-Century Science, Technology and Medicine', *History of Science,* no. 32, 1994, p. 113.

82 J. E. Gray, *Collection of Illustrations of Cetaceans and Sirenia Comprising Prints, Original Drawings, Tracings and Photographs Dating from the 18th Century*, two volumes, Natural History Museum, London, 88 f G.

83 Ibid. – see, for example, ff. 158 and 184 (letters and photographs sent from Gerard Krefft in 1865).

84 A. C. L. G. Gunther, 'A Series of Albums of Printed Illustrations, Drawings, Photographs and Printed Ephemera', 1870s–80s, 39 volumes, mss case 22, Linnean Society, London.

85 Gruber and Thackray, *Richard Owen Commemoration*, p. 161; Christopher Mills, *Images from Nature: Drawings and Paintings from the Library of the Natural History Museum, London*, Natural History Museum, London, 1998, p. 7.

86 Mills, *Images from Nature*, p. 6.

87 Jean M. Ingles and Frederick C. Sawyer, 'A Catalogue of the Richard Owen Collection of Palaeontogical and Zoological Drawings in the British Museum (Natural History)', *Bulletin of the British Museum (Natural History)* (historical series), vol. 6, no. 5, 1979, pp. 111–2.

88 Ibid., pp. 111–2; Nicolaas Rupke, *Richard Owen*, pp. 70–83.

89 Jim Endersby, *Imperial Nature*, p. 177.

90 See Jacob W. Gruber, 'The Richard Owen Correspondence: An Introductory Essay', unpublished ms, Natural History Museum, London – Owen Collection 87/1, p. 19.

91 Ingles and Sawyer, 'A Catalogue of the Richard Owen Collection', pp. 111–2.

92 William Bedford to Richard Owen, 7 February 1835: Owen Collection, Library, Royal College of Surgeons, London.

93 Ludwig Becker to Richard Owen, (1855): Owen Correspondence, Natural History Museum, London (NHM), 3, 22–3; A. G. L. Shaw, 'Ludwig Becker', *Dictionary of Australian Artists Online* – http://www.daao.org.au/bio/ludwig-becker/biography/ (accessed 1 October 2010, updated 2012).

94 William Epps Cormack to Richard Owen, (1850), Owen Correspondence, Natural History Museum, London (NHM), 8, 426/431–2 (4 letters).

95 Rupke, *Richard Owen*, p. 23; Rudwick, 'George Cuvier's Paper Museum of Fossil Bones', pp. 51–68.

96 William Broderip to Richard Owen, 10 March 1857: Owen Correspondence, Natural History Museum, London (NHM), 5, 309.

97 Edwards, *Raw Histories*, p. 52.

98 Richard Owen 'Exhibition of a Bone of an Unknown Struthious Bird of Large Size from New Zealand', *Proceedings of the Zoological Society of London*, vol. 7, 1839, pp. 169–71; Richard Owen, 'On Dinornis (Part XXV): Containing a Description of the Sternum of *Dinornis elephantopus*', *Transactions of the Zoological Society of London* vol. 12, no. 1, 1886, pp. 1–3.

99 Lawrence J. Prelli (ed.), *Rhetorics of Display*, University of South Carolina, Colombia, 2006, p. 17.

100 Richard Owen, *Palaeontology or, A Systematic Study of Extinct Animals and their Geographical Relations*, A. & C. Black, Edinburgh, 1861.

101 Rudwick, 'George Cuvier's Paper Museum of Fossil Bones', p. 53.

102 Ibid., p. 62.

103 Richard Owen to W. B. Clarke, 15 April 1872 in Ann Moyal, *The Web of Science: The Scientific Correspondence of the Rev. W. B. Clark, Australia's Pioneer Geologist*, vol. 2, Australian Scholarly Publishing, Melbourne, 2003, pp. 893–4.

104 Rudwick, 'George Cuvier's Paper Museum of Fossil Bones', p. 59.

105 Steven Shapin, 'The Invisible Technician', *American Scientist*, vol. 77, no. 66, 1989, pp. 554–63.

106 D. E. Allen, 'On Parallel Lines: Natural History and Biology from the Late Victorian Period', *Archives of Natural History*, vol. 25, no. 3, 1998, p. 365.

107 Susanne B. Keller, 'Sections and Views: Visual Representations in Eighteenth-century Earthquake Studies', *British Journal for the History of Science*, vol. 31, no. 2, p. 129; Michel Foucault, *The Order of Things*, Routledge, London, 2002.

108 Simon Knell, 'The Roller-Coaster of Museum Geology' in Susan Pearce (ed.), *Exploring Science in Museums*, Athlone, London and Atlantic Highlands N.J., 1996, p. 45.

109 British Museum: Index to Trustees' Minutes, 10 May 1856–C.8997; 'Messrs Colnaghi's Terms for the Sale of Photographs'; British Museum: Officers' Reports, 13 November 1857–f.219; British Museum: Officers' Reports, 12 December 1857, f.298.

110 British Museum: Index to Trustees' Minutes, 1 August 1857, C.9260.

111 British Museum: Original Papers, 19 January 1858–556.

112 British Museum: Index to Trustees' Minutes, 10 May 1856–C.8997, 'Messrs Colnaghi's Terms for the Sale of Photographs'; British Museum: Officers' Reports, 13 Nov 1857-f.219; British Museum: Officers' Reports, 12 Dec 1857, f.298.

113 *Art Journal*, 1857 – quoted in Colnaghi & Co, *Photography: The First Eighty Years*, introduction (n.p).

114 British Museum: Original Papers, 9 June 1859, f.317a; 'Photographs Taken for Government Institutions', *The Photographic News*, 30 September 1859, p. 4; 'Photography at South Kensington Museum', *The Photographic News*, 7 September 1860, pp. 224–5; *Select Committee Report on the South Kensington Museum*, 1860; Hamber, '*A Higher Branch of Art*', p. 384.

115 For a useful introduction to Lovell Reeve's promotion of stereoscopy in the context of his overall publishing career, see Amy Stark, 'Lovell Reeve (1814–1865), Publisher and Patron of the Stereograph', *History of Photography*, vol. 5, no. 1, 1981, p. 11.

116 *The Athenaeum,* 5 June 1858, p. 709.

117 Idem.

118 'Interior of the Museum of Practical Geology, Jermyn Street', *Stereoscopic Magazine*, no. 3: September 1858; 'The Succulent House Kew, Interior of the Aloe and Cactus House, Kew', *Stereoscopic Magazine*, no. 4: October 1858; 'Osteological Museum of the Royal College of Surgeons', *Stereoscopic Magazine*, no. 12: June 1859.

119 British Museum: Index to Trustees' Minutes, 9 October 1858-C.9473; British Museum: Index to Trustees' Minutes, 13 November 1858, C.9484; British Museum: Index to Trustees' Minutes, 13 March 1859-C.9542/3; British Museum: Index to Trustees' Minutes, 13 October 1860, C.9829.

120 Robert Hunt, 'The Stereoscope', *The Art-Journal*, 1856, p. 118.

121 Robert J. Silverman, 'The Stereoscope and Photographic Depiction in the 19th Century', *Technology and Culture*, vol. 34, no. 4, 1993, p. 730.

122 Lovell Reeve, 'The Stereoscopic Magazine', *Literary Gazette and Journal of Belle Lettres, and Art*, 24 April 1858, p. 408; Lovell Reeve, 'The Stereoscopic Magazine', *Notes and Queries*, 17 July 1858, (front matter).

123 Lovell Reeve, 'The Stereoscopic Magazine', *The Literary Gazette: A weekly journal of literature, science, and the fine arts*, 1858, no. 2153, 408; Lovell Reeve, 'Stereoscopic Magazine', *Notes and Queries*, 2nd ser., 6 (17 July 1858), front matter.
124 Lovell Reeve, 'The Stereoscopic Magazine', *The Literary Gazette: A weekly journal of literature, science, and the fine arts*, 1858, no. 2153, 408.
125 'The Stereoscopic Magazine', *The Guardian*, (n.d.): LRP/6/1 Press Cuttings 1851–1910, Archives, Royal Botanic Gardens, Kew.
126 'The Stereoscopic Magazine', *Literary Gazette*, (n.d.): LRP/6/1 Press Cuttings 1851–1910, Archives, Royal Botanic Gardens, Kew.
127 Lovell Reeve to Arthur Melhuish, 9 July and 20 September 1858; Lovell Reeve to Negretti & Zambra, 26 April and 14 January 1859: LRP/1/23 Indexed Letter Book, November 1847–September 1872, Archives, Royal Botanic Gardens, Kew.
128 European monuments and views appeared sporadically in the *Stereoscopic Magazine*.
129 Mary Warner Marien, *Photography: A Cultural Reader*, Laurence King, London, 2002, p. 84; Jonathan Crary, *Techniques of the Observer: On Vision and Modernity in the Nineteenth Century*, MIT Press, Cambridge (Mass.), 1990, pp. 122–3.
130 Crary, *Techniques of the Observer*, pp. 122–5.
131 Idem.; Asa Briggs, *Victorian Things*, Sutton, Thrupp, 2003.
132 Lynda Nead, *Victorian Babylon: People, Streets and Images in Nineteenth-Century London*, Yale University Press, New Haven and London, 2000, p. 86.
133 'The Megatherium', *The Stereoscopic Magazine*, vol. 1, no. 14, 1859.
134 British Museum: Officers' Reports, f.367, 23 April 1857.
135 See 'J. Wharton Jones, 'Observations on Single Vision with Two Eyes', communicated by Richard Owen to the Royal Society, London, 30 November 1839: mss. AP/23/33, Royal Society, London.
136 Dance, *A History of Shell Collecting*; S. Peter Dance, 'Reeve, Lovell Augustus (1814–1865)', *Oxford Dictionary of National Biography*, Oxford University Press, 2004, http://www.oxforddnb.com/view/article/23300 (accessed 30 November 2010).
137 Dance, 'Reeve, Lovell Augustus (1814–1865)'.
138 Richard Owen, 'Notes on the Stereograph of the Megatherium', *The Stereoscopic Magazine*, vol. 1, no. 14, 1859, pp. 193–8.
139 Richard Owen 'Notes on the Stereograph of the Giant Irish Deer', *The Stereoscopic Magazine*, vol. 2, no. 17, 1859, p. 3.
140 'The Stereoscopic Magazine', *The Leader*, (n.d.): LRP/6/1 Press Cuttings 1851–1910, Archives, Royal Botanic Gardens, Kew.
141 'The Stereoscopic Magazine', *Literary Gazette*, (n.d.): LRP/6/1 Press Cuttings 1851–1910, Archives, Royal Botanic Gardens, Kew.
142 'Reviews: The Stereoscopic Magazine', *The Art-Journal*, 1858, p. 255.
143 Idem.
144 'The Reptile Room, British Museum', *The Stereoscopic Magazine*, vol. 2, no. 20, 1860, p. 39.
145 Ibid., p. 40.
146 Ibid., p. 42.
147 The still-life study 'Fruit with Ivory and Silver Tankard', published in October 1860, was described in that issue as the 'First of a Series Taken for the Magazine by Mr. Fenton' – *The Stereoscopic Magazine*, vol. 2, no. 28, 1860, pp. 117–18.
148 'The Assyrian Gallery, British Museum', *The Stereoscopic Magazine*, vol. 2, no. 23, 1860.
149 'Group of Shells', *Stereoscopic Magazine,* vol. 3, no. 38, 1861, pp. 33–6; Svetlana Alpers, *The Art of Describing: Dutch Art in the Seventeenth Century*, University of Chicago Press, Chicago, 1983, p. 72.
150 'Group of Shells by Fenton', p. 35.
151 Lovell Reeve, *Conchologia Iconica, or, Illustrations of the Shells of Molluscous Animals*, 20 volumes, Reeve Brothers, London, 1843–78.
152 Jean-Louis Comolli, 'Machines of the Visible' in Teresa de Lauretis and Stephen Heath (eds.), *The Cinematic Apparatus*, Macmillan, London, 1980, pp. 122–3.
153 'General Progress in Art', *Sydney Morning Herald*, 13 May 1859, p. 9; see Warwick Reeder, 'The Stereograph and the Album Portrait in Colonial Sydney 1859–62', *History of Photography*, vol. 23, no. 2, 1999, p. 181.

154 Lovell Reeve to Roger Fenton, 6 August 1860: LRP/1/23 Indexed Letter Book November 1847–September 1872, Archives, Royal Botanic Gardens, Kew.
155 'Address', *The Stereoscopic Magazine*, no. 33, March 1861, (n.p.).
156 Pam Roberts, 'Roger Fenton and the Still-Life Tradition' in Gordon Baldwin, Malcolm Daniel and Sarah Greenough (eds), *All the Mighty World: The Photographs of Roger Fenton, 1852–1860*, Metropolitan Museum of Art, New York, 2004, p. 95.
157 Nicholas Thomas, *Entangled Objects: Exchange, Material Culture, and Colonialism in the Pacific*, Harvard University Press, Cambridge (Mass.) and London, 1991, p. 4.
158 Celeste Brusati, 'Natural Artifice and Material Values in Dutch Still Life' in Wayne E. Frantis, *Looking at Seventeenth-Century Dutch Art: Realism Reconsidered*, Cambridge University Press, Cambridge, 1997, p. 145.
159 See Marjorie Bull and Joseph Denfield, *Secure the Shadow: The Story of Cape Photography from its Beginnings to the End of 1870*, Terence McNally. Cape Town, 1970; Stephen Herbert, 'Frederick York (1823–1903)' in John Hannavy (ed.), *Encyclopedia of Nineteenth-Century Photography*, Routledge, London, 2007, p. 1520.
160 Frederick York to Sir Edward Bond , 6 March 1875: P. no. 1040, Central Archives, British Museum.
161 'Reviews from "The Times", "Daily Telegraph", and "Morning Chronicle"', newspaper clipping in the 'York' Album, c.1875: PS 345 182–345 238 and PS 345 263–345 318, Central Archives, British Museum.
162 *York & Son Collection*, National Monuments Records, Swindon, (n.d.); David Henry, 'York & Son', *New Magic Lantern Journal*, vol. 3, no. 1, 1984, p. 12 and vol. 3, no. 2, 1984, p. 13.
163 Explanatory note in the 'York' Album, c.1875: PS 345 18 –345 238 and PS 345 263–345 318, Central Archives, British Museum.
164 Album 8, no. 534 from A. C. L. G. Gunther, 'A Series of Albums of Printed Illustrations, Drawings, Photographs and Printed Ephemera', 39 volumes, mss case 22, Linnean Society, London.
165 Martha S. Vogeler, 'Harrison, Charles (1835–1897)', *Oxford Dictionary of National Biography*, Oxford University Press, 2004 – http://www.oxforddnb.com/view/article/38891 (accessed 6 May 2011); Hamber, '*A Higher Branch of Art*', pp. 385–6.
166 Charles Harrison, introduction in *Catalogue of a Series of Photographs by Stephen Thompson, from the Collection in the British Museum, Introduction by Charles Harrison*, W. A. Mansell, London, 1872, p. xliii.
167 Ibid., pp. xliii–xliv.
168 See Georgina Born, 'Public Museums, Museum Photography, and the Limits of Reflexivity', *Journal of Material Culture*, vol. 3, no. 2, 1998, p. 245; also Thomas Richards, *The Imperial Archive: Knowledge and the Fantasy of Empire*, Verso, London and New York, 1993; Tony Bennett, *Pasts Beyond Memory: Evolution, Museums, Colonialism*, Routledge, London and New York, 2004.
169 'Fine Arts', *The Athenaeum*, 7 September 1872, p. 309
170 See Susan Sheets-Pyenson, *Cathedrals of Science: The Development of Colonial Natural History Museums during the Late Nineteenth Century*, McGill-Queens University Press, Kingston and Montreal, 1988; John M. MacKenzie, *Museums and Empire: Natural History, Human Cultures and Colonial Identities*, Manchester University Press, Manchester and New York, 2009.
171 John Gascoigne, 'Science and the British Empire from its Beginnings to 1850' in Brett M. Bennett and Joseph M. Hodge (eds.), *Science and Empire: Knowledge and Networks of Science across the British Empire, 1800–1970*, Palgrave Macmillan, Basingstoke and New York, 2011, p. 56; see also MacLeod, Roy, 'On Visiting the "Moving Metropolis": Reflections on the Architecture of Imperial Science', *Historical Records of Australian Science*, vol. 5, no. 3, 1982, pp. 1–16.
172 Brett M. Bennett and Joseph M. Hodge (eds.), *Science and Empire: Knowledge and Networks of Science across the British Empire, 1800–1970*, Palgrave Macmillan, Basingstoke and New York, 2011, p. 3; see also Warwick Anderson, 'Postcolonial Technoscience', *Social Studies of Science*, vol. 32, nos. 5–6, 2002, pp. 643–58.

2 Museum traffic

Naturalist correspondents and the advent of photography in the colonial museum

Introduction

> [W]ith the best desires for accuracy and the greatest care, it is hardly possible for an author to so describe that his reader shall see seashells as he sees them.[1]

One of the most conspicuous features of natural history practice in the nineteenth century was the constant attention paid by naturalists to the comparative merits and meanings of visual and written accounts of the objects they studied. Two concerns were central to this: the importance of perfecting the way they communicated their own observations, and the difficulties in interpreting the perceptions of others. Of all the sciences, natural history seemed best suited to visual representation. This raised the question as to whether natural history could be more effectively disseminated by predominantly visual means rather than in the long, detailed descriptions that characterised much of scientific discourse and were considered essential for promoting an analytical approach to natural history. Joseph Hooker, for instance, director of the Royal Botanic Gardens at Kew and one of the most prominent naturalists of the empire, regarded those publications which omitted all illustrations to be 'more complete and philosophical', stressing this in a letter written in 1868 to James Hector, director of New Zealand's Colonial Museum in Wellington.[2] In contrast, John Edward Gray at the British Museum enthusiastically endorsed the view of William Sharp Macleay, every inch a philosophical naturalist, when he claimed that 'the modern art of describing is too long, often insufferably long, while human life remains as short as ever'.[3] Gray's pronouncement reflected the increasingly popular, but still contested, view that natural history was most usefully advanced by the visual arts. At the very least, he argued, it should be communicated in a concise description of a single example of an animal rather than hiding in 'a mass of words the characters you are looking for' by presenting long descriptions containing every detail 'of the genus or even the family or order to which the species belongs'.[4]

For naturalists practicing in the colonies and contributing to imperial science, the convention of describing or comparing new and uncommon species without illustrating them caused continual uncertainty. In the metropole, images that were derived from unproven sources or produced in unverified circumstances were often deemed to be dubious. In the decades following Gray's declaration, the situation appeared to be unchanged. In 1895, in a public and somewhat intemperate exchange with Edgar Smith, fellow conchologist and assistant keeper in the Zoological Department at the British Museum (Natural History), Charles Hedley, conchologist at the Australian

Museum, explained the challenges faced by naturalists in different locations in classifying species without recourse to images of the specimens described:

> Those who have had occasion to deal with the revision of species, either fauna by fauna or region by region, have observed that when an author has described a species without figuring it succeeding writers on the same subject have frequently renamed such species, but that when the first reporter also gave a figure of his species such usually escaped the infliction of synonymy.[5]

As Hedley argued, no positive identification can be made from mere description, 'even from one as good as Mr. Smith's invariably are'.[6] In a published response to Hedley's polemic, Smith concurred with his Australian colleague in regard to the importance of figures, although he also suggested that images were not always reliable and 'an inaccurate figure is almost worse than none'.[7] Even more to the point, Smith thought that illustrations of species were very often extraneous to the purpose of classification, since the British Museum had an ever-growing complement of specimens 'at hand for comparison'.[8] He then upped the ante by suggesting that Hedley's criticisms had more to do with colonial jealousy than with the proprieties of their profession: 'Some Americans are said to be very jealous of interference by Europeans with their fauna; and it seems almost as if the "green-eyed monster" were tripping in the Antipodes'.[9]

For naturalists operating on the periphery of the empire, metropolitan naturalists seemed to possess immense and seemingly incontestable authority. As Hedley responded to Smith's immodest reference to the infinite resources available to him at the British Museum:

> A London writer, who has at his command the ablest men, the wealthiest museums, and most complete libraries in the world, cannot appreciate the difficulties under which an American, and still more an Australian, student pursues his work in a city far from civilisation's centre, poorly equipped with books, specimens, or apparatus, and alone from fellow workers. If Mr Smith, who can identify almost any known shell by a glance at an authentic specimen in his official custody, could realize how the head of a student of his writings has ached in reading and re-reading one of his brief unfigured diagnoses and in endeavouring to match it with a specimen at hand, he would never, I believe, again issue an unfigured description.[10]

While limited resources in the colonies perpetuated the situation, the tendency of those at the centre of imperial science was to treat colonial scientific literature and contributions to the field in a 'flippant manner', as Hedley characterised it, and destined them to remain ignorant of the local, environmental information necessary for a proper appreciation of their subject.[11]

With the prestigious aspects of natural history, such as naming, being dominated by metropolitan savants, recognition of the crucial work undertaken by the empire's 'invisible technicians' was scarce: 'What credit may attach to the naming of species is appropriated by some Europeans, who leave the drier crust of classification, anatomy, distribution, &c. to be laboriously worked out by others'.[12] At the core of this long-standing dispute – and of significant concern for those located on the periphery as well as those at the centre – were three main issues: the reception and reliability of observations and images received from naturalists contributing to imperial science from unfamiliar and far-flung locations; the hierarchical and inherently problematic processes whereby various types of information were acknowledged and interpreted; and the diffusion of received information back out into the world as authoritative scientific knowledge that could provide a sound basis for further contributions. While colonial naturalists honed their working procedures and methodically recorded their observations and discoveries with greater fidelity towards making some headway in imperial science, metropolitan scientists were disinclined to relinquish any of their control over the more prominent aspects of their profession.

In this chapter, I examine the traffic in natural history between individuals and institutions at the centre and periphery of the British Empire as well as the complex meanings and expectations that were part of those exchanges. Arjun Appadurai has argued that 'We have to follow things themselves, for their meanings are inscribed in their forms, their uses, their trajectories. It is only through the analysis of these trajectories that we can interpret the human transactions and calculations that enliven things'.[13] In this chapter, that journey follows three routes, all of which pass through the Australian Museum in Sydney.

In this exploration of the role of photography in the colonial museum, I examine the Australian Museum under the stewardship of German-born Gerard Krefft and its relationship with the British Museum and other centres of authority during the 1860s and 1870s. The first section traces the rise and consolidation of natural history networks and their key function in the development of the colonial museum. Within this context, I address contemporary perceptions of the colonial masculine ideal and the Victorian 'self-made man', and, more particularly, their role in the ascendance and increasing legitimacy of colonial naturalists operating within an imperial class structure. The second section analyses the introduction of photography into the Australian Museum and its promotion by Krefft as not only a tool considered useful for the 'modern art of describing' but also as a means to reorient colonial natural history away from its traditional one-way trajectory based on the colonial–imperial binary of Sydney and London to a more cosmopolitan perspective. Specifically, these first two sections identify the different circumstances within which the exchange of specimens took place and the conventions governing this exchange. Next, protocols related to the correspondence between naturalists at the centre and periphery of the empire are evaluated, followed by a discussion of the way in which photography was employed to either initiate or complement exchanges of objects and information, to foster relationships between naturalists and to establish or enhance credentials. The third leg of the journey through this period of the Australian Museum follows several of Krefft's scientific expeditions to the Wellington Caves between 1866 and 1872. Krefft made a considerable effort to photograph the progress and findings of his explorations for the purpose of communicating new scientific knowledge to Richard Owen at the British Museum and to comply with the colonial government's reporting processes. His foray into expeditionary photography is considered in the context of the Du Chaillu affair, which played out in the London press and scientific societies during 1861–62 and was re-presented almost immediately through the colonial media. Both of these episodes drew the British Museum into the public mire of colonial and scientific credibility. They emphasise the mistrust with which colonial explorers' and naturalists' observations and material evidence were sometimes received by London's scientific elite due to their lack of credentials or social standing. Both cosmopolitan figures and self-made men, German-born Krefft and the French–American adventurer Paul

Du Chaillu originated from quite different backgrounds to those British-born savants who similarly moved around the empire, sometimes as expeditionary naturalists but increasingly in the newly established colonial museums. The presence of naturalists from various backgrounds – who travelled extensively, who were entrepreneurs and reformers and who took up positions of authority in significant numbers – shows that the practice of natural history during the nineteenth century did not represent a simple division between the advance of British interests and their impact on the colonial subject but indicates a more complex cultural terrain and introduced into the mix a set of interests in the form of traditions and loyalties that existed in parallel with those of the British Empire.

From German settler to cosmopolitan scholar: colonial masculinity and the rise of the self-made man

The significance of photography in the progress of natural history, and its role in colonial museums, was a subject that often appeared in nineteenth-century writings, usually generated by those naturalists travelling to and working in the colonies.[14] These reports presented a counterpoint to the state of affairs in metropolitan museums, including the British Museum, where – apart from Roger Fenton's brief and problematic association – photography had only been taken up in an ad hoc way and was usually prompted by commercial entrepreneurs, print dealers or publishers. In the colonial museums, on the other hand, more concerted programs – mostly officially sponsored or endorsed – began to appear over the following decade. Krefft, as curator and secretary of the Australian Museum, advised that 'a photographic establishment [was] one of the most essential parts of a modern museum' and was integral to most museological concerns, including display, research and collection development.[15] In a paper presented before the Royal Society of New South Wales in August 1868, Krefft observed that: 'The interest which all classes take in Natural History, has gradually changed the old fashioned curiosity shops of fifty years ago, into useful Museums – where rational amusement, combined with instruction, is offered to the mass of people, and where students have every opportunity to examine and study the specimens, of which the Museum consists.'[16] Like the learned societies that preceded them, colonial museums encouraged a public appreciation of science – especially the natural sciences – and embodied a mode of culture particularly appropriate to ensuring the progress of colonial society. As Krefft noted:

> With the British Museum for a model – we had, at first, adopted almost everything that is good and bad in that Institution, till experience showed plainly that there was much room for improvement, and, this it appears has also been felt by Dr. J. E. Gray. In a pamphlet sent to me by last mail, the Principal Keeper has suggested to the British Museum Trustees, certain alterations which I have advocated for years, and which, as far as it was possible, have long been adopted in our Museum.[17]

Improvements to the Australian Museum displays outlined in Krefft's paper included moveable cabinets, instructive labelling and large photographs accompanying the objects; and the presentation of exhibits in ways that would be attractive to the public: 'instead of each case, as of old, containing a single specimen, it should embrace a series of specimens, selected and arranged so as to present a special object for study'.[18] One inspiration for Krefft's museum reforms can found in Germany, where there was a general appreciation of popular science's important role that was 'inextricably linked to the spirit of the European Enlightenment' and the emergence of a mass reading public keen to engage with ideas and with an enthusiasm for rational entertainment.[19] Science was espoused as an agent of political liberalism that was capable of transforming culture and society.[20] The natural sciences, particularly, Kurt Bayertz argues, 'were considered to be the motive force for all areas in social life; whoever wanted to be "up with the times" had to be familiar with their success and method of thought'.[21] Consequently, the popularisation of science was a significant element in the way Germany constructed and conveyed its own self-image in this period.[22]

This concern for public instruction and entertainment – as part of the promotion of the museum as a socially useful institution – further distinguished the colonial museum from its metropolitan counterpart, the British Museum, which had done rather less to encourage public dissemination. It also highlighted a fundamental difference in the approaches to natural history taken in England and Australia, which was evident in other British colonies as well. The origins of this divergence can be traced to the transformations in the intellectual traditions underpinning scientific methodology in England and the practical exigencies that influenced cultural and intellectual life in colonial Australia. In Britain, the Baconian doctrine of empiricism, fact gathering and practical application that had long determined naturalists' approach to their discipline began to be challenged by a 'new ideology of research' championed by reformers such as T. H. Huxley, W. S. Jevons and John Tyndall, who promoted theory and innovation over method and industry.[23] In visual terms, this changing intellectual climate was exemplified in Ford Madox Brown's painting *Work*, 1852–65, which portrays the unmistakable figures of Thomas Carlyle and F. D. Maurice – described by Brown as 'brainworkers' – contemplating the labours of navvies and street sellers.[24] These instantly recognisable luminaries contrast with the anonymous labouring male figures, positioned in the centre of the composition – the latter depicted as bronzed muscular bodies with enormous arms and calloused hands – and thus represent the two contrasting corporeal motifs in the form of hand and head that symbolise the 'manual and mental forms of labour'.[25] Two divergent styles of masculinity are on display, therefore: the brutish masculinity of the manual worker, and the ascetic masculinity of the philosopher, whose manliness is founded on self-discipline.[26] Yet, the basis of the distinction between scientific practice and theory, between workers and theorists, was somewhat more complex, for, as Roy McLeod notes, it also 'recast science as a "reformed" enterprise which was universal, cosmopolitan, open, disinterested, and far-sighted. The image of the "practical man", by contrast, was particular, self-interested and in pursuit of immediate returns'.[27] Although these latter qualities were not necessarily regarded as a virtue in England, they were considered a necessity in the colonial setting, where a combination of initiative and industriousness was valued and, as in the case of the naturalist, demonstrated by 'the addition to knowledge of facts hard won from nature'.[28] Thus, while a division in hegemonic masculinity, represented by a separation between the cultivation and application of technical expertise, was a trend that was increasingly evident through the nineteenth century, the development of masculine identity in Australia and other British settler societies was different from those of Europe and was more closely linked to the Smilesian ethos of the self-made man.[29]

In accordance with this image of the colonial figure as a practical, hardworking fact-gatherer, colonial naturalists were commonly cast as merely transcribers of British science, and the study of natural history in the colonies was thought to involve little more than being a gatherer for metropolitan science.[30] Yet, if the creation of the British Empire was indeed 'a gendered exercise of great significance in the formation of Western masculinity', as Clive Moore has argued, the basis of work within colonial cultures, as one aspect of masculine claims to authority, was never so one-dimensional.[31] To proceed beyond the idea of outright colonial scientific subservience towards a more productive level of inquiry necessarily introduces questions regarding the complexity and contested nature of colonial identity and uncovers a range of associated issues relating to the value and recognition of work – both intellectual and physical – as well as leisure, class, gender, ethnicity and citizenship.[32] The mix of cultures in the British Empire was far more diverse during the nineteenth century than is often recognised, and thus the projection of these overlapping modes of identification was similarly complex.[33] Even within a single culture, masculine identities tended to be manifold and unfixed; they were even more complicated when cultures were brought together in a settler society like Australia.[34] Accordingly, there were multiple modes of scientific masculinity and, while the ideal of the self-made man also existed in nineteenth-century British science domestically, it had particular resonance for scientists in Australia and other settler colonies.[35]

Central to Australian colonists' self-image was the male heroic tradition. The category of work performed defined each individual in relation to specific types of masculinity. Yet, the image and persona of the colonial naturalist was indeterminate within that tradition. The fact that natural history was only starting to become professionalised meant that any reasonably educated man could teach himself to become proficient or even excel as a naturalist and thus join the ranks of the civic-minded and useful citizenry. Colonial natural history was an especially productive and physically active endeavour that contributed materially to the progress of the empire and was thus identified with masculine achievement. Furthermore, recognition in the world of science, either by one's peers or more broadly, conferred status and thus 'contributed to the construction of an autonomous masculinity closely related to contemporary ideas of citizenship'.[36] Yet, many aspects of nineteenth-century natural history involved, and indeed necessitated, practices and fields of inquiry that were also deemed to be acceptable areas of inquiry for women or shared qualities with various female craft traditions.[37] As Carol Harrison observes, natural history was typically 'detail-oriented and small-scale, and, ultimately, profoundly concerned with the aesthetics of presentation and display'.[38] Although women artists were commissioned by naturalists, recognition of their work was rare and usually qualified. For instance, in his 1869 publication of *Snakes of Australia*, Krefft was somewhat reserved in his acknowledgement of the contributing artists – the 'gifted daughters of A.W. Scott, Esq., Miss Scott and Mrs. Edward Forde – [who] had done everything in their power to give correct figures of the reptiles illustrated'.[39] In a letter Krefft wrote to Richard Owen later that year, he described the sisters as being unsuited to certain subjects as they did not have the proper scientific grounding: 'my artists are not quite up to bones & do not draw them yet with the necessary understanding'.[40] The role of women in the production of natural history illustrations and their perceived natural aptitude for such delicate pursuits blurred the boundaries of the ostensibly male domain of colonial natural history. Men of science in the public eye, both in colonial settings and in the metropole, attempted to demonstrate their prowess and authority through various formulations and performances of masculinity. As James Warren has argued, in the early to mid-Victorian period, when Britain's imperial power was being tested on a number of fronts, 'masculinity was the key conceptual guarantor of authority ... [B]eing white, Anglo-Saxon, Christian, and bourgeois [were] crucial to the construction of authoritative masculinity', which was personified in figures such as 'the muscular Evangelical', 'the colonial man of action' and 'the intellectual man'.[41]

Even within this narrow template, settler colonists comprised a range of cultural configurations; that is, not only British immigrants but also Europeans whose ethnicity and socio-political affiliations subtly distinguished them from policy-makers and other authorities in the metropole and from the British-born settlers. As such, settler colonisers were never inherently unified, nor did they necessarily share common interests and concerns; as Laura Stoler insists, 'their boundaries – always marked by whom those in power considered legitimate progeny and whom they did not – were never clear'.[42] There was a very specific ordering of masculinity in colonial society, which involved an imperial class dominated by 'the manly Englishman' and comprising senior British officials and social elites.[43] Other groups and classes occupied a place further down the hierarchy depending on the attributes they shared with the figure of the manly Englishman. Classifications of the relative 'manliness' of different races or ethnicities and their relative social status circulated widely throughout the British Empire and became so entrenched that they were received as part of general knowledge.[44]

Although non-British colonials were seldom accepted into the highest echelons of society, self-made men such as Krefft could move up the hierarchy of colonial masculinity through recognition of their accomplishments. However, they could also slide back down again if their perceived or actual authority – either intellectual or institutional – was challenged or diminished. For example, Harriet Scott, whose father, the entomologist Alexander Walker Scott, counted the Macleays as family friends, alluded to Krefft's rising reputation in her letter to Edward Ramsay: 'Fancy Mr Krefft introducing Papa to *a friend of his*'.[45] The comment underscores the entrenched divisions in colonial society based on class and ethnicity, and indicates that although the German was acknowledged for his professional position and scientific achievements, he nevertheless remained an outsider in Sydney's Anglo-Irish establishment.

German immigrants comprised the largest category of non-British settlers in the Australian colonies in this period.[46] Krefft's ethnicity – and others' preconceptions about his 'European temperament' – would always be an issue in Sydney society nonetheless. Yet, while his reforms at the museum and scientific accomplishments caused some resentment, the early trajectory of his career was not dissimilar to many colonial naturalists. Krefft arrived in Australia in 1852, having travelled from Germany to New York a year or so earlier.[47] After working on the Victorian goldfields for five years and having made some money, he applied himself to studying John Gould's *Mammals of Australia* in the Melbourne Public Library.[48] Around that time, Prussian–Polish-born William Blandowski, the Victorian government zoologist, was recruiting for an official expedition to the junction of the Murray and Darling Rivers, which was due to depart in December 1856. He engaged the young German as a naturalist and secretary, and, although the two men went on to develop a volatile relationship – owing to them both being extremely ambitious – their meeting marked the beginning of Krefft's career.[49] Krefft went on to produce an accomplished body of illustrative work during that expedition, comprising watercolours and drawings in pencil and ink (Figure 2.1).[50] The expedition also introduced him to

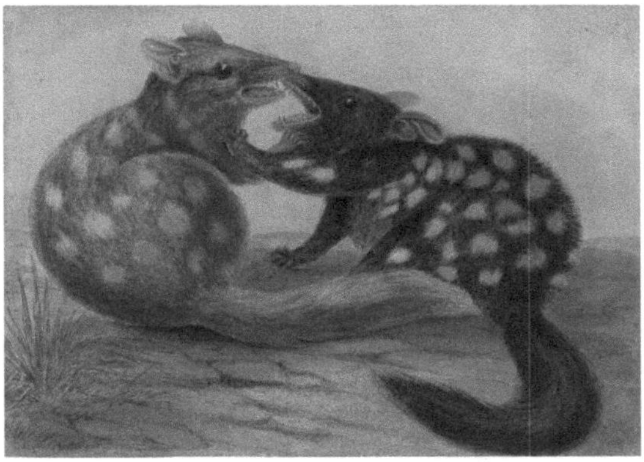

Figure 2.1 Gerard Krefft, *Dasyurus Viverrinus* (Eastern Quolls), 1860, watercolour, in Gerard Krefft album of watercolour drawings, 1857–66. Collection: Mitchell Library, State Library of New South Wales (63890).

Frederick McCoy, professor of natural sciences at the University of Melbourne and director of Victoria's National Museum, who invited Krefft to catalogue the expedition specimens presented to the museum. With these contacts firmly established, in 1858 Krefft briefly returned to Germany following the death of his father. En route, he had visited England, where he laid more of the groundwork for his future career in the Australian colonies by presenting a paper at the Zoological Society of London, visiting the British Museum and introducing himself to some of the major figures of imperial science.[51] He did the same in Germany and secured a contract to supply Australian specimens to the German Museums Commission. Upon returning to Australia, Krefft presented his, by then, many references from metropolitan and colonial figures to Governor William Denison in New South Wales and was subsequently offered the position of assistant curator at the Australian Museum in Sydney under the curator Simon Pittard, who was already infirmed with tuberculosis and died a year later.[52]

When Krefft took up his appointment in June 1860, the Australian Museum – which had opened to the public in 1857 – comprised a single exhibition gallery and several floors of staff living quarters.[53] Copies of Darwin's *Origin of Species* had been available in Sydney since February 1860, and his theory of evolution by natural selection was starting to attract local interest.[54] Krefft soon began to instigate major reforms at the museum.[55] After acting in the position for three years while the trustees wavered about his suitability, he was promoted to the position of curator and secretary of the Australian Museum in 1864 (Figure 2.2). It was a post that conferred a high level of status in the intellectual and social world of the colony of New South Wales. Yet, this status was dependent on him deferring to the authority of the museum trustees, who included Dr George Bennett, Rev. W. B. Clarke and William John Macleay, whose own standing within the London scientific establishment was inalienable.

One need look no further than the figure of Krefft and his identification as a cosmopolitan to recognize the flaws in any crude characterisation of the subservience

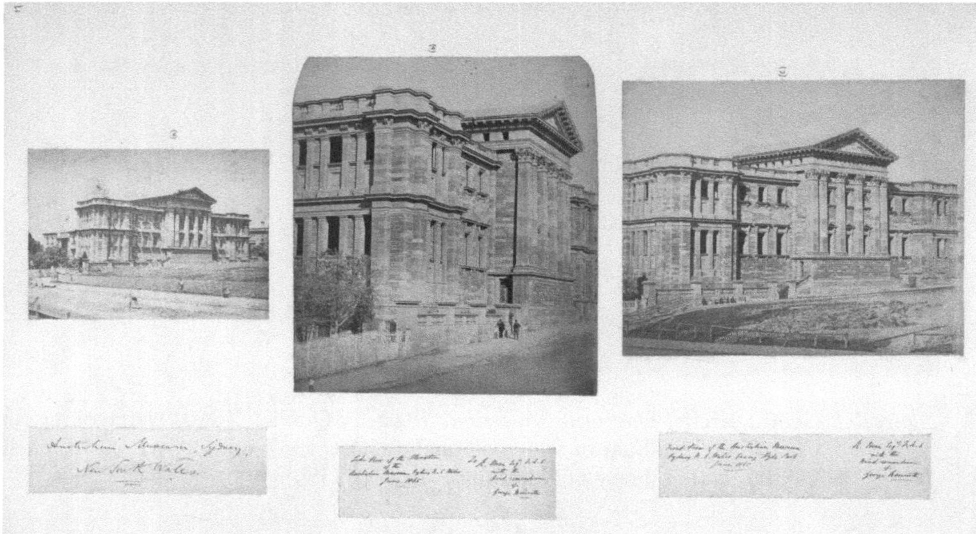

Figure 2.2 Henry Barnes (attrib.), *Australian Museum, Sydney New South Wales* (front and side views), June 1865, Richard Owen Collection of Palaeontological and Zoological Drawings, f.17 (albumen silver photographs sent by George Bennett to Richard Owen). Collection: Natural History Museum, London.

of colonial natural history to imperial authority. A series of photographic portraits of Krefft, produced in 1868, challenges the simplistic view of the colonial naturalist as mere apparatchik (Plate 5). The photographs appear deceptively informal despite Krefft adopting a range of poses and variations of attire as if rehearsing for different roles. Taken in a courtyard of the Australian Museum, the portraits include a potted plant and, in some views, also an empty jardinière, indicating the multiple functions of the museum and underscoring the complex relationship between labour and leisure and the practical and aesthetic aspects of Victorian natural history. Carefully positioned, they provide not only a formal counterpoint to Krefft within the composition but also blur the boundary between public areas and work places and allude to the domestic spaces of Krefft's own living quarters within the museum, thus conferring upon him a sense of ownership.[56] Most striking is Krefft's relaxed composure in front of the camera and the performative nature of the portrait sequence. Swapping between vest and shirt sleeves, and then top hat and coat, his changing dress and posture create a narrative in which his guise alternates between that of a confident flâneur or dandy – swaggering a little and leaning proprietarily against a wall with hands in pockets – and a more self-disciplined, reflective thinker. Occasionally addressing the viewer, he assumes control of the space in which he variously stands and sits. Yet, it is the handsome specimen of a huge manta ray, laid out before him and extending from wall to wall, that dominates the composition and gives the photographs much of their significance. Certain visual conventions have been adopted: the placement of the fish in a different orientation in each photograph to show its features from all sides; and the use of the human figure and architectural surroundings to give an indication of scale. Conspicuously, the series is also a tribute to the official visit of Prince Alfred in 1868, for whose entertainment Krefft staged a fight between a snake and a mongoose during a Friday afternoon tour of the museum, an extraordinary spectacle that went well beyond the bounds of Krefft's official role and exemplified the colonial museum as a site of complex, sometimes conflicted, identities encompassing a range of spontaneous and unconventional activities. As reported in the press, Krefft guided the royal visitor through the museum, pointing out various objects and collections of interest, including manuscript journals and other relics from the late lamented German explorer and naturalist Ludwig Leichhardt:

> Krefft then produced a case containing some live snakes, and one of these was taken out and placed on the floor; at the same time a mongoose, the property of Mr. [Henry] Parkes, [Colonial Secretary], was liberated. In a few seconds the mongoose observed the snake and at once attacked it. The reptile appeared perfectly powerless, and in a few seconds lay dead on the floor. Another smaller but not venomous snake was brought, and this shared the same fate. The little animal seized the reptile by the head, and with one bite placed it *hors de combat* ... His Royal Highness watched the experiments with the mongoose and snakes with much interest. Having inspected the whole of the Museum, and expressed his thanks to Mr. Parkes and Mr. Krefft, his Royal Highness drove back to Government House.[57]

This powerful, impromptu performance was clearly intended to impress Alfred and to position Krefft not simply as a loyal subject but to draw attention to a more tacit connection between them as both being men of the world in every sense. Having discovered the manta ray in that same year, Krefft declared his loyalty to the Duke of Edinburgh by naming this new species in his honour: *Ceratoptera alfredi* anonymous [Krefft]. However, Krefft's allegiance was not merely to the British throne but also in recognition of Alfred's heritage and ongoing connections with Germany as heir to the duchy of Saxe-Coburg Gotha. In Britain, Alfred's association with Germany was well known and openly discussed, often in unfavourable terms; yet it had never been a particularly discernible aspect of how Alfred defined or promoted his persona. During his royal visit to Australia, on the other hand, Alfred explicitly acknowledged – indeed spectacularly displayed – both his German and British imperial connections by wearing German military dress at local military reviews, including the uniform of a Prussian colonel at a review of troops in Adelaide in November 1867 (Figure 2.3) and that of a general officer of Saxe-Coburg Gotha at a review of military and volunteer forces in Sydney the following January.[58] He also portrayed himself as a manual worker – a symbol of the practical colonial man – when he posed in a mining costume at the Band of Hope Gold Mine in Ballarat in December 1867 (Figure 2.4).

By devising his portrait series as a homage to the prince, Krefft demonstrates his identification with Albert's German heritage and transnational identity as evident on his tour of the colonies. The series also celebrates Krefft's achievements in imperial science and his increased standing in colonial society. For Krefft, his rising status was due in part to the transnational dimensions of his own identity, which came to the fore as he navigated the networks of Victorian science and society. Central to these connections were the many ongoing exchanges of specimens that were a core function of virtually all colonial museums. As such, the cultivation of networks determined the longer-term strategic direction of most museum-based naturalists, influenced the focus of their scientific work and, for many, became an indicator of national loyalties and personal ambitions.

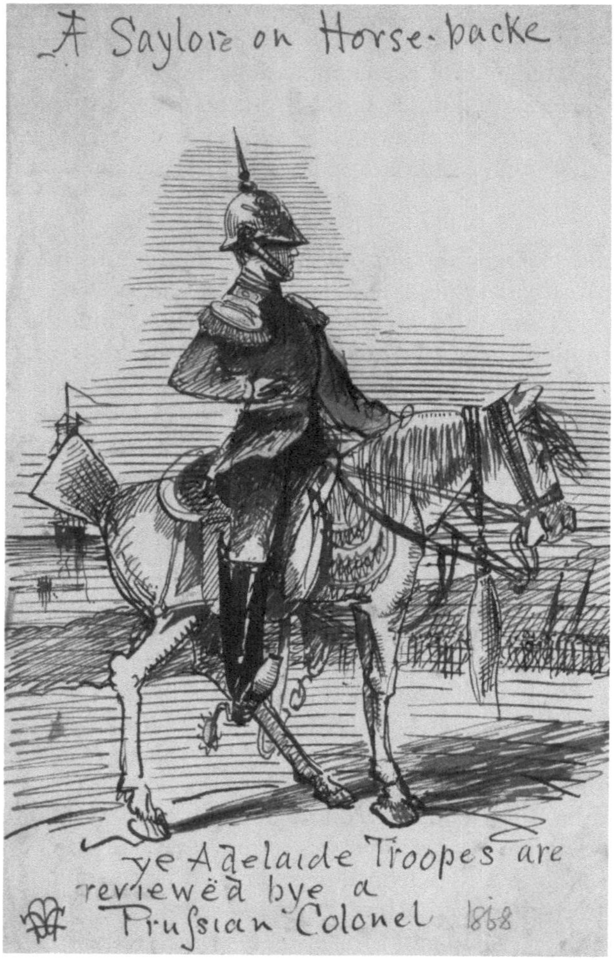

Figure 2.3 George Gordon McCrae, *A Saylore on Horse-backe. Ye Adelaide Troopes are reviewed by a Prussian Colonel*, 1868, pencil and watercolour. Collection: National Library of Australia, Canberra (nla.obj-139386514).

Naturalist correspondents, photography and the reframing of natural history

Transnational figures such as Krefft were subject to a range of different influences. A good deal of cultural transfer, both visible and invisible, occurred between Germany and Britain's colonies. These included specialist and popular publications, which gave German readers an overview of the most important expeditions into the inland regions of Australia; and Germany's state museums – established from the turn of the century, and which had acquired expedition collections of Australian species and associated records created by an earlier generation of German naturalists.[59] Moreover, the transnational network linking Germans scientists in different locations across the British Empire was significant. For instance, when Krefft visited London en route to Germany and read a paper before the Zoological Society in 1858, he made the acquaintance

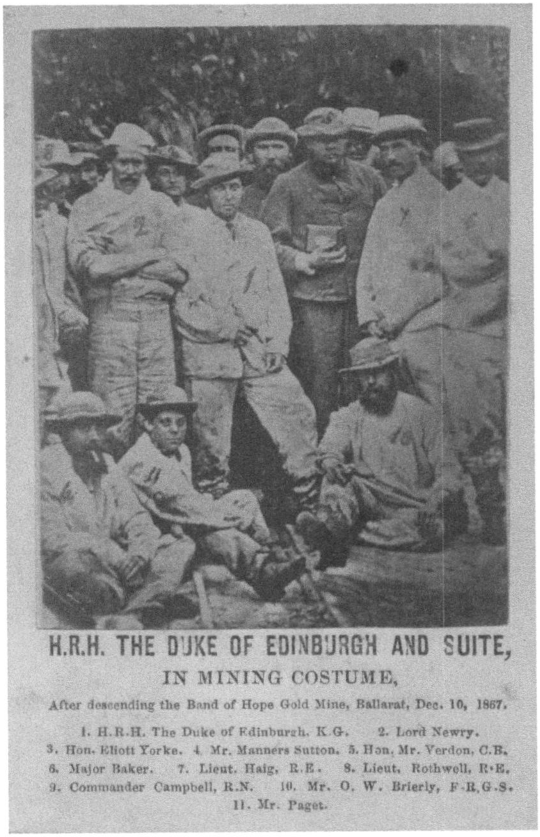

Figure 2.4 Roberts Brothers, photographers, *H. R. H. the Duke of Edinburgh and Suite in Mining Costume after Descending the Band of Hope Gold Mine, Ballarat, Australia*, 10 December 1867, *carte-de-visite* photograph. Pictures Collection, State Library Victoria (H84.439).

of Albert Gunther, among other figures, at the British Museum. Following this visit, Gunther became his confidant, with the two men exchanging photographs – including their own portraits – and many long letters, both in German and English, over the ensuing 10 years. The letters covered natural history matters and also Krefft's concerns regarding the management of the museum in Sydney. In 1864, when Krefft was promoted to curator at the Australian Museum, he sent a *carte-de-visite* photograph to Gunther (Figure 2.5). Posed holding a snake that trails up his left leg and across his lap, and whose head extends menacingly from his right hand towards the camera, Krefft no doubt aimed to demonstrate his growing expertise in herpetology, a field of study in which he would later distinguish himself for his expert knowledge of the Australian species.[60] The benefits that accrued to the Australian Museum from the informal correspondence between these two German compatriot naturalists were substantial. While the British Museum had received countless specimens from Australia prior to Krefft's friendship with Gunther, it was through Krefft's assiduous correspondence, Gunther notes, that the Australian Museum 'became the first to receive an exchange of specimens from the British Museum, a reward for its many contributions'.[61]

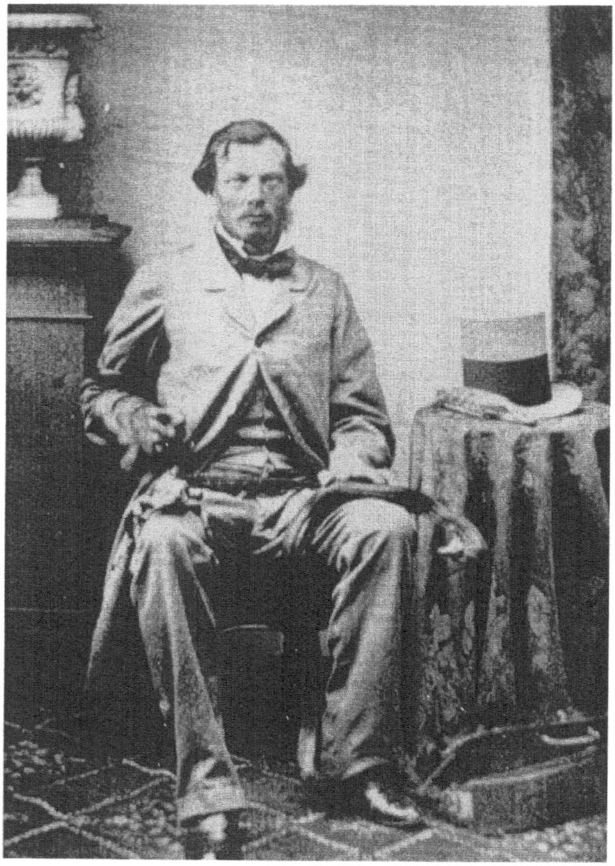

Figure 2.5 Photographer unknown, *Portrait of Gerard Krefft sent to Albert Gunther at the British Museum in 1864*, reproduction of an albumen silver photograph. Collection: Australian Museum, Sydney.

In the past, Royal Navy expeditions were a major source for the empire's principal collections such as those of the British Museum and the Royal College of Surgeons. From the early decades of the nineteenth century, metropolitan savants began to seek out different sorts of specimens and new areas of knowledge by engaging the services of inland travellers, settlers, colonial governors and other civil servants, corresponding with them either directly or through senior officials or local figures they knew or trusted by reputation.[62] As Richard Owen declared of the British Museum, 'The annual additions of specimens continued to increase in number and in value year by year. I embraced every opportunity to excite the interest of lovers of natural history travelling abroad, and of intelligent settlers in our several colonies'.[63] Albert Gunther, in the British Museum's zoological department, was just as assiduous in his own of field of ichthyology; he cultivated his own network of German-born contacts and also 'missed no opportunity of making contact with anyone in any part of the world ... [who] might be persuaded to assist in collecting'.[64] From the colonial perspective, Peter Cunningham wrote that, 'a number of individuals in Sydney earn a good livelihood by collecting our beasts, birds and insects; stuffing, preparing and arranging them in cases; and disposing of them to individuals leaving the colony'.[65] Amateur gentlemen–naturalists, including William Macleay and Owen's long-standing friend, Dr. George Bennett, personally contributed to England's burgeoning collections, initially in a private capacity and then also through the Australian Museum, from where they coordinated collecting activities across New South Wales and in other Australian colonies.[66] With the formation of colonial institutions such as the Australian Museum – whose early trustees had personal connections with London's scientific establishment – this role continued automatically and initially without demur.[67] As Gunther noted, the curators of the 'newly established museums in the growing colonies were an early source of specimens ... [and] correspondence over the next thirty years with these men assumed enormous proportions'.[68] The many German scientists in colonial institutions represented a useful network predisposed to assist Gunther in this way.

Of course, there were benefits to be had on both sides. Many exchanges were clearly as advantageous to those providing the specimens as they were to their recipients. This was especially the case in the early years of colonial museums, when resources and expertise were limited. Thus, while a typical exchange of letters might have contained details and clarification about the latest shipment of specimens, it might also have included requests for specialist books and scholarly papers. These transactions, however, were not always equitable. The Australian Museum, for instance, ensured an almost continuous traffic of specimens bound for London collections while at the same time purchasing cases of books chosen by Owen on their behalf.[69] With the selection of these publications often left to the judgement of the sender, metropolitan savants could thus ensure that specialist knowledge was cultivated in such a way as to complement their own collecting requirements.[70] Alternatively, Gunther insists that 'not all potential collectors gave as good as they got. Pro-consuls of the empire anxious to make a "contribution to science" could be a nuisance ... [H]undreds of written questions took more time to answer than to ask. Nor was there [always] a reward in specimens'.[71] Gunther's remonstration highlights the expectations that metropolitan savants had of colonial collectors and their exasperation if the balance of those relationships was not to their advantage.

The exchanges that flowed from these global connections were fundamental to both the naturalists' research and their institutions' progress. To interpret these imperial and transnational relationships solely in terms of the transfer of actual specimens, however, would mean neglecting other significant factors that defined the practice of colonial natural history, including scientific illustration and the multifarious uses of photography. From the first days of settlement, the two activities of specimen collection and natural history illustration were almost synonymous. While traditional illustrations functioned primarily as scientific data, they were often also impressive works of art, appreciated as much for their exquisite aesthetic effects and extraordinary detail.[72] As prized objects, the movement of such illustrations through colonial life was usually quite distinct compared to photographs, which tended to be treated as interim records. Yet, the fact that it is not unusual to find photographs being requested, referred to and circulated amongst naturalists from the 1850–60s onwards indicates something of a turning point in the way in which the observations and speculations of correspondents were communicated, shared or authenticated.

When the Australian Museum commenced its photography program around 1864, it was introduced with the intention of fostering a new form of communication that would promote, but in some respects also replace, the traffic in natural history specimens. Late that year, Krefft arranged for a number of images to be taken with the express purpose of sending them to Owen at the British Museum. They were the first of many that the young German naturalist would present to the professor over the next decade or so. On that first occasion, just three small albumen prints were sent and, while they all depict animal skeletons, the representation of natural history specimens in those three images could not be more dissimilar or more calculated to impress (Plate 6). In two of the photographs, a carefully articulated specimen – shown from above and side on – has been placed against a neutral background to delineate its complex structure.[73] The museum's taxidermist, articulator and, recently, museum photographer, Henry Barnes, was almost certainly behind the camera, probably having already helped Krefft to compose each scene. In these images, the adherence to scientific and aesthetic conventions faithfully follows the protocols of natural history illustration and was surely intended to demonstrate not only the expertise of the museum's articulator but the efficacy of photography for making this kind of scientific record, which could be interpreted by practitioners anywhere.[74]

The third photograph sent from Krefft to Owen reveals a chaotic group of skeletons and stands in stark contrast to the first two images. It is perhaps also indicative of the realities of practicing natural history on a global scale. As Krefft points out in his accompanying letter, a shipment of specimens selected by Owen to furnish the Australian Museum had arrived 'in such a wretched condition that it [would] cost a considerable sum to repair them'.

> [Eight] of the smaller animals ... are completely destroyed, partly by being crushed by some of the larger & heaviest bones & by the action of the sea water, the ordinary precaution of having the packing cases lined with tin being omitted. You will perhaps scarcely believe it Sir but it is a fact that Mr. Gerrard did not consider it worth his while ... to pack the smaller skeletons into separate cases, no Sir he fitted a Cape Hyrax & a Hare into the same case with a camel & a bear not protecting these small skeletons otherwise than by wrapping some paper round the bones – the consequence is that some of the larger bones, becoming detached, crushed the small skeletons & the action of the sea water did the rest.[75]

Presenting a direct view into the interior of the shipping case, the image is clearly intended to corroborate Krefft's claims. A shattered jumble of fragile bones betrays the careless work of the British Museum's taxidermist, who packed the specimens for transport to New South Wales. Krefft's use of photography here thus had a dual purpose, for while he took the trouble to arrange a photograph showing the poor condition of the skeletons sent from London, he was just as careful to ensure that he also enclosed 'for [the professor's] inspection [some] Photographs of specimens prepared at the Australian Museum'.[76] Whereas the subject of these latter two photographs would probably have seemed unremarkable to Owen, the image conjured by the other photograph would previously have been conveyed only through written exchanges, as in Krefft's accompanying letter. Although naturalists depended heavily on correspondence as a means for gathering information and procuring or presenting specimens, issues about reliability surfaced when the moral standing of correspondents was uncertain or the manner of communication was unconventional. Anne Secord argues that 'the early nineteenth-century community of naturalists largely overcame these problems, by developing an almost formulaic mode of correspondence, which served to establish the trustworthiness of a writer proposing an exchange of specimens, books or ideas'.[77] Having corresponded just once or twice during the preceding two years, Krefft and Owen had barely established a relationship, and yet this letter and many of those that followed, although quite innocuous on the surface, signalled Krefft's determination to participate in the theoretical and speculative side of science rather than just attend to its practical aspects. Anticipating a new trend in professional assertiveness in Australia, Krefft was one of the first to advocate colonial independent expertise by contesting the supremacy of British mentors as the undisputed regulators of all aspects of imperial science.[78]

With a greater recognition of natural history as a vocation that came with the increasing prominence of colonial museums, Australia's hitherto peripheral role in imperial science became more complex. As the colonies began to establish their own scientific institutions, in Sydney, Hobart, Melbourne and Christchurch, it was often German naturalists who tended to have the requisite experience to fill the professional posts. Throughout most the nineteenth century, Germany was much more advanced than Britain in formal scientific education. In contrast to many of the British-born immigrants who took up scientific pursuits, the majority of the Germans arriving in Australia who subsequently became known for their scientific work came already trained for careers as professional scientists.[79] In other cases, expatriate connections enabled self-taught figures such as Krefft to join scientific expeditions and then rise through the ranks of the profession. Crucially, it was the naturalists from these traditions who seemed to move most effectively within and across the global natural history networks. In this respect, some early studies of colonial science tended to generalise the settler societies as culturally homogenous and failed to account for the cultural complexity introduced by cosmopolitan figures such as Krefft. Like many scientists working in Australia, England was never 'home' for Krefft as it was for the majority of colonists. Nor was it necessarily regarded as providing the sole intellectual influence for his investigations. For many, however, the peak scientific bodies in London remained pre-eminent on the world stage.[80] Acknowledgment of one's discoveries and achievements in London was the equivalent of world-wide recognition. Consequently, it was as much pragmatism as impartial deference that compelled German naturalists to suppress their national loyalties and, instead, to court the favour of powerful London-based institutions and individuals.

With an expanding network of correspondents, Owen, Gray and Gunther at the British Museum communicated daily with numerous naturalists and other collectors in distant locations. The main benefit for metropolitan scientists in nurturing these correspondents was access to specimens. Therefore, while many naturalists in the colonies were developing unique and specialized knowledge of their own or adopted regions, their status in imperial science remained for the most part as field collectors and was based on the quantity and quality of their collections. Their role was vital for metropolitan scientific communities but largely invisible and almost always underutilised in the process of settling and formalising scientific knowledge. On those occasions when colonial naturalists attempted to engage in the more speculative aspects of scientific discovery by offering their own analysis of specimens, the response from their metropolitan confreres usually ranged from disdainful to politely dismissive. Of course, the response depended very much on the social status and credentials of the colonial naturalist concerned. Yet, generally, observations and images offered by colonial figures were simply incorporated into the scholarly findings of metropolitan scientists without due recognition.

Expeditionary photography and the 'New Traveller's Tales'

There was always some doubt surrounding the material evidence and observations received from naturalists and explorers in the colonies, as they did not have the same credentials or social standing as those located at the centre of imperial science.[81] This was especially the case for those correspondents and collectors who were from non-British cultural backgrounds and with no significant connections in London. For Owen and Gray at the British Museum, these issues came to the fore in 1861, when they became embroiled in one of the most public disputes of their respective careers. At the heart of this dispute was the validation of scientific evidence and challenges to the reliability of photography for that purpose. The year-long debate, played out in *The Athenaeum*, *The Times* and several science journals, pertained to the explorations of French–American adventurer Paul du Chaillu in Equatorial Africa and involved not only Owen and Gray but also photographer Roger Fenton and prominent publisher John Murray. Du Chaillu's authority was challenged based on his lack of formal education, low social standing and possible mixed-race background.[82] Exchanges took place at the peak scientific societies in London and in the pages of the metropolitan press. Reverberating across the empire, the substance of the debate was zealously reported in articles reprinted in colonial papers such as the *Sydney Morning Herald*.[83] For colonial naturalists, explorers and other contributors to imperial natural history, the controversy – which arose less than a year after Krefft's appointment to the Australian Museum – was an object lesson in the perils and consequences of overstepping the bounds of one's position and attempting to communicate new knowledge to scientific authorities at the centre of imperial science.

The Du Chaillu affair started with a mostly positive, but unattributed, review and a response from Gray that questioned the accurateness of facts described in the traveller's recently published *Explorations and Adventures in Equatorial Africa*.[84] The issue escalated with Owen's public support for the explorer and proposed purchase from him of gorilla skins and skeletons for display in the British Museum. Owen had much invested in settling the scientific classification of the gorilla and the question of its relationship to man, as he had been publicly debating the matter with Thomas Huxley since the previous year, arguing that humankind belonged to a distinctly different order than the great apes.[85] These exhibits would help Owen support his case against Huxley. Furthermore, they had the cachet of being associated with a high-profile and intrepid explorer whose thrilling escapades, recounted in his popular book and reflected in his public persona, would only enhance the public's reception of them.[86] As Owen wrote to Murchison just prior to the eruption of the affair, Du Chaillu was an 'adventurous traveller' who reminded him 'of Livingstone [more] than any other of the African wanderers', and it was exhilarating just to converse with 'a man who had stood face to face with the living gorilla in his native forest'.[87] The controversy gained momentum. Gray's initial article was followed by a succession of commentaries written by Gray and other contributors that continued to identify inconsistencies, inaccuracies and fabrications in the adventurer's account.

A vital piece of evidence in Gray's damning critique was an illustration comparing human and gorilla skeletons that Du Chaillu presented as his own but which, as Gray argued, bore an unmistakable resemblance to a well-known photograph by Roger Fenton. As another anatomist concurred, Du Chaillu's illustration was clearly 'an exaggerated copy of Fenton's photograph' (Figure 2.6).[88] Questions were also raised about discrepancies in the book's stated chronology and locations of events. One correspondent pointed out that certain aspects of Du Chaillu's adventure 'were wholly irreconcilable' and, as Gray added, must have been partly or entirely fabricated.[89] Such inconsistencies might just as easily be attributed to a poorly maintained traveller's journal. On the other hand, the explorer's obvious appropriation of a published photograph seemed to provide incontrovertible proof of the dubiousness of some or all of his claims and was a fact that could be discerned by anyone who cared to compare the images regardless of their scientific knowledge or first-hand experience.

To Gray, the appropriation of Fenton's photograph was obvious, and he objected to Du Chaillu having copied the plate without acknowledging its source. The skeleton of a gorilla that was already in the British Museum collection had been photographed

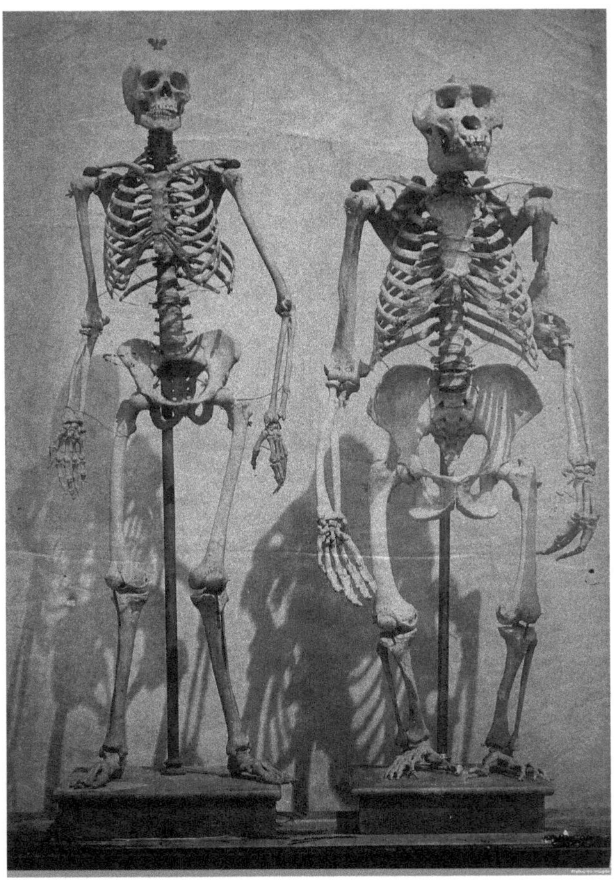

Figure 2.6 Roger Fenton, *Skeleton of Man and of the Male Gorilla (Troglodytes Gorilla) I*, 1854–58, salted paper print. Collection: Wellcome Library, London.

by Fenton in the spring of 1858.[90] The actual photograph used by Du Chaillu, as Gray pointed out, was taken from this group of studies made by Fenton for the trustees of the British Museum. These images had recently been transferred to the South Kensington Museum, where they were readily available for purchase by anyone 'for a few pence'.[91] Any careful observer could not fail to notice the similarities:

> In the specimen in the British Museum, the upper bone of the left arm had been broken, and, consequently shortened, and this is represented in Mr. Fenton's photograph. In Mr. Du Chaillu's copy … the bone is represented as perfect, but the copyist has forgotten to lengthen it so as to correspond with the arm of the opposite side … This will be evident to anyone who will examine the two with an accurate eye, and who will take a pair of compasses and measure the relative length of bones. The latter process is, however, quite unnecessary, as the position of the skeleton itself and of its several parts is quite sufficient to determine the question, and to show how little confidence can be placed in the traveller's most positive assertions.[92]

In addition to this, Gray argued that Du Chaillu had, on more than one occasion, copied an illustration representing one kind of animal for the purpose of denoting another that he called 'new and undescribed'. 'He evidently uses these words not in the ordinary sense', Gray explained sarcastically, 'for he calls an animal new and undescribed though he applies to it the scientific name under which it was figured and described in the year 1836, at least twenty-five years ago, and which has been noticed by several African travellers'.[93] A week later, Gray's stance had become even more resolute and called into question every aspect of the foreigner's reports.[94]

The controversy did incalculable damage to Du Chaillu's reputation. Owen, however, went on to reproduce the controversial image in 1866, including it in his own studies of the gorilla (Figure 2.7).[95] In the wake of the controversy, Du Chaillu took lessons in photography in Paris as part of his preparation for a subsequent African expedition. Funded from the sale of specimens to the British Museum, the purpose of this expedition was to obtain evidence for some of his disputed observations in *Explorations and Adventures in Equatorial Africa*.[96] In his subsequent 1872 book, *Country of the Dwarfs*, he recounted his discovery of a large female gorilla, mortally wounded from a hunter's bullet and with a baby suckling at her breast (Figure 2.8):

> [T]he terrific screams of the mother kept me awake … The next morning I immediately prepared my photographic apparatus, and took an excellent photograph of the wounded mother with the child on her lap … The night after I had taken the photograph … her moanings were more frequent and in the morning they gradually became weaker as her life ebbed out, and about 10 o'clock she died. Her death was painfully like that of a human being and her child clung to her to the last.[97]

Rather than reproducing the photograph he took, he included an illustration of himself in the process of photographing the gorilla and, while obviously more intent on demonstrating the veracity of his first-hand account, he failed to capture the utter pathos of the episode as recounted in his narrative. Du Chaillu went on to publish a number of similar popular accounts of his journeys in Africa. Surprisingly, Gray later appeared to relent from his previously uncompromising view.[98] Nonetheless, virtually every aspect of the Du Chaillu controversy, and the way it was reported, drew attention to the

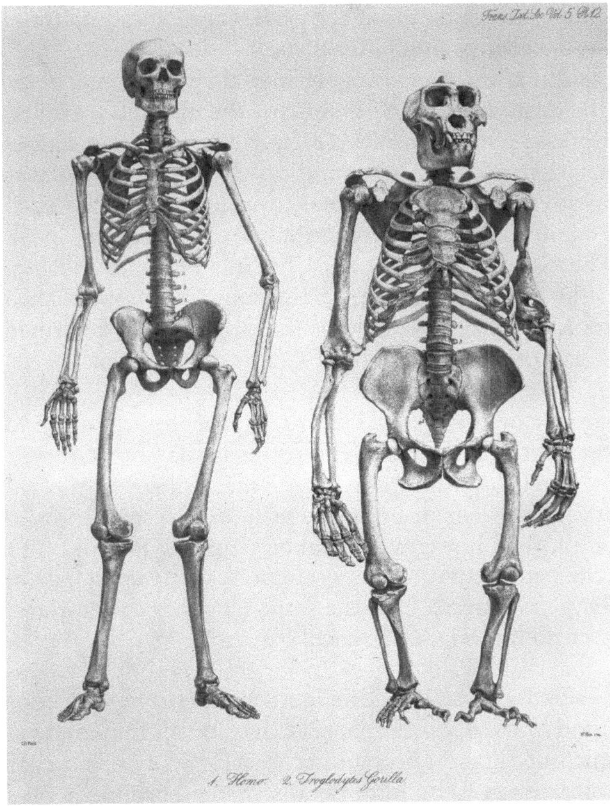

Figure 2.7 *Homo sapiens, human and Gorilla gorilla, western gorilla* [after Roger Fenton] in Richard Owen, 'Osteological Contributions to the Natural History of Anthropoid Apes – No. 7', *Transactions of the Zoological Society of London*, vol. 5, p. 1, 1866. ©The Trustees of the Natural History Museum, London.

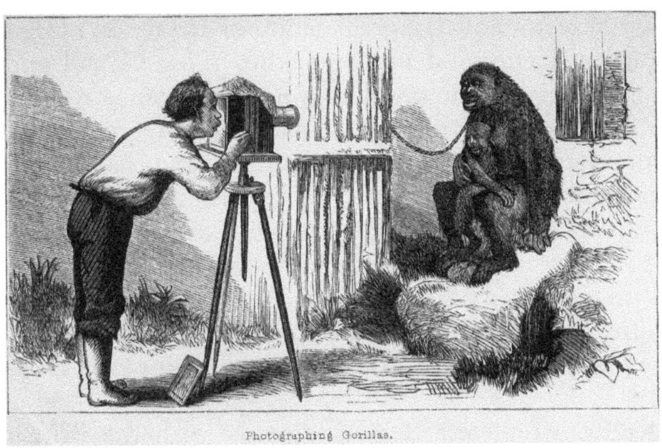

Figure 2.8 'Photographing Gorillas' from Paul Du Chaillu, *The Country of the Dwarfs*, Harpers & Brothers, New York, 1872, p. 83. Source: University of California – courtesy of HathiTrust.

underlying scepticism of metropolitan scientists about the observations and material evidence they received from colonial explorers and naturalists.

In New South Wales, the Du Chaillu affair was keenly followed by a colonial audience undoubtedly spellbound by the dramatic turn of events for the intrepid traveller. Numerous articles in the Sydney press – often syndicated from metropolitan dailies and journals – tracked the course of Du Chaillu's reputation in England as it was transformed from that of a daring adventurer to a colonial traveller of dubious aptitude and integrity and then back again to an invaluable supplier of impressive specimens to the British Museum.[99] 'The Gorilla War', by Du Chaillu's British publisher, John Murray, which appeared in the *Athenaeum* in July 1861 and was republished in the *Sydney Morning Herald*, took the defensive side of the dispute and described Murray's forensic study of Fenton's photograph in light of Gray's allegations:

> I have made a careful comparison of the two skeletons in the photograph and woodcut, aided by compasses and tracing paper. The result is that Dr. Gray ... is utterly mistaken ... With the English photograph before me, I pledge my word that the two skeletons are totally different in position, proportions, and in individual parts ... But more than all, the photograph is taken from a different point of sight, which entirely alters the perspective in the two figures, while the left arm, instead of being shorter, as Dr. Gray asserts, is of the same length with the right, so that the compasses can detect no material difference![100]

To support his conclusions, Murray had also referred the matter to Fenton, who concurred that the two images were fundamentally different and that 'no skill in adaptation could have made the woodcut out of the photograph'.[101] Murray insisted that Gray's profession involved the comparison of minute details in the specimens presented to him; consequently, it was inexcusable for him, 'either wilfully or through want of care, to try to make the world believe that black is white' and, furthermore, it seemed peculiarly ungenerous of him to so virulently persecute a foreign traveller without persuasive evidence.[102] By late October 1861, Du Chaillu's career seems to have been saved by Owen's interventions, with a report stating that 'the African traveller has gained a triumph' with his two gorillas and other animals having been purchased by the British Museum, specimens that were 'now the great centre of attraction in that national exhibition'.[103] As the London correspondent noted, this 'must be gall and wormwood to Dr. Gray, who, as an official of the museum, must be doomed to pass M. Du Chaillu's specimens every day on his way to the department over which he presides'.[104] By December, the same correspondent was reporting that the charismatic Baptist preacher Charles Spurgeon had entered into the debate with a popular lecture accompanied by music, singing and a magic lantern show with dissolving views. Spurgeon's lecture focussed on the veracity of Du Chaillu's claims and the genuineness of the pictures. The columnist was unconvinced of the evangelist's argument, however, recounting that Spurgeon's 'chief ground for thinking the adventure true seems to be that he did not believe that any traveller would have imagination or genius enough to invent them out of his own head'.[105]

For colonial figures such as Krefft, who were dealing directly with the British Museum and other centres of authority, Murray's assertion that Gray was guilty of trying to 'make the world believe that black is white' was surely cause for concern. The challenges for colonial naturalists seeking acknowledgement by their metropolitan colleagues encompassed so many shades of grey, especially when a figure such as Spurgeon completely disregarded the resourcefulness of those working in the field. Other concerns were also evident throughout the dispute, usually conveyed through pointed references to Du Chaillu's foreignness and reports of his flighty temper.[106] By 1863, colonial readers would most likely have noticed that London reviewers had become very wary of travellers' tales and were much more reluctant to unquestioningly accept colonial correspondents' accounts of their exploits and observations.[107]

Prior to his appointment to the Australian Museum, Krefft had had some experience with the problems and uncertainties of photography as a member of William Blandowski's expedition to the junction of the Darling and Murray Rivers in 1857. He thought it was a mission for which they were ill-prepared as they carried throughout the entire journey 'a useless photographic apparatus and chemicals' which they could not get to work.[108] The frustrations involved in photographing on inland expeditions during those early years were always a matter of concern. Initially, the problems were mostly logistical and technical.[109] Despite that, the Royal Geographical Society actively promoted photography. Dr. Pole's 'Photography for Travellers and Tourists', 1871, re-published in the society's journal, endorsed the new dry-plate process and attempted to disabuse prospective travellers of the commonly held notion that photographic processes involved 'a fearful array of dark rooms, huge instruments, chemical paraphernalia, water, and mess, which no sane person, out of the professional photographic guild, would think of burdening himself'.[110] Nonetheless, the introduction of dry-plate photography was initially slow and only came into general use in the Australian colonies from the 1880s.[111]

In 1866, Krefft set off to explore the Wellington Caves, situated 370 km northwest of Sydney, carrying the museum's photographic equipment with him. Discovered during the 1820s, the site was the earliest known source in Australia for the fossil remains of giant, extinct marsupials. The fossils were first noted in 1830 by Bathurst magistrate George Rankin and in the same year by the New South Wales surveyor-general, Thomas Mitchell, who sent some teeth and a fragment of lower jaw to Owen at the Royal College of Surgeons, who would use them to slowly build up a profile of the extinct *Diprotodon*, a rhinoceros-sized ancestor of the wombat.[112] A full report of Mitchell's exploration of the caves – comprising a long evocative account of the physical environment and the difficulties in accessing it – eventually appeared in his two-volume expedition narrative, *Three Expeditions into the Interior of Eastern Australia*, 1838.[113] When Krefft decided to explore the Wellington Caves more than 30 years later, their location and some of their contents were therefore known within elite scientific circles and already had significant cachet.

Colonial naturalists needed to be especially assiduous in recording their working procedures and observations in order to make any progress in imperial science. Prior to Krefft's expeditions, a thorough and systematic investigation of the contents of the caves had never been undertaken.[114] The site proved to be a prolific source for fossil specimens. Krefft's reports from the Wellington Caves expeditions, which were occasionally published in the press, were enlightening documents with regard to the ordeals he and his colleagues endured as part of their explorations as well as the issues facing all colonial naturalists as they attempted to have their findings recognised by metropolitan savants. In this respect, expectations relating to field expeditions revealed as much about the attitude of the scientific elites as they did about the discovered objects themselves.[115] That is, metropolitan scientists were reticent to devolve any of their authority, while those operating on the ground in the colonies came to increasingly depend on their visual and written records as testimony to the reliability of their methodologies, observations and, ultimately, their discoveries.

One of Krefft's more substantial reports, which described one of his first expeditions to the Wellington Caves, was published in the *Sydney Morning Herald* in 1866 and provided particularly thorough detail about the practices of museum naturalists who sought to not only secure specimens but also to document their surroundings and verify their processes. As Jennifer Tucker notes, such dedication to 'accuracy of observation, repetition of small acts, attention to detail, method, punctuality, diligence, perseverance in difficult situations, and devotion to duty' was the mark of a disciplined, self-made man and a basic tenet of Samuel Smiles' *Self-Help* manual'.[116] Krefft's report took the form of a field journal, which involved the reader in his activities and observations on almost a daily basis.[117] His evocative, immersive account calls to mind Alexander von Humboldt's 'intuited connection between the sensuous and the intellectual' as an exemplar of the German Romantic influence in scientific inquiry.[118] As well as bringing science to the public in an accessible way, the purposes of this and other popular articles by Krefft were to establish a thoroughgoing process of authentication of his scientific procedures and to gain broad approval for his work. Departing from the museum on Monday 12 November 1866 with a buggy, two horses and three men, the party took a week to travel the rough roads and tracks to their destination. Krefft then exhaustively recounted his movement through and examination of different caverns section by section, describing the narrow passages that had to be negotiated to access larger vaults but which were so full of rubble and fine dust that it took considerable time and effort to clear them sufficiently to inch through. In these less accessible caverns, the remains of bones and teeth were more abundant. As Krefft was keen to emphasise, however, the 'tedious task of examining great heaps of red dust is difficult to conceive, as the fine particles of the deposit, rising in clouds at every movement of the body, often extinguish the candles, and make breathing difficult'.[119] Nearly all of the deposits had been disturbed, and only one small hollow appeared to be untouched. This space was examined methodically, the dirt sieved and the bones and teeth retrieved, producing an excellent collection of fossil remains. Yet, the party's explorations were soon cut short as heavy rain had set in and they needed to start back for Sydney immediately.

> I left on Thursday, 22nd November, making small stages on account of the heavy roads. That afternoon a thunderstorm came on, and though the tent was immediately erected to protect a large quantity of loose breccia which I carried with me, I regret to say that everything was soon saturated, and a large portion of the precious freight washed into the adjoining Bell River. I do not think that many specimens were lost, as I had been busy when the rain commenced to pick out whatever bones were visible, and obtained a good many the next morning. The roads were very heavy, the horses tired, and the carriage chattered so that but little progress could be made, and Saturday, 25th, found me still within twenty miles of Orange, where I came to a dead halt. I despaired of ever reaching Sydney with the conveyance, and took passage in the passing mail coach bringing all the specimens, with me.[120]

Before abandoning their exploration of the caves, Krefft and his assistants had attempted to obtain a photographic view of the interior of the largest cavern by the aid of magnesium lighting. Unfortunately, this failed, and they had to content themselves with views of the valley and the limestone rocks around the caves: 'Five fine pictures were thus secured, many more being spoiled through the myriads of flies which penetrated everywhere'.[121]

In contrast to earlier images of the area, the photographs reveal the nondescript appearance of the limestone outcrop where the caves are located. Apart from Cathedral Cave, the largest structure, most of the caverns were found to have inconspicuous entrances rather than grand soaring portals. It was in these lesser spaces with their low narrow thresholds, rather than in the imposing caverns set into the face of the limestone, that the richest deposits of breccia and fossilized bones were found. Mitchell's book mentioned some of these smaller cave entrances. Yet, the explorer's main concern in his illustrations had been to capitalise on the dramatic appeal of the larger spaces, with their vaulted ceilings and spectacular shimmering stalagmites accentuating the eerie darkness of their depths (Figure 2.9). Similarly, the artist Augustus Earle, who travelled to the caves in 1826, used the diminutive scale of human figures holding flickering torches to emphasise the massive size of the most impressive of the cave

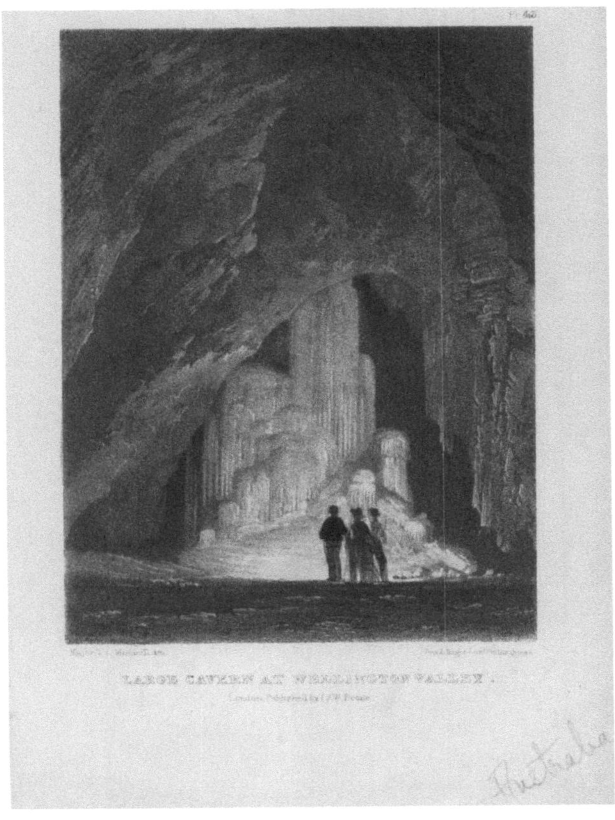

Figure 2.9 Major T. L. Mitchell (Day and Haghe, Lith.), 'Large Cavern at Wellington Valley' in Thomas L. Mitchell, *Three Expeditions into the Interior of Eastern Australia*, vol. 2, T & W Boone, London, 1839. Collection: National Library of Australia, Canberra (nla.obj-136107184).

structures and the labyrinthine system of passages and vertical shafts (Figure 2.10).[122] In these earlier representations of the Wellington Caves, the contrast in scale between man and nature is amplified and the allusion to danger ever present, situating the images unmistakably within the aesthetic tradition of the sublime. Mitchell, in particular, equated the awe-inspiring sight of Cathedral Cave with an almost religious experience, provoked by its soaring interior spaces and astonishing, majestic formations.[123]

The fact that Krefft had not only hoped to photograph the principal cavern but had adopted the observational style of the explorer's journal as his preferred mode of written documentation – both of which had featured so prominently in *Three Expeditions* – is not surprising. He undoubtedly saw himself as following in the footsteps of Mitchell and made a point of reminding readers of the significance of that continuity. Consequently, while Krefft – assisted by Barnes – was unable to record the soaring spaces inside Cathedral Cave using the camera, he did produce a series of contiguous views to construct a seamless panorama of the surrounding scenery, thereby re-creating the first illustration from Mitchell's series of the Wellington area. Drawn

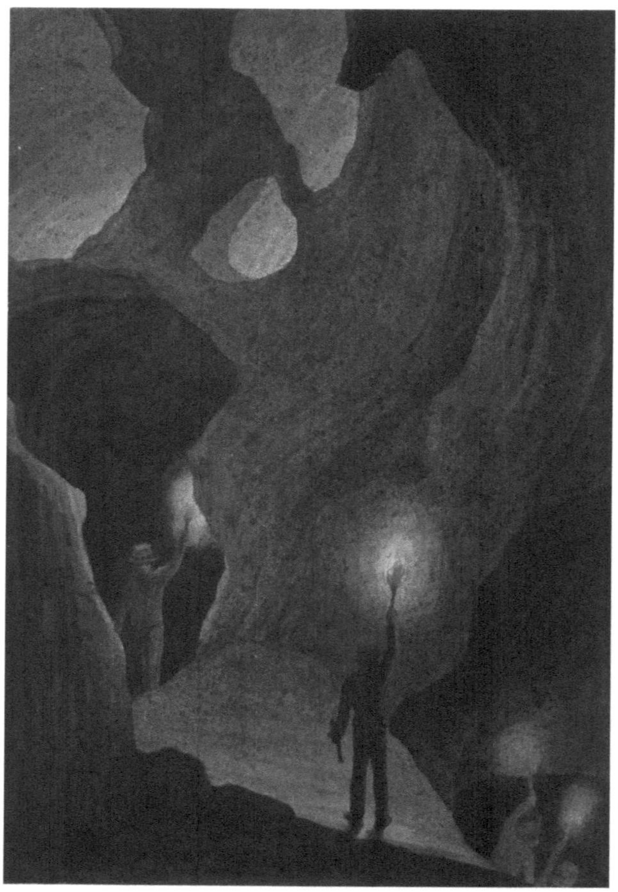

Figure 2.10 Augustus Earle, *Mosman's Cave, Wellington Valley, New South Wales, no. 3*, c.1826, watercolour. Collection: National Library of Australia, Canberra.

by fellow surveyor William Govett, this opening image depicts Mitchell in the process of sketching the entrance to the caves. Although Govett did not adopt the panorama format per se, he clearly intended to give a sense of the land unfolding and extending out and around the explorer. The placement of Mitchell seated below the viewer's eye level and looking across to the caves – rather than standing over and commanding the view – as well as the framing of the composition as a vignette calls attention to the fact that this is only one fragment of a much larger landscape that awaits the explorer's intervention.[124] Just as important is the representation of Mitchell sketching, alluding to the explorer's role in not only surveying the view but in more fully apprehending and accurately reproducing it for audiences 'at home'. As Mitchell states in the preface to the first edition of his published narrative, the sole merit that he claimed with respect to the expeditions was 'that of having faithfully described what he attentively observed; neither his pencil nor his pen has been allowed to pass the bounds of truth'.[125] In constructing his photographic panorama of the Wellington Caves area, Krefft sought to represent the site with the same fidelity but also to reveal the explorer's expanded view of the site; that is, not as a picturesque landscape – constructed and finite – but as an extended field of vision that the viewer could also experience as if they themselves were immersed in the scene.

Krefft's explorations of the cave interiors started in precisely the same location as Mitchell's had ended. As Krefft explained, he located a left lower incisor of the much-talked about marsupial lion, *Thylacoleo carnifex*:

> This tooth, was found within a few feet of the spot where Sir Thomas Mitchell discovered a similar tooth ... thirty years ago, figures of which are given in his *Three expeditions into the interior of Eastern Australia* ... but curious to say without name; nor does Professor Owen in his report on the fossils of the Wellington caves allude to them.[126]

In concluding his report, Krefft expressed the hope that a few hundred pounds would be granted by the legislature to the museum trustees to enable him to continue his research of the Wellington Caves. Based on the collection amassed from that single foray, he anticipated that a thorough investigation of those and other caves of the colony would not only contribute substantially to the knowledge of extinct animals but would be a means of attracting more scientific men to the colonies, 'by whose presence', he reasoned, 'the country would benefit in various ways'.[127]

Krefft's program of photography for the Australian Museum came into its own when he began to use it during the Wellington Caves expeditions, commencing with that early expedition in 1866 and becoming far more prominent during the colonial government-sponsored explorations in 1869 and 1870. With Barnes assisting, Krefft and Dr Alexander Thomson, reader in geology at the University of Sydney, sifted through enormous quantities of loose breccia and secured a considerable amount of fossil material, which, Krefft reported, included 'many valuable and rare specimens, some quite new to science'.[128] Numerous photographic plates were exposed in challenging conditions during that exploration. Further photographs were produced back at the museum and became records of the most important finds, resulting in more than 100 prints being presented by the colonial secretary to the British Museum over the following months.[129] The involvement of the British Museum, and Owen in particular, was pivotal in securing funding from the New South Wales parliament and in gaining

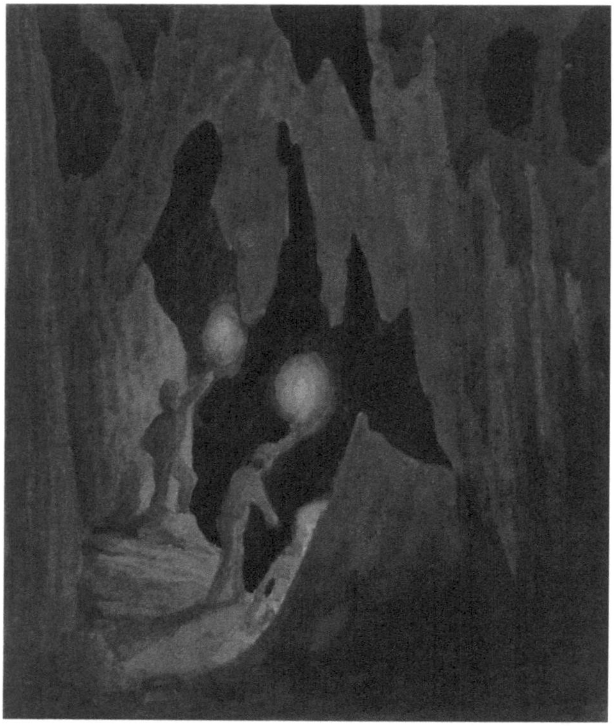

Figure 2.11 Augustus Earle, *Mosman's Cave, Wellington Valley, New South Wales no. 5*, c.1826, watercolour. Collection: National Library of Australia, Canberra.

the museum trustees' ongoing support for the venture. As Owen promptly pointed out to the trustees following a report in the *Times* of the expedition's 'astonishing and unexpected results', he had been corresponding with the colonial secretary since early 1867. He stressed that he had offered to 'devote time to the description of such specimens [as] might be transmitted for deposition in the British Museum' for the purpose of completing his major work on the *Fossil Mammals of Australia* and for which he had been collecting materials since receiving Mitchell's Wellington Caves fossils in 1832.[130] To this request, Owen advised the trustees, the colonial secretary had favoured him with a reply in June 1869, announcing 'that the sum of £200 had been voted by the local Parliament for carrying out your suggestion of a careful and systematic exploration of the Caves and Wellington Valley'.[131] Although modest in scope, the expedition had both imperial and colonial consequences. From Krefft's perspective, it was very much a local enterprise but one that had the potential to bring his own work imperial, and perhaps also international, attention; and to enhance the status of the Australian Museum.

Krefft's extensive report on the expedition included Thomson's account of the caves' geology and a selection of 44 photographs that accorded with the specific requirements of Owen's research. Yet, nowhere among any of these official records do the museum's photographs of the actual caves and environs appear. As Krefft wrote to

Owen in 1866, he had been investigating the site long before the formal exploration commenced, ostensibly to secure some of the more significant caverns against looters but also to take photographs of the 'neighbourhood' before returning to Sydney with 'a considerable number of specimens', which he planned to exhibit at the Exposition Universelle in Paris the following year.[132] Such a prestigious event would enable him to present his discoveries to a global audience. In many ways, however, the bold speculations Krefft broached in his communications with Owen and others at the centre of imperial science were more critical to his reputation than the exposure he gained at international exhibitions and were therefore riskier and likely to have a far greater impact on his career. To have ventured forward with photographs showing the local area or the expedition in progress might well have diminished the status of Krefft's scientific discoveries by casting him as a mere practical man of science or an overly ambitious foreign traveller, as Du Chaillu had previously been characterised by some in the scientific establishment. Photography's value to elite science during the nineteenth century was rarely to show nature 'unmediated' within a particular context – geographers, for instance, tended to use photography to reproduce maps rather than taking it into the field. Rather, it was a means of empowering natural history's own discourse by providing accurate, reproducible records of specimens that could be interpreted by any knowledgeable naturalist from Melbourne to Munich.[133]

The results of the Wellington Caves exploration, as given in the report and illustrated by the photographs, surpassed Owen's expectations 'in regard to novelty, instructiveness, and number of specimens'.[134] Yet, Krefft had withheld from that official account one of the most intriguing of the fossil fragments, which he believed to be a portion of a prehistoric human molar tooth. Almost a year after the expedition, he cautiously raised his 'most important discovery' in a letter to Owen, venturing 'to trouble [him] with a photograph again'. On this occasion, the fossilised molar was flanked by two recent human teeth for comparison: 'I do not think that any animal lower than Man wears its teeth in this peculiar way'.[135] Although very much puzzled by what else the specimen could be, Krefft never explicitly asked Owen's opinion on the matter and quite deliberately sent him a photograph along with his own hypothesis rather than the actual tooth. Owen initially concurred with Krefft's identification but later revised his findings and determined that it was not human. This incident illustrates how Krefft was increasingly turning to photography as a means of demonstrating his scientific abilities and putting his knowledge to the test while also keeping hold of valuable specimens, especially as he was starting to embrace Darwin's theory of evolution. Having discovered the supposedly human tooth among remains of the extinct *Diprotodon*, his conclusion was that it came from an earlier stage in human evolution. In declaring his findings, he was one of the first scientists working in the Australian colonies to openly make such an observation and thus to unequivocally embrace Darwin's revolutionary ideas. For the colonial naturalist, of course, there was always a risk of missing out on public recognition for one's hard work and discoveries if the raw data were merely shipped off to others who would announce the findings at meetings and publish them in the proceedings of metropolitan scientific societies as their own.

Krefft's strategy of sending photographs of specimens to the British Museum instead of the actual objects was nevertheless fraught with complications. The correspondence of metropolitan savants noted the challenges in discerning from photographs all the necessary particulars to satisfactorily describe the specimen in detail. For example, in response to a photograph sent by Krefft purportedly depicting the jaw bone of a new species of *Nototherium*, Owen declared how 'puzzled' he was by what the image he was sent actually showed, although he nevertheless used it to masterfully segue into his own matters and reiterate an earlier request he had made to Krefft to locate certain specimens for him:

> Is the mandibular ramus broken across in front of the 3rd molar, counting forward? And, if so, is the anterior piece with the base of the incisor rightly fitted, or does it really belong, to the hinder portion? You see the difficulty of dealing with Photographs only. However, if [you] carry out your liberal intentions, I feel confident in coming at, or very nearly at, the truth of these matters.[136]

Sometimes, photographs that were sent were unwittingly or conveniently overlooked and simply not acknowledged. On other occasions, the exchange of photographs prompted expansive and thoughtful responses, and thus for colonial naturalists this remained a practicable, if not always dependable, way of extracting information from correspondents and complementing their knowledge without forfeiting their own collections. By sending images along with their written observations or queries pertaining to local conditions of which those at the centre had little or no direct knowledge, colonial naturalists showed themselves to be unwilling to continue as invisible technicians deferentially sending raw data but instead increasingly inclined to more assertively incorporate their own expertise into the process of knowledge formation. This circulation of photographs raised wider questions about the images' status as proxy specimens and thus their reliability as representations able to be interpreted in the same way by different viewers over space and time.[137] The complex social and political framework within which these transactions occurred was also crucial as to how – and, indeed, whether – new knowledge was endorsed.[138] Regardless of the medium, there was always some doubt about the speculations and observations of naturalists in the colonies, who were often self-taught or came from diverse cultural backgrounds.[139] While not dissimilar to their metropolitan colleagues in this regard, these naturalists had fewer avenues through which to make themselves known and to establish their credentials on the imperial stage.

As Owen had argued during the Du Chaillu affair, 'a photograph does not always give the precise testimony required by him who uses it: it may tell more than is wanted; it may represent an object in a position which conveys an erroneous idea of the facts at issue; it may show a manipulation and not a natural structure'.[140] Even Gray could not wholly disagree with Owen's summation of some of the limitations of the medium.[141] Yet, in their key positions at the British Museum, both men found themselves increasingly on the receiving end of photographs intended as substitutes for actual specimens or presented as an alternative mode of natural history illustration. In contrast to the experience of these metropolitan naturalists, whose knowledge was developed and enhanced primarily through their capacity to reflect on, compare and draw together an array of information gathered from different locations, the expertise of colonial naturalists encompassed not only cognitive and theoretical aspects of scientific exploration but the practical and phenomenological aspects as well.[142] The precise recording of observations was therefore critical to the credibility of their accounts, which were entered into journals or produced as visual records usually without delay, often very rapidly and therefore at the moment of greatest intensity for the individuals experiencing them. Hence, colonial naturalists' travel reports negotiated a fine line between the affective and the intellectual. Metropolitan scientists reading their reports were sometimes sceptical, nevertheless, and did not necessarily hold with Humboldt's premise of the intuited connection between these two modes of encounter. Always cautious of hyperbole and self-aggrandisement, philosopher scientists were particularly wary of self-made men whose cultural and educational backgrounds were very different to their own and who seemed overly enthusiastic in their willingness to venture into unfamiliar locations. 'Travellers' tales' was one of the various disparaging terms that had been used not only in the press but also specifically by Gray to describe Du Chaillu's accounts of his adventures. It was an expression usually directed from the centre to the periphery and was of concern to all colonial explorers who wished to have their discoveries accepted. That is, the perils of scientific exploration were not confined to the physical and psychological hardships of field expeditions but extended to the uncertainty surrounding the recognition of one's speculations and findings in an environment of mounting distrust of scientific interlopers, pretenders and showmen in a period when natural history was becoming increasingly professionalised.

Conclusion

In contesting the pre-eminence of the metropolitan savants, Krefft made a compelling case for a greater role for colonial expertise in imperial science. His efforts to transform the Australian Museum into a modern institution that was also relevant to the broader public was crucial in increasing his standing in the colony, but it was also aimed at ensuring that the institution and its associated professionals would enjoy a reputation of international stature. Possibly inspired by Germany's strong record of supporting science, coupled with its purpose to engage the general populace, Krefft published around 200 popular and scholarly articles in the local press and in international scientific journals, addressing subjects ranging from the description of new species and reviews of Owen's publications on palaeontology to the treatment of snake bites. As well as publishing many announcements of his discoveries, expeditions and museum reforms in the *Sydney Morning Herald*, Krefft had a regular column entitled 'Natural History' in the *Sydney Mail*, which became the vehicle for him to publicly expound his support for Darwin's theory of evolution. Krefft's museum reforms included building up and retaining the museum collections as a distinctive reference collection. To achieve this, he began to reduce the number of specimens simply sent off by colonial operatives to be catalogued and classified at the British Museum and other centres of authority. Moreover, he insisted that only by maintaining such a unique and important collection would colonial naturalists have the opportunity to make substantial contributions to the intellectual and analytical work of the modern sciences in a global context. As he wrote to the premier of New South Wales, Henry Parkes, in 1873: 'a thorough history of our Animals can only be *written in this Country* and in this Colony'.[143]

This view stood in stark contrast to the way the trustees had not only established but wished to continue administering the Australian Museum, which had long served their own interests and maintained their social status in a class system that had been transplanted from Britain to colonial New South Wales. The distinction between Krefft's ambitions and those of his masters was as much an issue of who – or, more specifically, what type of person – should be in charge of the museum and how it should be run. Following the theft of a gold nugget from a display cabinet, which

Krefft reported on Christmas Eve 1873, the museum came under the closest public scrutiny it had received since its establishment more than a decade earlier. The New South Wales Legislative Assembly formed a select committee in March 1874 to report on the system of management and state of the Australian Museum. That process revealed a dysfunctional institution, and Krefft's reputation, along with his career as museum curator, went into a sharp and irreversible decline.[144] Much of the negative opinion against Krefft stemmed from the trustees' self-serving conduct and entrenched club-like attitude to the museum, which, despite Krefft's disapproval and his practical reforms, continued to be used by the trustees as a resource to enhance their own scientific collections and activities.[145] This state of affairs was not uncommon in British and colonial institutions and learned societies, most of which were founded by avocational scientists.[146] With the emergence of professional museum scientists – increasingly recruited from the highly trained and well-connected German scientific network – the clandestine practices and informal agreements that previously were condoned in museums as a mutually beneficial arrangement came to be regarded as an undesirable conflict of interests. As an article in *Nature* quipped: 'The authorities of the British Museum may congratulate themselves on their not being the only governing body which is considered to be on an antiquated and improvable foundation, which calls for a radical and speedy change'.[147] With two of the Australian Museum trustees also members of parliament and part of the select committee investigating the museum, a strong feeling of antagonism towards Krefft by some on the museum's board was not only exposed but explicitly introduced and skewed evidence towards a negative judgement of the curator's character and work.[148]

A list of charges was made against Krefft, allegations that appeared to be without foundation in most cases. The evidence suggested to some committee members that there was a conspiracy against him by the museum trustees who accused him of various misdemeanours, including – most damningly for his professional standing – maliciously breaking open a box containing a fossilized jawbone of a *Diprotodon*, a specimen belonging to Dr. Bennett and intended for transmission to Owen in London: 'It was packed at the museum by the working men, who stated that Mr. Krefft took a hammer, and smashed the box open, scattered the contents upon the floor, and stamped upon them. They said that Mr. Krefft then filled the box with rubbish, and that he declared that he would take care [that] Professor Owen should not get the *Diprotodon*'.[149] Yet, there was also proof that the incident never occurred. In a letter Bennett subsequently received from the British Museum, Owen thanked him for the *Diprotodon* specimen, and said that it had been received in excellent condition: 'The Museum will owe much to [Krefft's] energy and devotion. As soon as our new Museum [at South Kensington] is ready to enable us to sort out our accumulated stores, Sydney shall have my first collection of duplicates'.[150] This confirmed a view within the standing committee that some museum trustees and officers were trying to destroy Krefft's reputation. The controversy was also lampooned in *Sydney Punch* (Figure 2.12).[151] Representing Krefft in far less favourable terms than the committee's final estimation of him, a caricature showed him in the guise of a confidence trickster deviously making a fossil magically re-materialise from under a white handkerchief right before the eyes of the unsuspecting figure of Owen, who appeared as little more than an apparition. Dressed in an evening suit and stepping up onto a weighty publication, the implication was that Krefft was trying to artificially elevate both his social and professional status. In dubbing him 'the Great Wizard of the North', *Punch*

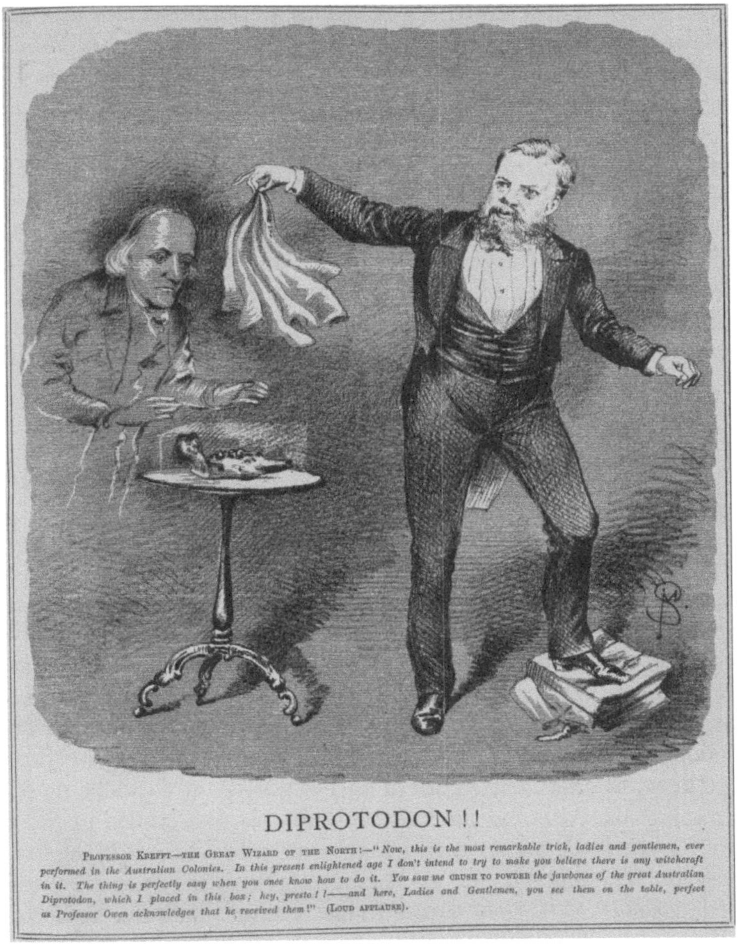

Figure 2.12 'Diprotodon!!', *Sydney Punch*, 26 June 1874, p. 209. La Trobe Collection, State Library Victoria.

likened Krefft to the famous Scottish magician John Anderson, who had died earlier that year and had been as famous for his showmanship and self-promotion as he was for his spectacular conjuring acts.

In spite of this condemnation by the press, the committee's evidence revealed that, as far as the museum's problems were concerned, the trustees had also been culpable in many respects. Almost as a counter-claim to the trustees' indictment of Krefft, the draft report mentioned a number of charges against the board members.[152] As each accusation came to light, the assessment of the museum's trustees became only more damning in their rejection of Krefft's attempts to reform the institution and improve the standing of colonial science: 'The members of the managing body have shown little desire to advance the welfare of the institution committed to their charge [and, instead,] ... have given too much consideration to trivial matters of complaint against their curator'.[153] On the other hand, it was noted from the statements of some trustees, including Dr. Bennett, that Krefft was a talented man of science and a most valuable, hard-working public servant.[154] Yet, with so much negative commentary provided by those with an interest in restoring the *status quo ante* at the museum, the inquiry had elicited a number of facts that made it difficult to entirely exonerate him, and it would have been impossible to allow a public servant – let alone one in such a prestigious position – to remain in office unless he cleared himself of all the charges made against him. During the months of the select committee's inquiry, further dubious accusations against Krefft's character arose. In particular, it was alleged that he was conspiring with Barnes to use the museum's premises and equipment to produce and distribute pornographic photographs.[155] This seemed to be perfectly calculated to permanently tarnish his standing as a man of good character and a responsible public figure. Although the select committee made no such recommendation, the trustees soon thereafter informed the government of their intention to dismiss the curator and, after he was locked out of his residence in the museum, Krefft and his family were evicted from the premises in September 1874.[156]

Stigmatized and shunned by Sydney society, Krefft was financially destitute for the remaining few years of his life and turned to popular science in earnest. In 1877, he published the first and only issue of *Krefft's Nature in Australia*, a journal 'for the discussion of questions on natural history', and he advertised the imminent opening of 'Krefft's Australian Museum'. Beyond those first notices, however, nothing further eventuated from either enterprise, which was perhaps to be expected since a substantial part of his magazine consisted of a long, rhyming diatribe against the museum trustees.[157] In his final months, he returned to his early occupation as an artist to earn a small income from the government printer by hand-colouring large-format photographs for the Melbourne Exhibition of 1880–81.[158] Yet, even in this relatively menial work, Krefft believed he was being cheated out of part of his fees and was being passed over for further commissions. Although he had become a British citizen in the same year he was promoted to the position of curator at the Australian Museum, he felt that Germans in Sydney were 'very much hated' and he had few options other than to turn to his compatriots for assistance.[159] As he wrote to his brother, William, in Germany: 'I am totally ruined because of the untrue and semi-official reports of the Museum Trustees, and I have no choice but to beg the German Consul to clarify my point, that the Government, who hasn't protected me, is refusing to give me work'.[160]

The traducement of Krefft's reputation following the controversy, and his subsequent dismissal, inevitably raised questions about his professional judgement and, for some time, put a pall over all his scientific work and his reforms at the Australian Museum. The year after Krefft's death in 1881, Edward Ramsay, his successor at the Australian Museum, decided to restart explorations at the Wellington Caves, testing the validity of the German's findings for himself and on behalf of imperial science.[161] Barnes commenced a new series of photographs of the caves using magnesium light and finally captured the views of the spectacular formations in Cathedral Cave that Krefft had hoped to attain on one of his very first expeditions. Receiving many 'very instructive letters' from Owen, Ramsay confirmed in his correspondence over the following years a number of details from Krefft's accounts of the Wellington Caves but adopted a much more submissive role in his communications with the aging professor at the British Museum.[162] The laborious field work was soon devolved to Barnes and representatives from the New South Wales Mines Department.[163] Consequently,

Ramsay, like the savants in London, ultimately became reliant on second-hand intelligence rather than his own observations as the source for his scientific reports. In his reports, there is no suggestion of the physical effort or the sense of discovery that Krefft had been so keen to communicate. Indeed, apart from four interior views of the caves, the few extant photographs from those later expeditions give the impression of a relaxed outing in the countryside. With little indication of the digging that the Mines Department had been undertaking, Barnes had minimal investment in the expeditions' scientific outcomes and was clearly more intent on exploring the aesthetic potential of photography as evidenced in his picturesque views of the much less austere surroundings of the nearby Bell River.[164] In these latter photographs, the environment that previously had provided a setting for Krefft's scientific aspirations is transformed from an inhospitable setting of physical danger and masculine prowess to a feminised landscape of abundance, conviviality and reverie.

Through his process of retrieval and meticulous documentation – which entailed the securing of specimens and the recording of physical activities and surroundings – Krefft sought acknowledgment for the arduous fieldwork involved in scientific endeavour and attempted to lay the groundwork that would allow him to ultimately legitimise the scientific analysis he offered. His first-hand experience of the land and the hazards involved in accruing scientific evidence were the crucial factors that separated the fieldworker's knowledge from that of their correspondents, who were the ultimate recipients of specimens and arbiters of discoveries. At the outset, Krefft's detailed observations and regular supply of important specimens would have labelled him as a model contributor to imperial science. However, when he began to speculate more profoundly about the objects he had discovered and sought recognition for his own analysis and identification of fossil specimens, his position and the reputation he had built up within the conservative structure of imperial science and colonial society became untenable. Krefft's field-journal style of reporting his experiences, which replicated his daily procedures, was designed to satisfy the savants in London of the veracity of his observations. Yet, although he desired to be recognised in international, not just local, science, his method could never entirely circumvent the long, complicated process of scientific legitimation that wound its way through the elite scientific establishment. If anything, it exacerbated the precariousness of his status as an outsider in New South Wales's predominantly British colonial society. With the Australian Museum's scientific trustees dominated by amateur practitioners whose main interests and expertise were focussed on amassing their own collections rather than the more analytical aspects of the natural sciences, Krefft increasingly found himself in conflict with the guardians of his own institution. Where the trustees desired only to maintain the status quo of colonial science by simply collecting specimens and collating them – a role that perfectly complemented the objectives of imperial science – Krefft sought recognition and a level of independence for the intellectual work of the Australian Museum.

Nonetheless, Krefft could claim significant scientific achievements and was in fact successful in his aspiration for international recognition. As the influential London journal *Nature* remarked in its reports on the curator's plight and the parlous state of the Australian Museum under the trustees, Krefft's scientific standing was much more secure at the centre of imperial science than many in the New South Wales establishment realised:

> That institution seems to be in the hands of a few collectors of the old school, who treat it as a plaything of theirs, rather than a public institution, supported by

> public funds. They have a curator, Mr. Gerrard [sic] Krefft, of whose very high scientific position in the mother country they cannot be fully aware, or they would be more liberal to him and give him more opportunities for the employment of his abilities.[165]

Moreover, the unidentified author criticised the prevailing dismissive attitude towards the practical work and local specialist knowledge of naturalists. Addressing the need for 'energy and progressive spirit', the writer argued, was as vital to the enhancement of the British Museum as the 'national collection' as it was to the progress of colonial institutions and necessitated scientific independence on the periphery and greater recognition of colonial 'scientific attainments'.[166] Scientific independence was a complex notion, however, and could only be achieved to a certain degree. As one of the earliest colonial supporters of Darwinian evolutionary theory, Krefft flew in the face of conventional opinion and challenged not only the Australian Museum trustees but also Owen at the British Museum, Darwin's foremost ideological opponent and the most powerful figure of imperial natural history.[167] Any individual or institution that broke ranks with the key figures and peak bodies of imperial science so as to embark on an autonomous path was in danger of becoming marginalised and irrelevant. The route to scientific independence, therefore, was more like a gradual emergence from a subservient role to a level of self-sufficiency with regard to the recognised knowledge and experience of colonial scientists and the authority and resources available to them in their own museums.[168] While Krefft's role in the ongoing process he had started became untenable with the abrupt termination of his position at the Australian Museum, which precipitated the end of his scientific career, his reforms and discoveries were recognised by many of his contemporaries and remain on the record as important contributions to imperial and global science.

In the following chapter, I continue this investigation of the professional credibility and public reputation of Victorian naturalists by shifting my focus back to the centre of empire. The foundation of Victorian scientific culture – and thus scientists' own reputations – relied on them accruing, substantiating and communicating verifiable facts. As I have demonstrated, reliability very often centred on the figure of the naturalist himself, identified and portrayed as a model of moral standing, practical application and intellectual rigour. This raises questions about the specific ways in which prominent scientists sought to present themselves, both to their colleagues and to the wider public. In the next part of my inquiry, I will examine the role of portraiture and the visual strategies employed by metropolitan naturalists. I will explain how these reflected not only their station in Victorian public life but also their influence and celebrity in the eyes of the general public and the importance of their work to the civic and professional institutions they represented and their many connections throughout the empire.

Notes

1 P. P. Carpenter quoted in Charles Hedley, 'A Query as to the Synonymy of *Rhysota Armiti*, Smith', *The Annals and Magazine of Natural History*, series 6, vol. 16, no. 92, 1895, p. 201.

2 J. D. Hooker to James Hector, undated (after 23 December 1868) in John Yaldwyn and Juliet Hobbs (eds.), *My Dear Hector: Letters from Joseph Dalton Hooker to James Hector, 1862–1893*, Museum of New Zealand Te Papa Tongarewa, Wellington, 1998, p. 107.

3 William S. Macleay quoted in J. E. Gray, *Catalogue of Shield Reptiles in the Collection of the British Museum*, Pt. II, British Museum, London, 1872, p. 2.

4 Gray, *Catalogue of Shield Reptiles*, p. 2.
5 Hedley, 'A Query as to the Synonymy of *Rhysota Armiti*, Smith', p. 201.
6 Idem.
7 Edgar A. Smith, 'Notes on *Rhysota Armiti*', *The Annals and Magazine of Natural History*, series 6, vol. 16, no. 94, 1895, p. 352.
8 Idem.
9 Idem.
10 Charles Hedley, 'The Imputed Jealousy of European Workers on Australasian Faunas by Local Writers', *The Annals and Magazine of Natural History*, series 6, vol. 17, no. 99, 1896, pp. 258–60. I am indebted to Ian Loch, curator of malacology at the Australian Museum, for bringing this article to my attention.
11 Ibid., p. 259.
12 Ibid., pp.258–9; Steven Shapin, 'The Invisible Technician', *American Scientist*, vol. 77, December 1989, pp. 554–63.
13 Arjun Appadurai (ed.), *The Social Life of Things: Commodities in Cultural Perspective*, Cambridge University Press, Cambridge, 1986, p. 5.
14 See F. A. Bather, 'Some Colonial Museums', *Report of the Museums Association*, 1894, pp. 193–239.
15 Gerard Krefft, 'The Improvements Effected in Modern Museums in Europe and Australia', *Transactions of the Royal Society of New South Wales for the Year 1868*, Sydney, 1869, p. 24.
16 Ibid, p. 15.
17 Idem.
18 Ibid, p. 17.
19 Alfred Kelly, *The Descent of Darwin: The Popularization of Darwin in Germany, 1860–1914*, University of North Carolina Press, Chapel Hill,1981, p. 10.
20 Peter J. Bowler, 'Popular Science' in Peter J. Bowler and John V. Pickstone (eds.), *The Cambridge History of Science Volume 6: The Modern Biological and Earth Sciences*, Cambridge University Press, Cambridge, 2009, p. 626.
21 Kurt Bayertz, 'Spreading the Spirit of Science: Social Determinants of the Popularization of Science in Nineteenth-Century Germany' in Terry Shinn, and Richard Whitley (eds.), *Expository Science: Forms and Functions of Popularisation*, D. Reidel, Dordrecht, the Netherlands, 1985, p. 209.
22 Idem.
23 Roy MacLeod, 'The "Practical Man" Myth and Metaphor in Anglo-Australian Science', *Australian Cultural History*, vol. 8, 1989, p. 30.
24 Ford Madox Brown, *Work*, 1852–65, oil on canvas. Collection: Manchester City Art Galleries. [Ford Madox Brown], *The Exhibition of* WORK *and other Paintings, by Ford Madox Brown, at the Gallery, 191, Piccadilly*, London, 1865.
25 Tim Barringer, *Men at Work: Art and Labour in Victorian Britain*, Yale University Press, New Haven and London, 2005, p. 69.
26 See James Eli Adams, *Dandies and Desert Saints: Styles of Victorian Manhood*, Cornell University Press, Ithaca and London, 1995.
27 MacLeod, 'The "Practical Man" Myth and Metaphor', p. 30.
28 Ibid., p. 35.
29 R. W. Connell, *Masculinities*, Allen & Unwin, St. Leonards, 1995, p. 194; Samuel Smiles, *Self Help*, Ward & Lock, London, 1859.
30 See George Basalla, 'The Spread of Western Science' in William K. Storey (ed.), *Scientific Aspects of European Expansion*, Variorum, Brookfield, Vt., 1996, p. 6; also Ian Inkster, 'Scientific Enterprise and the Colonial "Model": Observations on Australian Experience in Historical Context', *Social Studies of Science*, vol. 15, 1985, p. 684; Colin Finney, *Paradise Revealed: Natural History in Nineteenth-century Australia*, Museum of Victoria, Melbourne, 1993, p. 2.
31 Clive Moore, 'Australian Masculinities', *Journal of Australian Studies*, no. 56, 1998, p. 4.
32 See Roy MacLeod, 'Introduction. Nature and Empire: Science and the Colonial Enterprise', *Osiris*, 2nd series, vol. 15, 2000, pp. 1–13.
33 Laura Stoler, 'Rethinking Colonial Categories: European Communities and the Boundaries of Rule', *Comparative Studies in Society and History*, vol. 31, no.1, 1989, pp. 136–7.

34 Moore, 'Australian Masculinities', p. 8; see also Robert Hogg, 'Performing Manliness: "Unmanly" Men on British Frontiers in the Mid-Nineteenth-Century', *Journal of Australian Studies*, vol. 35, no. 3, 2011, pp. 355–372.
35 See Paul White, *Thomas Huxley: Making the 'Man of Science'*, Cambridge University Press, Cambridge, 2003.
36 Carol E. Harrison, 'Citizens and Scientists: Toward a Gendered History of Scientific Practice in Post-revolutionary France', *Gender & History*, vol. 13, no. 3, 2001, p. 445.
37 Ibid., pp. 447 and 460.
38 Ibid., p. 460.
39 Gerard Krefft, *The Snakes of Australia: An Illustrated and Descriptive Catalogue of all Known Species*, Thomas Richards, Government Printer, Sydney, 1869, preface.
40 Gerard Krefft to Richard Owen, 21 December 1869: 27, f.22–6, Owen correspondence, Natural History Museum, London.
41 James H. Warren, 'Contesting Colonial Masculinity/Constituting Imperial Authority: Ceylon in Mid-nineteenth-century British Public Debate', *New Zealand Journal of Asian Studies*, vol. 6, no. 2, 2004, p. 45.
42 Stoler, 'Rethinking Colonial Categories', p. 137.
43 Mrinalini Sinha, *Colonial Masculinity: The 'Manly Englishman' and the 'Effeminate Bengali' in the Late Nineteenth Century*, Manchester University Press, Manchester and New York, 1995, p. 2.
44 Margaret Allen, '"A Fine Type of Hindoo" Meets "The Australian Type": British Indians in Australia and Diverse Masculinities' in Desley Deacon, Penny Russell and Angela Woollacott (eds.), *Transnational Ties: Australian Lives in the World*, ANU E-press, Canberra, 2008, p. 45; Catherine Hall, Keith McClelland and Jane Rendall, *Defining the Victorian Nation: Class, Race, Gender and the Reform Act of 1867*, Cambridge University Press, Cambridge, 2000, p. 55.
45 Harriet and Helena Scott to Edward Ramsay, 22 October 1862 and 31 December 1865: E. P. Ramsay Correspondence, ML, SLNSW (emphasis in original).
46 Beverly Kingston, *The Oxford History of Australia, vol. 3: Glad, Confident Morning, 1860–1900*, Oxford University Press, Melbourne and Oxford, 2001, p. 128.
47 Born in Brunswick, Germany, on 17 February 1830, Krefft left school at the age of 15 and took satisfaction throughout his life on being a self-educated man – Jenny Nancarrow, 'Gerard Krefft: A Singular Man', *Proceedings and Transactions of the Royal Society of Victoria*, vol. 121, no. 1, 2008, pp. 146–54; G. P. Whitley and Martha Rutledge, 'Krefft, Johann Ludwig Gerard (Louis) (1830–1881)', Australian Dictionary of Biography, Volume 5, Melbourne University Press, 1974, pp. 42–44; Joan Kerr (ed.), 'Johann Ludwig Gerard (Louis) Krefft (1830–1881)', *The Dictionary of Australian Artists: Painters, Sketchers, Photographers and Engravers to 1870*, Oxford University Press, Melbourne, 1992, p. 437. At the same time, he also admired formally trained scientists who had made substantial contributions to new knowledge; for instance, publishing a biographical article on the explorer and naturalist, Ludwig Leichardt – Gerard Krefft, 'A Biographical Sketch of the Youth and Early Manhood of Dr. Ludwig Leichardt', *The Maitland Mercury and Hunter River Advertiser*, 3 February 1866, p. 3. After working as a clerk in Halberstadt, the earliest indication of Krefft's developing interests in the natural sciences was when he travelled to New York at age 20 and spent a year studying and copying the French-American ornithologist and artist John James Audubon's *Birds of America*, 1827–38, in the Mercantile Library. He sold these illustrations to pay for his passage to Australia – Gerard Krefft to Charles Darwin, 6 September 1873: Australian Museum Archives, Sydney; Ronald Strahan, *Rare and Curious Specimens: An Illustrated History of the Australian Museum 1827–1979*, Australian Museum, Sydney, 1979, p. 28.
48 Krefft's notes for official inquiry into the Australian Museum, 1874 – quoted in Nancarrow, 'Gerard Krefft', p. 146.
49 Gerard Krefft, unpublished mss: Krefft papers, Mitchell Library, microfilm no.930, A265 and A267.
50 Krefft's drawings included highly detailed studies of fauna; scenes of members of the expeditionary party hunting for specimens and negotiating the rugged landscape; and portraits of Aborigines that contained insightful observations of their customs and surroundings.

51 Gerard Krefft, 'A Few Remarks on the Habit and Economy of the Brown-capped Potamorhinus (P. ruficeps, Hartlaub)', *Proceedings of the Zoological Society of London*, vol. 23, pp. 352–3.
52 *A Few Letters and Testimonials from Distinguished Men of Science Addressed to Mr. Gerard Krefft, Curator and Secretary of the Australian Museum, From 1858 – 1874*, 1874: A26; P1, Gerard Krefft Papers, Mitchell Library; Unpublished mss, Gerard Krefft papers, Mitchell Library, microfilm no.930, A265 and A267. Prior to his appointment, Krefft had in fact applied to join the ill-fated Burke and Wills expedition.
53 *Annual Report*, Australian Museum, 1858; Strahan, *Rare and Curious Specimens*, p. 28.
54 'New Books, ex Mail Steamer', *Sydney Morning Herald,* 13 February1860, p. 5.
55 'The Australian Museum', *Sydney Morning Herald,* 13 May 1864, p. 2.
56 Letter, 4 March 1859, Australian Museum 1842–66: 4332 (2/567), Archives Office of NSW – cited in Public Works Department of NSW, *The Australian Museum: A Conservation Analysis of the Complex and its Site with a Statement of its Significance*, 1984, p. 12.
57 'The Duke of Edinburgh in New South Wales', *Sydney Morning Herald*, 15 February 1868, p. 7.
58 'The Duke of Edinburgh in Adelaide', *Sydney Morning Herald*, 15 November 1867, p. 6; 'The Duke of Edinburgh in New South Wales: The Review', *Sydney Morning Herald*, 25 January 1868, p. 7; see also Cindy McCreery, 'A British Prince and a Transnational Life: Alfred, Duke of Edinburgh's Visit to Australia, 1867–8' in Desley Deacon, Penny Russell and Angela Woollacott (eds.), *Transnational Ties: Australian Lives in the World*, ANU E-press, Canberra, 2008, pp. 60–62.
59 Peter Bernhard, 'Australia in the German Geographical Literature between 1775 and 1840' in H. Lamping and M. Linke (eds.), *Australia: Studies on the History of Discovery and Exploration*, Institutes für Wirtschafts und Sozialgeographische der Johann Wolfgang Goethe-Univeristät, Frankfurt am Main, 1994, p. 47; Ilse Jahn, 'A Reflection of German Investigations about Australia in Pictures and Written Form Documents of the Museum für Naturkunde (Museum of Natural History) at the Humboldt-University in Berlin' in Lamping and Linke, *Australia*, pp. 61–63.
60 See Colin McCarthy, 'Paper Snakes', *Nature*, vol. 306, 1 December 1983, p. 524.
61 Gunther, *A Century of Zoology*, pp. 320–1.
62 Finney, *Paradise Revealed*; Roy McLeod, 'On Visiting the "Moving Metropolis": Reflections on the Architecture of Imperial Science', *Historical Records of Australian Science*, vol. 5, no. 3, 1982, p. 6.
63 Quoted in Rev Richard Owen, *The Life of Richard Owen*, vol. II, John Murray, London, 1894, p. 45.
64 Albert E. Gunther, *A Century of Zoology at the British Museum: Through the Lives of Two Keepers, 1815–1914*, Dawsons, London, 1975, p. 281.
65 Peter Cunningham, *Two Years in New South Wales*, Angus & Roberson, Sydney, 1966 (1827), p. 159.
66 See Peter Stanbury and Julian Holland, *Mr. Macleay's Cabinet: The History of the Macleays and their Museum*, The Macleay Museum, University of Sydney, Sydney, 1988; also Strahan, *Rare and Curious Specimens.*
67 Australian Museum, *Report of the Trustees of the Australian Museum of the Year ending 1860.*
68 Gunther, *A Century of Zoology*, pp. 281–2.
69 Gerard Krefft to Richard Owen, 20 January 1862: Owen correspondence, Natural History Museum (NHM), London, 27, f.1–2.
70 Idem.
71 Gunther, *A Century of Zoology*, p. 282.
72 Caroline Jordan, *Picturesque Pursuits: Colonial Women Artists and the Amateur Tradition*, Melbourne University Press, Melbourne, 2005.
73 Owen specialised in the osteology of Australian mammals, and he would have recognised it immediately as an echidna skeleton. I am indebted to Ross Sadlier at the Australian Museum for his assistance in identifying the skeleton.
74 See David N. Livingstone, *Putting Science in its Place: Geographies of Scientific Knowledge*, University of Chicago, Chicago and London, 2003.
75 Krefft to Owen, 21 October 1864: Owen correspondence, NHM, London, 27, f.3–5.

76 Idem.
77 Anne Secord, 'Corresponding Interests: Artisans and Gentlemen in Nineteenth-century Natural History', *British Journal for the History of Science*, no. 27, December 1994, p. 384.
78 Ann Moyal, 'Sir Richard Owen and his Influence on Australian Zoological and Palaeontological Science', *Historical Records of Australian Science*, vol. 3, no. 2, 1975, pp. 41–56.
79 R. W. Home, 'Georg von Neumayer and the Flagstaff Observatory, Melbourne' in David Walker and Jürgen Tampke (eds.), *From Berlin to the Burdekin: The German Contribution to the Development of Australian Science, Exploration and the Arts*, New South Wales University Press, Sydney, 1991, pp. 40–41.
80 Marie Boas Hall, *All Scientists Now: The Royal Society in the Nineteenth Century*, Cambridge University Press, London, 1984.
81 Felix Driver, 'Distance and Disturbance: Travel, Exploration and Knowledge in the Nineteenth Century', *Transactions of the Royal Society*, sixth series, vol. 14, 2004, pp. 81–2.
82 Stuart McCook, 'Paul du Chaillu and the Legitimation of Evidence in the Field Sciences', *Osiris*, 2nd ser., vol. 11, 1996, pp. 179-80.
83 Numerous articles also appeared in Sydney's *Empire*, Melbourne's *Argus*, Hobart's *Mercury*, the Launceston *Examiner*, Brisbane's *Courier* and the *South Australian Register*; the controversy made at least one or two appearances in virtually every regional newspaper across the Australian colonies.
84 'Literature', *The Athenaeum*, 11 May 1861, pp. 621–3; John Edward Gray, 'The New Traveller's Tales', *The Athenaeum*, 18 May 1861, pp. 662–3.
85 See Nicolaas Rupke, *Richard Owen: Victorian Naturalist*, Yale University Press, New Haven and London, 1994; also Joel Mandelstam, 'Du Chaillu's Stuffed Gorillas and the Savants from the British Museum', *Notes and Records of the Royal Society of London*, vol. 48, no. 2, 1994, pp. 228–230.
86 See K. David Paterson, 'Paul B. Chaillu and the Exploration of Gabon, 1851–1865', *The International Journal of African Historical Studies*, vol. 7, no. 4, 1974, p. 654.
87 Owen to Murchison, 4 February 1861: Owen-R-31, Geological Society, London.
88 John Edward Gray, 'The Gorilla War', *The Athenaeum*, 13 July 1861, p. 51.
89 R. B. S., 'The New Traveller's Tales', *The Athenaeum*, 25 May 1861, p. 695; John Edward Gray, 'The New Traveller's Tales', *The Athenaeum*, 1 June 1861, p. 728.
90 British Museum: Original Papers, 18 March 1858, f.243.
91 John Edward Gray, 'The New Traveller's Tales', *The Athenaeum*, 25 May 1861, p. 695; British Museum: Original Papers, 18 May 1859, f.159 and 9 June 1859, f.317a.
92 John Edward Gray, 'The New Traveller's Tales', *The Athenaeum*, 15 June 1861, p. 798.
93 John Edward Gray, 'The New Traveller's Tales', *The Athenaeum*, 25 May 1861, p. 695.
94 John Edward Gray, 'The New Traveller's Tales', *The Athenaeum*, 1 June 1861, p. 728.
95 Richard Owen, 'Osteological Contributions to the Natural History of the Anthropoid Apes – No. 7', *Transactions of the Zoological Society of London*, vol. 5, pt. 1, 1866, pp. 1–32.
96 Paterson 'Paul B. Chaillu and the Exploration of Gabon', p. 654; McCook, 'Paul du Chaillu', pp. 195–6.
97 Paul Du Chaillu, *The Country of the Dwarfs*, Harpers & Brothers, New York, 1872, pp. 80–85.
98 See, for example, P. Du Chaillu and Dr. J. E. Gray, 'Notice of a Skull of a New Species of Bush-goat (Cephalophus longiceps), Sent from the Gaboon [sic]', *Proceedings of the Zoological Society of London*, vol. 33, issue 1, 1865, pp. 204–6.
99 See Francis Buckland, 'The Gorilla at Home', *Sydney Morning Herald*, 2 August 1861, p. 2; 'The Gorilla Controversy', *Sydney Morning Herald*, 9 September 1861, p.8; 'Gorilla Warfare', *Sydney Morning Herald*, 11 October 1861, p. 8; 'The Great Gorilla Controversy', *Sydney Morning Herald*, 3 December 1861, p. 8. Du Chaillu's *Explorations and Adventures in Equatorial Africa* was available from a number of Sydney booksellers from mid-July 1861, having been described in lurid detail over two pages in the *Sydney Morning Herald* – 'The Discoveries in Africa, by M. Du Chaillu. (From *The Times*, May 20)', *Sydney Morning Herald*, 19 July 1861, pp. 2–3.
100 John Murray, 'The Gorilla War', *Sydney Morning Herald*, 4 October, 1861, p. 8.

101 Idem.
102 Idem.
103 'From Our London Correspondent', *Sydney Morning Herald*, 14 October 1861, p. 8.
104 Idem.
105 'European News', *Sydney Morning Herald*, 20 December 1861, p. 2.
106 John Murray, 'The Gorilla War', *Sydney Morning Herald*, 4 October, 1861, p. 8; 'Miscellaneous News', *Sydney Morning Herald*, 21 September 1861, p. 13.
107 See W. Winwood Reade, 'News from the Gorilla Country', *Sydney Morning Herald*, 6 March 1863, p. 8.
108 Unpublished mss, Gerard Krefft papers, Mitchell Library, microfilm no. 930, A265 and A267.
109 Unpublished mss, Gerard Krefft papers, Mitchell Library, microfilm no. 930, A268, f.2; see also Kerry Heckenberg, 'Photography and Exploration: Issues Related to the Uptake of Photography in Australian Exploration', *Journal of Australian Colonial History*, vol. 9, 2007, p. 161.
110 Dr. Pole, 'Photography for Travellers and Tourists', *Proceedings of the Royal Geographical Society*, 1871, p. 47.
111 See Gael Newton, *Shades of Light: Photography and Australia 1839–1988*, Australian National Gallery and Collins, Canberra and Sydney, 1988, p. 208; Anne-Marie Willis, *Picturing Australia: A History of Photography*, Angus and Robertson, Sydney, 1988, p. 74; Alan Davies and Peter Stanbury, *The Mechanical Eye in Australia: Photography 1841–-1900*, Oxford University Press, Melbourne, 1985, p. 16.
112 See Lyndall Dawson, 'Marsupial Fossils from Wellington Caves, New South Wales: The Historic and Scientific Significance of the Collections in the Australian Museum, Sydney', *Records of the Australian Museum*, vol. 37, no. 2, 1985, p. 57; Joan Starr and Doug McMillan, *The Wellington Caves*, Macquarie Publications, Dubbo, 1985.
113 Thomas L. Mitchell, *Three Expeditions into the Interior of Eastern Australia: With Descriptions of the Recently Explored Region of Australia Felix and of the Present Colony of New South Wales*, T. W. Boone, London, 1838 (rev.1839).
114 The Secretary and Curator, Australian Museum to the Principal Under Secretary, 4 July 1869, in *Papers Relating to the Exploration of Caves and Rivers (Fossil Remains at Wellington Caves, NSW)*, 1869–1882, Australian Museum Archives, Sydney; Starr and McMillan, *The Wellington Caves*.
115 Anne Secord, 'Science in the Pub: Artisan Botanists in Early Nineteenth Century Lancashire', *History of Science*, no. 32, 1994, p. 269.
116 Jennifer Tucker, 'Voyages of Discovery on Oceans of Air: Scientific Observation in the Age of "Balloonacy"', *Osiris*, 2nd series, vol. 11, 1996, pp. 164.
117 'Mr. Krefft's Report on the Fossil Remains Found in the Caves of Wellington Valley', *Sydney Morning Herald*, 18 December 1866, p. 3.
118 Alexander von Humboldt, *Kosmos* – quoted in Robert J. Richards, 'The Impact of German Romanticism on Biology in the Nineteenth Century' in Karl Ameriks (ed.), *The Impact of Idealism: The Legacy of Post-Kantian German Thought*, Cambridge University Press, Cambridge, 2013, pp. 105–33.
119 Mr. Krefft's Report on the Fossil Remains', p. 3.
120 Idem.
121 Idem.
122 See Jocelyn Hackforth-Jones, *Augustus Earle, Travel Artist: Paintings and Drawings in the Rex Nan Kivell Collection*, National Library of Australia, Canberra, 1980, pp. 8 and 23.
123 See Martin Terry, 'Curious Caverns: Picturesque Geology in Australian Art', *The Australian Antique Collector*, no. 33, 1986, pp. 46–47; also Mary Mackay, *The Geological Sublime: A New Paradigm*, PhD dissertation, University of Sydney, 1990.
124 See Simon Ryan, *The Cartographic Eye: How Explorers Saw Australia*, Cambridge University Press, Cambridge, p. 91; also Charles Rosen and Henri Zerner, *Romanticism and Realism: The Mythology of Nineteenth-Century Art*, Faber and Faber, London, 1984, p. 25.
125 Mitchell, *Three Expeditions into the Interior of Eastern Australia*, Preface to the first edition, n.p.
126 'Mr. Krefft's Report on the Fossil Remains', p. 3.

127 Idem.
128 Report of Gerard Krefft to the Trustees of the Australian Museum, 7 October 1869, in *Papers Relating to the Exploration of Caves and Rivers.*
129 Krefft to colonial secretary under the direction of Australian Museum trustees, 4 November 1869; 'List of Photographs of Australian Fossils, for Transmission to Professor Owen F. R. S.'; Owen to colonial secretary, 21 March 1870; Colonial secretary to Owen, 28 May 1870 in *Papers Relating to the Exploration of Caves and Rivers.*
130 'New South Wales (From Our Own Correspondent) Sydney, October 9', *The Times*, 1 December 1869, p. 10.
131 Colonial secretary to Owen, 16 June 1869 in *Correspondence Relative to Exploration of Wellington Caves, Presented to the Houses of Parliament, New South Wales, 1870*; Owen to Australian Museum trustees, 1 December 1869, in *Papers Relating to the Exploration of Caves and Rivers.*
132 Krefft to Owen, 22 December 1866: Owen correspondence, NHM, London, 27, f.10–2; *Catalogue of the Natural and Industrial Products of New South Wales Forwarded to the Paris Universal Exhibition Commissioners*, Thomas Richards, Government Printer, Sydney, 1867, pp. 91–124.
133 See Paul Carter, 'Invisible Journeys' in Paul Foss (ed), *Island in the Stream: Myths of Place in Australian Culture*, Pluto Press, Leichhardt, 1988, p. 49.
134 Owen to colonial secretary, n.d. (reply to a letter of 21 March 1870) in *Papers Relating to the Exploration of Caves and Rivers.*
135 Krefft to Owen, 15 July 1870: Owen correspondence, NHM, London, 27, f.50–4.
136 Owen to Krefft, 26 February 1870, f.30–32 and 22 April 1870: 27, f.39–41, Owen correspondence, Natural History Museum, London.
137 See Bruno Latour, *Science in Action: How to Follow Scientists and Engineers through Society*, Harvard University Press, Cambridge (Mass.), 1987.
138 See Susan L. Star and James R. Griesemer, 'Institutional Ecology, "Translations" and Boundary Objects: Amateurs and Professionals in Berkeley's Museum of Vertebrate Zoology, 1907–39', *Social Studies of Science*, vol. 19, 1989, pp. 387–420.
139 On photography and the issue of scientific credibility, see Jennifer Tucker, *Nature Exposed: Photography as Eyewitness to Science*, John Hopkins University Press, Baltimore, 2005.
140 Richard Owen, 'The Gorilla', *The Athenaeum*, 5 October 1861, pp. 445–6.
141 J. E. Gray, 'The Gorilla', *The Athenaeum*, 12 October 1861, p. 479.
142 Driver, 'Distance and Disturbance', p. 82.
143 Krefft to Parkes, 23 September 1873, quoted in Moyal, 'Sir Richard Owen and his Influence', p. 53 [emphasis in Moyal].
144 'Australian Museum', *Sydney Morning Herald*, 17 April 1875, p. 9.
145 See Strahan, *Rare and Curious Specimens*; Nancarrow, 'Gerard Krefft'.
146 Sally Gregory Kohlstedt, 'Australian Museums of Natural History: Public Priorities and Scientific Initiatives in the 19th Century', *Historical Records of Australian Science*, vol. 5. 1983, pp. 8–9.
147 'The Australian Museum', *Nature*, vol. 10, 4 June 1874, pp. 81–2.
148 'Legislative Assembly: The Sydney Museum', *Sydney Morning Herald*, 17 June 1874, pp. 2–3.
149 Idem.
150 Idem.
151 'Diprotodon!!', *Sydney Punch*, 26 June 1874, p. 209.
152 'Legislative Assembly: The Sydney Museum', *Sydney Morning Herald*, 17 June 1874, pp. 2–3.
153 Idem.
154 Idem.
155 'Australian Museum', *Sydney Morning Herald*, 17 April 1875, p. 9.
156 Idem.
157 'A British Colony Ruled by Trustees', *Krefft's Nature in Australia*, vol. 1, no. 1, pp. 2–4.
158 Gerard Krefft to William Krefft, 23 November 1880: Krefft family letters, Australian Museum Archives, Sydney – translated by Bodo Matziche from originals held by the Krefft family, Hamburg.

159 Ibid; Register no. 3, p. 56, item 4/1202, reel 130, State Records Authority of NSW.
160 Gerard Krefft to William Krefft, 23 November 1880.
161 E. P. Ramsay to Owen, 18 October 1882: f.54–7, Owen correspondence, Natural History Museum, London.
162 E. P. Ramsay to Owen, 29 March 1885: f.67–70, Owen correspondence, Natural History Museum, London.
163 E. P. Ramsay to Owen, 14 September 1885–8 August 1888: f.71–98, Owen correspondence, Natural History Museum, London.
164 'Photographs by the Staff of the Australian Museum, Sydney, NSW', 1880–90: album VA232, Australian Museum Archives, Sydney.
165 'The Australian Museum', *Nature*, vol. 10, 4 June 1874, p. 82.
166 Ibid, p. 81.
167 Gerard Krefft letters from Charles Darwin 1872–76, ML mss 5828, Mitchell Library, Sydney.
168 R. W. Home and Sally Kohlstedt (ed.), *International Science and National Scientific Identity: Australian between Britain and America*, Kluwer Academic Publishers, Dordrecht, 1991, p. 3.

3 The rhetoric of exemplarity

Portraiture and the naturalist as celebrity

Introduction

In 1849, the *Art Journal* announced the recent publication of a photographic portrait of 'the able experimental philosopher Dr. Faraday, whose researches on electricity and magnetism have placed him in the first rank of the investigators of nature'. Lithographed from a daguerreotype taken by Antoine Claudet, Michael Faraday's 'likeness' was the first in a series of similar portraits of 'eminent scientific men' to be issued for general purchase, prompting what was perhaps the earliest mention of a photographic enterprise that would become synonymous with the rising celebrity culture in the Victorian era: 'The reality of these productions is such, that the public will be certain of obtaining possession of the most faithful resemblances which can be secured of those whose genius and industry have secured them the world's honour'.[1]

Although the image itself is an unassuming head-and-shoulders study chosen from several versions photographed by Claudet, the selection of Faraday to launch the series was without doubt a most apposite choice (Figure 3.1). Having not only announced the invention of 'photogenic drawing' at the Royal Institution exactly ten years earlier, as part of his public Friday Evening Discourses, Faraday was also one of the era's most widely admired and dynamic popularisers of science. His enthusiasm for the medium, and his willingness to sit before successive studio photographers, meant that every detail of his face, if not necessarily the finer points of his scientific discoveries, was familiar to Victorian consumers. With advances in paper-based photographs, especially the introduction of the *carte de visite*, the production of photographic portraits soon became more than just an occasional endeavour and proliferated into a vast industry.

Metropolitan men of science such as Faraday, among other contemporary 'worthies', were persistently sought by commercial studios for sittings, prompting the circulation of large quantities of celebrity photographs over subsequent decades. Consequently, where enterprising publishers had previously issued volumes of scientific biographies and engraved portraits that projected the image of science as a national tradition inherited from iconic figures of the past, these new photographic albums were unapologetically contemporary.[2] 'It is curious to reflect how little in past ages mankind knew of their contemporaries', remarked Stephen Thompson in 1862. 'The form and features of those who "moved the world" were as utterly unknown to those who lived in the same period as they are to us. [Yet] the men of mark of our own age are as familiar to us as "household words."'[3] As one of the most visible scientists

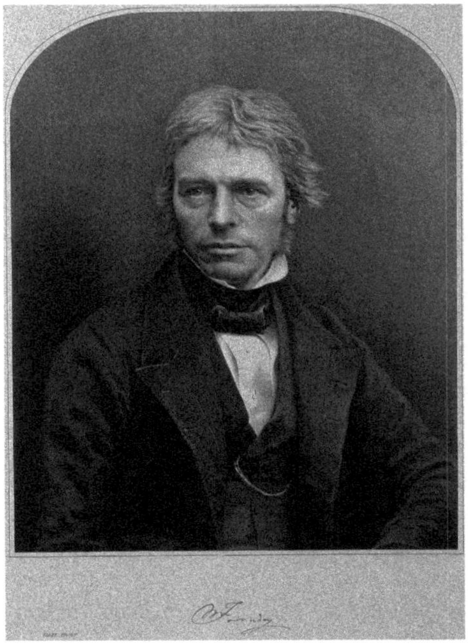

Figure 3.1 After Antoine Claudet, *Michael Faraday*, c.1849, lithograph after a daguerreotype. Collection: Wellcome Library, London.

of his day, Michael Faraday was himself an ardent collector of portraits.[4] He was also an exemplary scientific discoverer and promoted photography's potential for recording observations from nature.[5] In an era that saw an increased emphasis on visuality and the empirical sciences, the scientific observer had himself become the object of scrutiny and emulation.

It was perhaps inevitable, then, that in the field of natural history – the preeminent realm of Victorian science – the new generation of influential figures such as Richard Owen, T. H. Huxley and Joseph Hooker, and even the reclusive Charles Darwin, used the medium to their advantage at a time when the public identity of the scientist was transitioning from an amateur vocation to that of a civic and professional occupation. Furthermore, the imperial and global influence of many of these modern men of science was not only tacitly understood by viewers but was often explicitly signalled in identifiable attributes within their portraits. Personifying the pursuit of imperial natural history as often as the spirit of disinterested scientific inquiry, the Victorian naturalist had various roles and projected multiple, complex identities. Joining the innumerable portraits of royalty, parliamentarians, clergymen, writers, actors, singers, artists and other figures of public interest, images of naturalists soon predominated amongst the commercial photographs of scientists. Their portraits appeared in a range of presentational formats – from lavish publications celebrating the achievements, cultural status and moral probity of public identities to informal albums created by individuals who collected portrait photographs according to their own particular interests and allegiances.

In this chapter, my inquiry into role of photography in Victorian natural history focuses on portraiture through an analysis of contemporary notions of exemplarity. By looking at the contexts within which images of metropolitan naturalists appeared – how and by whom portraits were used – I consider the way that photographic portraits portrayed both the practitioner and the practice of natural history and represented the individual and shared accumulation of knowledge and kudos. As such, they effectively addressed a range of audiences. In the first section, I analyse the rise of studio portraiture in relation to the different kinds of nomenclature used by naturalists as part of an ongoing process of self-identification. In published portrait anthologies, two descriptive forms – visual and textual – were brought together, highlighting the ways that photographic portraits and accompanying biographies presented Victorian middle-class society as cohesive and exemplary. However, they also illustrated subtle distinctions between different descriptions and images of men of science. This first section also examines contemporary commentary about *cartes de visite* and the anxiety at the time about the amount and types of detail they revealed. Some critics of the format were concerned to ensure that photographic portraits embraced the aesthetics of traditional portraiture by observing formal principals of art, which, rather than rendering the sitter's features in minute detail, would engender a more idealised conception of the subject. Yet, for others, the production of detail was considered to be photography's strongest point and, hence, the opportunity to delineate every feature and object was irresistible. Victorian naturalists and other men of science responded to these conflicting viewpoints in different ways. These issues were pertinent to them as the very basis of Victorian scientific culture – and thus their own reputations – relied on them accruing, substantiating and communicating all sorts of details as verifiable facts.

In the second part of this chapter, I focus mainly, although not exclusively, on the figures of Charles Darwin and Richard Owen and examine the production and reception of portraits of these two prominent contemporaries of Victorian natural history. As England's most renowned amateur naturalist, Darwin was nevertheless a reticent sitter for portraits and a reluctant public celebrity. Yet, Darwin's hesitancy to submit to portrait sittings, and his subdued appearance when he did acquiesce, was not merely due to his apprehension about the process of visiting a studio or his distaste for public attention. Rather, these factors encapsulated his general approach to self-fashioning, whereby he purposely downplayed his subjectivity in order to assert himself as a poised and dispassionate scientific investigator and theorist. As a counterpoint to Darwin, I examine several characteristic examples from the extensive cycle of portraits of Owen produced over the course of his career. Ever conscious of the impact on the popular imagination of visually impressive museum displays and specimens – particularly the immense extinct vertebrates that had distinguished his career – he chose to actively exploit these attributes in his portraits in several ways. In the first instance, the portraits commemorated Owen's role as the discoverer of many intriguing and spectacular species, collected on his behalf from locations all across the empire and brought together by him at the British Museum. They also conveyed information of a practical nature about the specimens, including their size and structure, as well as how the anatomist imaginatively and literally reconstructed them.[6] The representation of specific, identifiable objects and practices of natural history as part of the portraiture genre performed two separate tasks: to explicitly identify sitters as part of the increasingly diverse and multifarious community of naturalists; and to engage viewers and familiarise them – through the intricacies and minutiae of natural history – with various kinds of knowledge. 'Any attempt to represent nature in visual form', as Bernard Lightman suggests, 'was unavoidably a claim to possess knowledge and an assertion of scientific expertise'.[7] The use of portraits thus performed a similar rhetorical function to other natural history illustrations and represents a sophisticated amalgamation of genres that presented viewers with a more complex pictorial scheme than immediately evident.

Gentleman amateurs and professional bodies: the social formation of Victorian science

As the most pervasive visual form to arise in the nineteenth century, studio portrait photography coincided with an emergent middle class that was keen for objects that symbolised their social status.[8] Prior to the popularisation of photography, very few in society had the means to commission portraits, and even fewer managed to acquire copies of their portraits to send to or exchange with others.[9] 'By having one's portrait done', Gisèle Freund argues, 'an individual of the ascending classes could visually affirm his new social status both to himself and to the world at large'.[10] Another aspect of this aspiration was the appetite for images of contemporary 'worthies' produced in the same mode. As such, there has been much scholarship in recent years regarding the role of portraiture in commodity cultures. These studies have tended to focus on the visual ordering of broad social strata.[11] Less attention has been given, however, to the production and circulation of portrait photography within distinct sectors, including knowledge- or skill-based groups such as specific disciplines and their associated occupations. Portraits of individuals were not necessarily acquired in order to slavishly model one's life on the famous but rather to provide a more subtle point of comparison between one's own life and that of colleagues, supporters and superiors. Accordingly, the prominent role of portraiture in the social construction of identity provides a salient starting point for observing fundamental changes within certain occupations and avocations at that time, particularly those that had to guarantee, both to themselves and others, their efficacy and reliability.[12]

As natural history gained in popularity and became the most widely practiced nineteenth-century science, it represented one such occupation, comprising as it did an increasingly diverse and geographically dispersed community of enthusiasts and operatives ranging from the working class to the elite. Active across a range of institutional sites and geographical locations, naturalists did not share a ready-made identity. For this reason, they were inclined to occupy themselves with various processes of self-fashioning as an important aspect of their professional and scientific practice.[13] With natural history and other gentlemanly intellectual pursuits becoming far more accessible while at the same time progressively more structured and professionalised, naturalists as a group became increasingly regulated by testimonials, formal instruction and educational credentials. Photography assisted in this process by representing in visual form not only the individual but their place within these collective identities.[14] The production of photographic portraits and their circulation – by both formal and informal means – was a way of establishing or consolidating professional affiliations, similar to naturalists' exchange of specimens. Portraiture was a very immediate way to consolidate reputations and was useful in gaining publicity for particular causes or projects within a broader intellectual or public arena.

The ways in which Victorian naturalists chose to define themselves were complex. Scientific practice itself was becoming increasingly specialised, and the cultural landscape of Victorian natural history was similarly changing and becoming more diverse. A precise picture of the social formation of science during that period is difficult to discern and all but impossible to interpret if our understanding is based on strict demarcations between amateur and professional practitioners.[15] For instance, some historians view the nineteenth century as being dominated by the ideal of the amateur gentleman, with science at the time being transformed very slowly, and not necessarily completely, away from an interest pursued by mostly wealthy individuals towards a regular vocation.[16] Naturalists increasingly sought both financial security and formal recognition for their scientific endeavours, nevertheless, and by the 1840s–1850s, figures such as Joseph Hooker already had a definite idea about the appropriate qualities of a professional man of science.[17] Furthermore, analysis of provincial natural history in the late Victorian era demonstrates that there were many different amateur identities and practices at play, which, rather than being diminished or undermined by the growth of the new professional community, facilitated its rise as these amateurs continued to work alongside the professional naturalists. These observations suggest a general refashioning of the roles of both amateur and professional naturalists throughout this period.[18] Indeed, naturalists belonged to an increasingly varied sector that was active in a range of occupational contexts and physical locations. Neither their professional situation and standing nor the knowledge they produced was solely attributable to their immediate circumstances; instead, they pieced together their identities using the different cultural resources that accrued from the environments in which they found themselves.[19] Nor could their work be described as a solitary activity.[20] There was a general acknowledgement of the existence of a scientific community, with corresponding ideas of inclusion and exclusion.[21]

A nuanced language of self-description – which appeared in both public rhetoric and private correspondence – was one way in which affiliates, prospective participants and onlookers could recognise and differentiate between members of the scientific community.[22] For instance, particular expressions were used with varying frequency and had different connotations. Through her comparison of terms in common usage to refer to the nineteenth-century scientist, Ruth Barton has demonstrated that these designations were applied quite precisely to draw attention to specific attributes.[23] Certain qualities of mind and character, for example, were usually implied by the characterisation *man of science* or *scientific man*: that is, 'their self-sacrificing and public-spirited labours gave them moral authority, [and] their long experience of drawing conclusions from complex and even contradictory evidence trained their reasoning powers and made their judgements particularly reliable'.[24] Accordingly, these titles were assigned to a group of individuals who were deemed worthy of such connotations. In contrast, the terms *cultivator of science*, *student of science* and *scientific worker* or *scientific labourer* drew attention to the painstaking and performative aspects of scientific practice, the first two suggesting a devotion to and an active promotion of science, along with a spirit of inquiry; and the latter emphasising the laborious aspects of scientific work. *Philosopher*, on the other hand, tended to be more exclusive in its application and described only the loftiest investigators: those members of the scientific elite who took the minutiae – the 'mere facts' – of practical science and transformed them into broad theoretical explanations.[25] This title was used to designate towering figures such as Faraday, Owen and Darwin – for example, the *Art Journal*'s description of Faraday as 'the able experimental philosopher'.[26] Similar ranges of terms tended to be used for men within a diversity of vocations, but in portrait anthologies, unsurprisingly, publishers included only the most prominent figures.

These various descriptions – *cultivator of science*, *student of science*, *scientific worker*, *scientific labourer* and *philosopher* – clearly established a division between the practical work and the brainwork of science and were all commonly used conventions in a variety of publications, including popular scientific pamphlets and articles. While these terms were never used in the titles or prefaces of portrait anthologies, it was not unusual for them to appear in the individual biographies that accompanied the photographic portraits of individuals. Indeed, different designations were sometimes applied to the same person across various publications. Biographies allowed a flexible and highly nuanced interpretation of the persona, status and role of the Victorian scientist, which might change with the context of the publication and in anticipation of a specific audience. Crucially though, within this typology of scientific nomenclature, there is no indication of formal occupational status. The labels 'amateur' and 'professional' were rarely used and, when they did appear, their meaning could be ambiguous and inconsistent.[27] Instead, it is the subjects' moral qualities and their practical or theoretical contribution to science that are conveyed. While the social formation of Victorian science was in fact complex, such rhetoric constructed an identity for all scientific men as perceptive, respectable, trustworthy figures who were worthy of emulation, and elevated the generic man of science into the upper echelons of the social hierarchy, a position previously held only by the leisured elite.[28]

The language of exemplarity pervaded science in its various settings, from presidential addresses at the Royal Society and the British Association to popular science journalism. With science's growing profile and status in the nineteenth century depending heavily on the kinds of people it attracted, there seemed to be a genuine awareness of the temperament, motivations and virtues of those involved in formulating new knowledge. In this way, the scientific enterprise was socially sanctified.[29] Consequently, while towards the end of the century, scientific knowledge was increasingly being presented as the product of accepted and shared practices and institutions, for the major part of the Victorian era, the individual was perceived to be the basis of scientific authority.[30]

The Victorians' fascination with new visual modes and technologies meant that the burgeoning business of commercial portraiture became just as important for the social construction of identity as it was for the construction of individual identity, and it could be used with similar rhetorical effect. Published portrait anthologies maintained the hegemonic order, which was further endorsed by titles – such as *Men of Eminence* and *Men of Mark* – that drew on the language of exemplarity.[31] Thus, men of science came to be incorporated within that pantheon of contemporary Englishmen that included explorers, military heroes, artists, industrialists, politicians, clergy and men of letters. Collectively, they represented an idealised group portrait of the accomplishments and potential of the British nation itself.

Because the professional status of subjects was generally not clearly articulated, it is difficult to discern between those men of means whose discoveries and achievements had afforded them honorary positions and those who had worked their way up through the ranks within paid employment. Similarly, the occupation and interests of most sitters were seldom identified, except in accompanying biographies. In many studio portraits, the subject is portrayed simply holding an open but unidentified book or poised beside a bureau with a document and pen in hand. Occasionally, sitters identify themselves as scholars by wearing academic robes. In particular instances, motifs directly pertaining to the sitters' work and achievements are featured. Such identifying attributes were only used in relation to certain occupations and, considering the great volume of portraits produced, the incidence of such details tended to be the exception rather than rule. Crucially, the inclusion of specific scientific objects as identifying attributes was a significant feature distinguishing very many naturalists' portraits. Almost as often, however, these Victorian notables posed against typical studio backdrops emulating domestic interiors and thus were portrayed in the same mode as every other sitter of the period. *Photographic Portraits of Men of Eminence in Literature, Science, and Art*, 1863–67, typifies this, with all the sitters' backdrops being virtually interchangeable.[32] Photographed by Ernest Edwards in one of London's leading studios, the series exemplifies the 'plain background with a very limited number of accessories ... regarded as the standard style' for almost all such portrait sittings.[33]

From the photographer's side of the transaction, the pragmatic use of a set range of studio accessories, although often sumptuous in themselves, was symptomatic of the exigencies of commercial studio portraiture. The opposite perspective is equally revealing, however. For the sitter, and ultimately for the viewer, the repetition visually unites these sitters within each album or series and implies a social connection between them. It links them likewise to the modestly comfortable world of the middle-class consumer as viewer. With the same array of studio backdrops, furniture and other domestic accessories recycled continually across the series – a feature criticised by some commentators as repetitive and undesirable – these individuals were seen to reside in a comfortable world that was familiar not only to the sitters but to the majority of their viewers.[34] At the same time, however, the artifice of the studio was always clearly identifiable.

The unconcealed artifice of the studio setting, reproducing the values of middle-class domesticity, was a constant reminder for sitters that the photographic portrait was very much about performance and was therefore more a means of self-fashioning rather than of revealing the individual. This performance or self-fashioning allowed sitters not only to distinguish themselves but also to establish how and where they belonged in the bigger picture. To a great degree, the repetition across numerous *cartes de visite* reflected the desire of sitters to present a persona that, in many cases, appeared exactly the same as, and indeed was modelled on, that of their peers and their contemporaries. Hence, Edwards' portraits of Sir Charles Lyell, William Hooker and Richard Owen, for example, are almost interchangeable in setting and pose, as are the photographs of Sir Roderick Murchison, Thomas Huxley and Michael Faraday (Figures 3.2–3.4). The way these figures approached the portrait process was also mirrored and mimicked by scores of other sitters who visited Edwards' studio or the many other firms like it. These figures presented themselves to their contemporaries, Roy Porter suggests, primarily as 'sober, steady, successful men of affairs'.[35] However, the sorts of visual information offered to Victorians across a range of portraits of 'men of science' suggest that they were actually involved in a rather more complex and historically nuanced process of self-fashioning.

Photographs offered seemingly endless and unmediated detail and promised to reveal much about the subjects portrayed. At the same time, critics of photography were concerned to ensure that photographic portraits embraced the aesthetics of

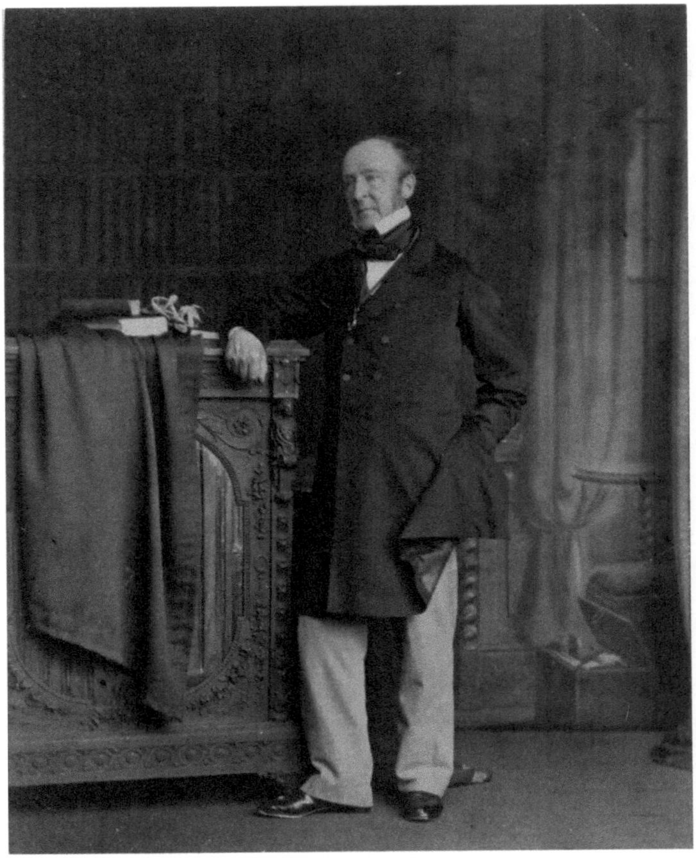

Figure 3.2 Sir Roderick Impey Murchison, p. 23 in Lovell Reeve (ed.), *Portraits of Men of Eminence in Literature, Science and Art with Biographical Memoirs. The Photographs from Life by Ernest Edwards*, Lovell Reeve & Co., London, 1863. Source: The Getty Research Institute – courtesy of HathiTrust.

traditional portraiture by observing certain conventions of the genre. An article that appeared in the *Photographic News* in April 1860, as part of a larger series by William Lake Price, suggested that photographers had a great deal to learn from studying the compositions and use of chiaroscuro by the great masters Velázquez, Titian and, especially, Sir Joshua Reynolds:

> We find much variety in their treatment, but, on the whole, the portraits which impress the spectator … are those which are exceedingly *simple* and *natural* in their attitude and expression … [Nevertheless, Reynolds] has often introduced many accessories in his portraits, analogous to the pursuits of the person represented.[36]

As Lake Price argues, in the hands of a lesser artist, attention to mere detail in a portrait 'is not only insufficient to please, but carried to excess, is positively damaging to

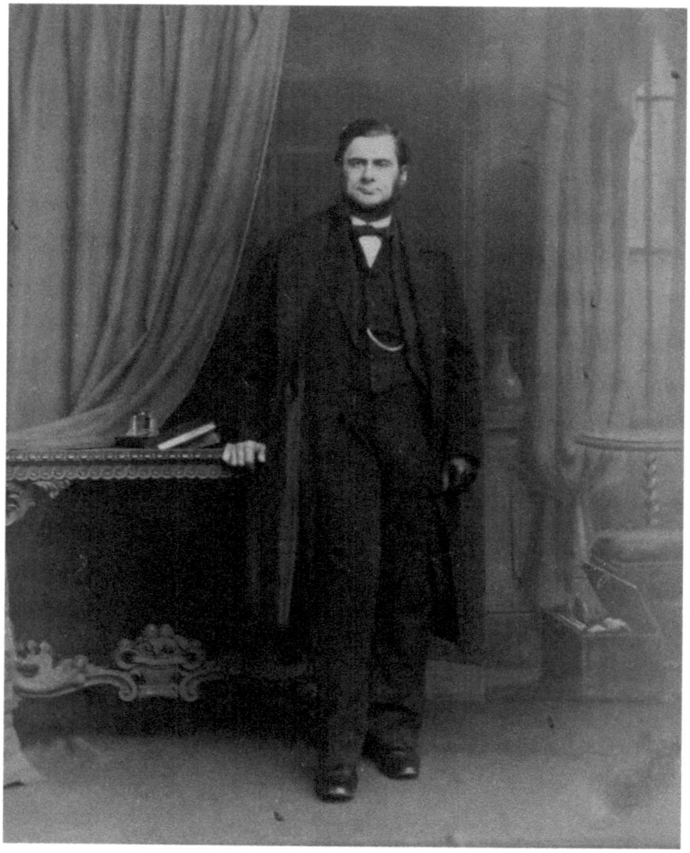

Figure 3.3 Thomas Henry Huxley, p. 127 in Lovell Reeve (ed.), *Portraits of Men of Eminence in Literature, Science and Art with Biographical Memoirs. The Photographs from Life by Ernest Edwards*, Lovell Reeve & Co., London, 1863. Source: The Getty Research Institute – courtesy of HathiTrust.

the work'.[37] To illustrate his argument, he presented a study after Reynolds' 1786 portrait of the great anatomist and surgeon John Hunter – surrounded by the objects of his endeavours – as an example of the artist's masterly approach, which, he conceived, might be 'successful in photography, since the lens executes such accessories with admirable facility' (Figure 3.5).[38] His formal analysis of the work is worth quoting at length as he carefully articulates how the skilful use of light and shade is most important not only in assisting the lines of the composition but also for suppressing or moderating 'commonplace' details; that is, in censoring the same attributes that made photography so promising for science and so appealing to many of its practitioners.

> The effect is broad and masterly. Observe the manner in which the head is relieved, the high light on the forehead being opposed to the depth of the background, whilst the shadow side comes off in dark rather from the half-tones, thus giving the greatest possible relief to the head. The catching *lights* on the glass vessels,

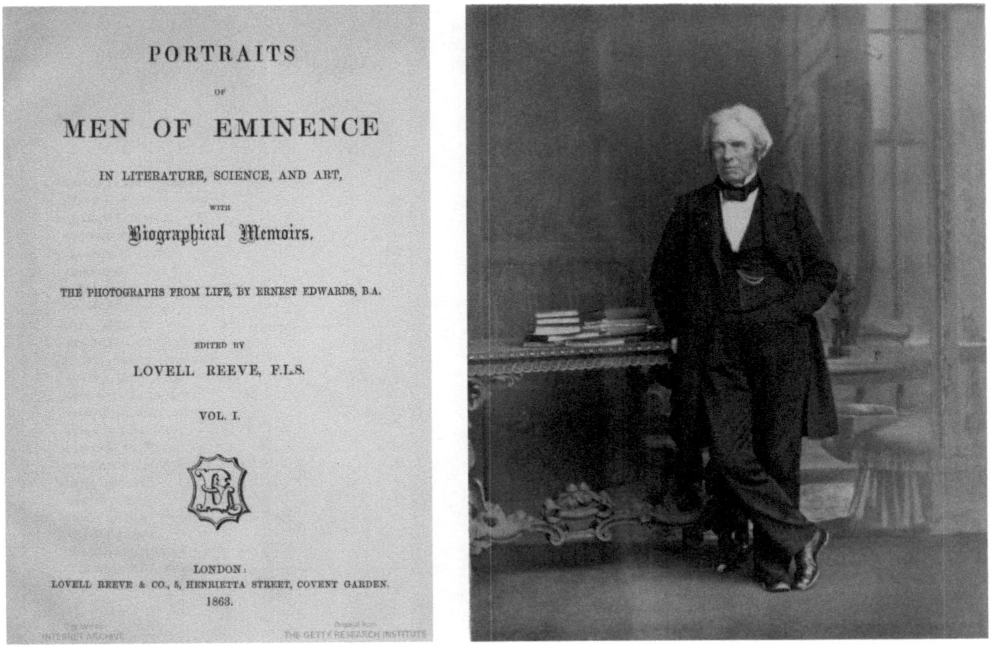

Figure 3.4 Michael Faraday, p. 147 in Lovell Reeve (ed.), *Portraits of Men of Eminence in Literature, Science and Art with Biographical Memoirs. The Photographs from Life by Ernest Edwards*, Lovell Reeve & Co., London, 1863. Source: The Getty Research Institute – courtesy of HathiTrust.

> intensify, by contrast, the depth of the background, and are most useful. We observe comparatively a small portion of the figure in salient relief, and that only in the vicinity of the point of principal interest, the head; the other forms being nearly lost in the background, with the depth of whose tones they assimilate. Use is made of the books and papers on the table, to spread the mass of the principal light on the head, and to blend it *gradually* with the half-tones, waning off into darks. Thus, the highest light in the picture, to which all other are subordinate, is on the head; and it is kept from being a spot … by its diffusion, and being led out of the picture.[39]

Note how Lake Price's description calls attention to the aesthetic effect of the accessories but is not concerned with their specific purpose and meaning. By referring to them in this way, he disregards the historical significance and practical function of the glass spirit and bell jars and the open book, which clearly illustrates examples of comparative anatomy. Similarly, the human skeleton which dominates the right-hand corner of the composition – the Irish giant, Charles Byrne – is entirely passed over. According to Lake Price, these objects are to be viewed not for their intrinsic interest or in any detail but purely as a function of the portrait itself. Indeed, he may have been inspired by a significantly altered detail of the Reynolds portrait that had appeared in the *Illustrated London News* the previous year and which had reduced the great anatomist to just

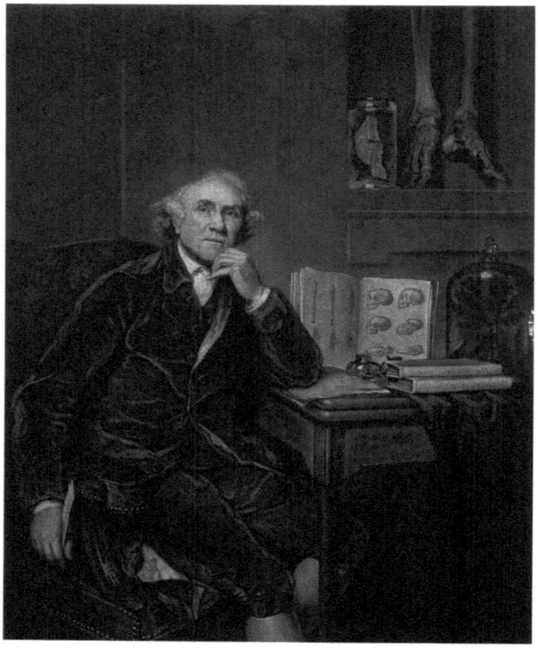

Figure 3.5 W. O. Geller after Joshua Reynolds, *Portrait of John Hunter*, 1836, mezzotint. Collection: Wellcome Library, London.

his head and a hand (Figure 3.6).[40] As Lake Price concluded, 'The camera is capable of producing heads of the highest quality, and only requires judgement and taste in its management by the photographer'.[41]

Studio portrait photographers were often not so discriminating. Many considered the production of detail to be photography's strongest point, and the opportunity to delineate every feature and object remained irresistible, albeit to the detriment of good taste. As Edward Dunmore argued in the *British Journal of Photography*, 'There naturally exists a disinclination to sacrifice one iota of this quality, and to this reluctance may be attributed the frequent failures that are met with in small composition pictures'.[42] The photographic *cartes de visite* and celebrity portraits, which had gained great popularity from the late 1850s, epitomised the small composition pictures to which Dunmore referred.[43] In 1862, at the very height of their popularity, Stephen Thompson could not help but observe that 'portraits by photography are generally considered *uncertain*'.[44] This is not surprising, he suggested, considering how 'chameleon-like are the flitting aspects of the human face. A passing thought – an "airy nothing" – will cast a shadow over the countenance at the critical moment and alter the whole expression'.[45] The propensity of the camera to document every detail, rather than provide a selective interpretation of the subject's appearance and manner, only highlighted the staged nature of studio portraiture by betraying the self-consciousness of the sitter. Thus, where we would expect to see the very likeness and essence of the subject revealed detail by detail, we see instead, as André-Adolphe-Eugène Disdéri put

JOHN HUNTER.

The remains of this distinguished anatomist, after a repose of upwards of half a century under the Church of St. Martin in the Fields, have been disentombed, in consequence of the sanitary movements now going on throughout this vast and overcrowded metropolis for the purpose of securing the health of the living. The Royal College of Surgeons of England have nobly stepped forward to secure for them a fitting resting-place. Having obtained the consent of the Dean and Chapter of Westminster Abbey, they determined to deposit Hunter's ashes in that splendid old pile, which is still allowed to retain undisturbed possession of our illustrious dead. On Saturday last Hunter's remains were removed with all care and reverence to the Abbey, and on Monday were deposited in the north aisle between the tombs of Ben Jonson and Sir R. Wilson. The ceremonial took place at four o'clock, and long before that hour the Jerusalem Chamber—in which room the members of the medical profession were to assemble—was densely crowded. Among the company were the Earl Ducie, the Dean of Westminster, Dr. Clark, Professor Owen. Mr. Buckland, the Presidents of the Colleges of Physicians and Surgeons, and most of the *élite* of the profession. The College of Surgeons intend to complete their work by erecting near this spot a monument to Hunter's memory, and £600 has been already been scribed towards it.

It was as an anatomist John Hunter began his bright career; and, long before its close, he had acquired a popularity as a surgeon which had never before been equalled, and has never since been surpassed.

John Hunter was the son of John and Agnes Hunter, of Kilbride, in the county of Lanark. He was born at Long Calderwood, on the 13th or 14th of February, 1728. The parish register bears the date of the 13th of February; and on the 14th of that month the Royal College of Surgeons of London celebrate the anniversary of his birth.

The late Sir Everard Home remarks that anatomy seems to have been a pursuit for which Mr. Hunter's mind was peculiarly fitted; and he applied himself to it with a perseverance of which there is hardly any example. He laboured for ten years in this branch of science, during which period he not only became acquainted with what was already known, but made considerable additions to that knowledge.

So eagerly did Mr. Hunter attach himself to the study of comparative anatomy, that he left no means unemployed to obtain possession of the rarer kinds of animals, with the view of examining into their peculiarities. For this end he applied to those who had the charge of the Royal Menagerie at the Tower for the bodies of the animals that died there; and he made similar applications to all those who made a business of collecting and exhibiting wild beasts to the public. He also purchased any of the rarer animals that came in his way; and these, with such others as were presented to him by his friends, he intrusted to the showmen to keep till they died, the more to encourage them to assist him in his labours.

In February, 1767, Mr. Hunter was chosen a Fellow of the Royal Society, an earnest of those many honours which were subsequently heaped upon him. The year 1771 was signalised by the publication of Mr. Hunter's treatise "On the Natural History of the Teeth," a work which led the way to many others in rapid succession. Mr. Hunter was now advancing rapidly into public celebrity; and in 1776 he had the honour of being appointed Surgeon Extraordinary to the King; and ten years afterwards was gazetted as Deputy Surgeon-General to the Army. Professional success, however, was never the ultimate aim of his ambition, and he valued it mainly as affording the necessary means for the attainment of those scientific objects in which his soul was centred.

In 1787 the Royal Society conferred upon him its Copleyan medal. About this time honours were showered thick upon him, and he was elected Fellow of the Royal Society of Science at Gottenburg, and of the American Philosophical Society, and also of both the Royal Society of Medicine, and of the Royal Academy of Surgery, at Paris.

His museum continued to enlarge with increasing rapidity, for which he was in no small degree indebted to the friendship of Sir Joseph Banks, who not only allowed him to take any of his own specimens, but procured him every curious animal production in his power, and afterwards divided betwixt him and the British Museum all the specimens of the animals he had collected in his voyage round the world.

Mr. Hunter's death was sudden. He went to St. George's Hospital on the 16th of October, 1793, in his usual state of health. Meeting with things which irritated him, he went into the next room, where, turning round to Dr. Robertson, one of the physicians of the hospital, he gave a deep groan, dropped down, and expired. Mr. Hunter was in his sixty-fourth year. After his death his museum was purchased by the British Government for the public benefit for the sum of £15,000; and subsequent grants were voted for a building for its suitable accommodation. At the time of Mr. Hunter's death the collection contained upwards of 14,009 preparations and drawings, and also a quantity of manuscripts, the great majority of which related to the museum, being voluminous notes in Mr. Hunter's own handwriting, and also the labours of several amanuenses, who for many years had been in the habit of writing under his direction.

Of this noble collection, over which the greatest genius of the age has so long and ably presided and become the exponent of, Professor Owen has well said:—"The preparations and dissections display astounding industry and genius. Hunter's genius, indeed, had penetrated far back into the zoology of other days; and in his museum he left behind him the germ of all the surpassing discoveries since made in palæontology: he left enough to prove that his acute mind had embraced even this vast subject, and had appreciated the stores of matter to be derived from its study. . . . . . Whether viewed in a professional light or as a man of science, his zeal for his profession on the one hand, and for natural history on the other, richly entitles him to the gratitude of posterity."

John Hunter had an elder brother, William, scarcely less distinguished than himself as an anatomist, whose valuable museum is in the University of Glasgow.

Our Engraving is from the celebrated painting, by Sir Joshua Reynolds, in the possession of the College of Surgeons.

LITERATURE.

The Arts Connected with Architecture, Illustrated by Examples in Central Italy, from the Thirteenth to the Fifteenth Century. By J. B. Waring. Lithographed and published by Vincent Brook.

This splendid volume forms a new and important chapter in the history of early Italian art. The works of Rosini, Ottley, D'Agincourt, and others, and, later still, the labours of the Arundel Society, have done much to make known the excellence of the pioneers of art in Italy, as far as composition, design, and expression are concerned, but supplied little or nothing in respect to colour. The art of colour-printing, but recently brought to a practical bearing, affords the means of supplying this desideratum, and of producing the chefs-d'œuvre of those great men on a reduced scale, but with the complete ensemble of effect which they exhibit on the walls of the venerable edifices to which they belong. And it may be observed that this new process is peculiarly adapted to works of this class, which do not involve qualities which prove so serious an impediment and drawback to all attempts at mechanical reproduction.

In glass-painting, mosaic, and wood and stone inlaying, the pieces of various-coloured materials being placed simply in juxtaposition, without any attempt at blending by graduated tints, the work may be absolutely and exactly copied in all its details by the printing process, with a sufficient number of workings; and even in fresco, though slightly modified, a broad and bold treatment of colours was employed, which proves easily capable of textile treatment. This we consider very fortunate, for, whilst we would prefer a good engraving of a Correggio or a Titian to any attempt at chromatic imitation, there are points of essential usefulness and interest in the works of the earlier artists, as involving the application of colour to architectural decoration, which cannot be too widely explained and illustrated, and which it would be impossible to make appreciable by any amount of description unassisted by the colours themselves.

It was long the habit to underrate the works in mosaic, glass painting, and other analogous processes, as mechanical, and unworthy of art; but later experience and a juster appreciation have recognised their important influence upon painting in its grandest characteristics and application. The vastness of the scale upon which these works are produced, the boldness, and withal the simplicity, of design required in them, gave that mastery of hand and freedom of outline which we admire in, and which formed the chief speciality of, earlier Italian art; and these severe but grand characteristics began gradually to disappear as the charms of graduated tints, and reflected lights and hues, and the graces of chiaroscuro began to be recognised, and to be attainable almost without limit, through the medium of oil colours. With the necessity for distinctness and firmness of outline went the power of producing it; drawing became more and more neglected; and the arts, depending rather upon outward effect than the substance that should be within, declined, as all simply imitative processes inevitably must. It is not too much to add, therefore, that, both as a guide and example in art-culture generally, and as a type for decorative art in particular, these remains of early Italian art are of inestimable value; and every attempt to make them patent to us is deserving of our thanks and encouragement.

The literary contents of the volume, though not so full as one would wish, supply many interesting observations, and are for the most part written with intelligence. As glass staining or painting is again coming into vogue amongst us, the remarks upon the principles which should guide the use of this art will be read with advantage and interest. Upon the *rationale* of window-painting, and the intended effect, for instance, we find it pertinently remarked, in a quotation from Mr. Winston:—

The remaining objection, that it is wrong to represent a receding picture

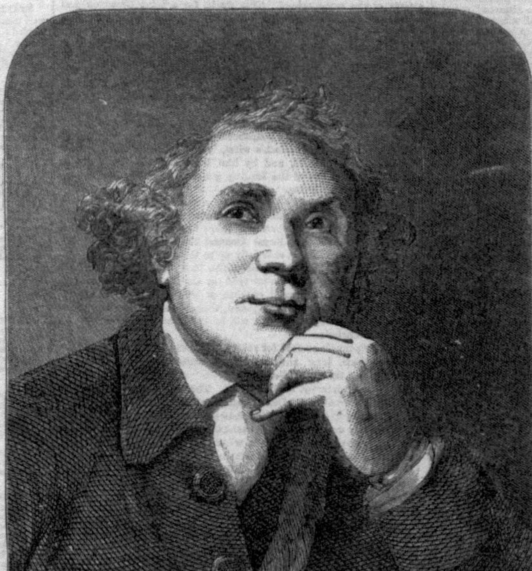

JOHN HUNTER.—FROM THE PAINTING BY SIR JOSHUA REYNOLDS.

Figure 3.6 'John Hunter – From a Painting by Sir Joshua Reynolds', *Illustrated London News*, 2 April 1859, p. 332.

it, the relentless fidelity of the photographic image, which 'shows us the actor where we believed we put the man, the theatrical action where we attempted to place the natural action'.[46]

The naturalist refashioned

The different ways that naturalists and other men of science responded to conflicting opinions in relation to their portraits is crucial to understanding the complex role of photography in natural history during the nineteenth century, especially as the foundation of Victorian scientific culture – and thus scientists' own reputations – relied on them accruing, substantiating and communicating verifiable facts. The very basis of that reliability was the figure of the naturalist themselves, identified and portrayed as a model of moral standing, practical application and intellectual rigour. This raises questions about the specific ways naturalists sought to present themselves, both to their colleagues and to the wider public. For instance, how did Charles Darwin – a private, relatively reserved 'gentlemanly amateur' but also a prominent intellectual and a 'scientific man' who aroused immense peer and popular interest – approach the many commercial portraits he sat for over the years? What sorts of choices were available to him and how did he envisage the process and outcome? Conversely, what were the visual strategies employed by a professional, museum-based naturalist such as Richard Owen, who sought to promote not only his moral standing and the trustworthiness of his scientific findings but also the museum he represented and the many connections throughout the empire that his public role involved?

Apart from his travel narrative *Journal of Researches*, 1839, based on his *Beagle* diary, the general public remained largely unaware of Darwin's scientific work until the *Origin of Species* was published in 1859 and caused a sensation. Even following its publication, however, as Alvar Ellegård warns, 'it would be a mistake to think that the whole of Britain knew about a theory which was to revolutionize our outlook on man'.[47] The general reading public's interest in Darwin's theory of natural selection only peaked during the Du Chaillu affair in 1861–62, although it was reignited at other times with the publication of related works such as Thomas Huxley's *Man's Place in Nature* and Charles Lyell's *The Antiquity of Man*, both of which appeared in 1863, as well as Darwin's later study *The Descent of Man*, 1871.[48] Darwin himself rapidly became public property and came to serve as the personification of evolutionary theory for the general population. Before long, he was viewed as the quintessential celebrity scientist. Darwin contributed to the popularisation of his species theory and aimed to influence public debate, with his supporters doing as much as possible to promote the man as well as his work. The production of commercial portraits provides a vivid illustration of this, as it conveys the process of representation from his standpoint as subject and demonstrates the way he presented himself to his peers and to the public through successive studio sittings.[49]

Darwin's first commercial sitting was for one of the earliest publications of celebrity portrait photographs, Maull and Polyblank's *Literary and Scientific Portrait Club*, 1855 (Figure 3.7). A subsequent sitting took place in 1865 when his photograph was taken for an album of biographical memoires by Edward Walford (Figure 3.8).[50] He posed for this latter portrait in the studio of Ernest Edwards, possibly during his visit to London between 8 and 20 November that year; he sat again for Edwards for a

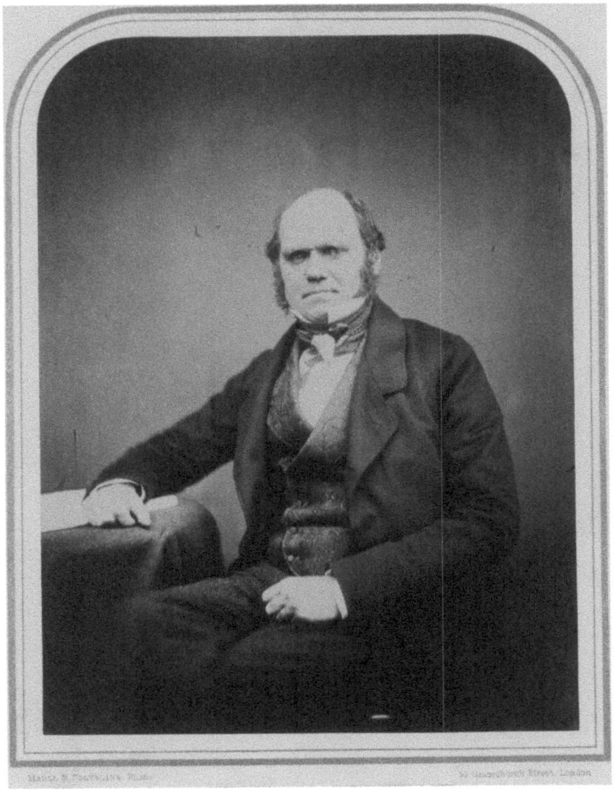

Figure 3.7 Maull and Polyblank, *Charles Darwin*, c.1855, albumen silver photograph, from the *Literary and Scientific Portrait Club*. Collection: National Portrait Gallery, London.

subsequent album by Walford entitled *Representative Men in Literature, Science and Art*, published in 1868.[51] Darwin did not recount the circumstances of either of these Edwards sittings in his personal papers. However, his family recalled that the anticipation following the first sitting for Edwards had been almost too much to bear. His brother Erasmus impatiently wrote to Emma Darwin just a few days after the event advising that he had 'not yet received the Photos' and, expecting the worst, confided: 'My present impression is that they are too hideous to be borne. I think you had better write to the man and say that C[harles] will try again when he comes up to town after Christmas, & you may quote your feminine feelings ... as a reason'.[52] Unusually for a sitter, Darwin did not initially order any copies for his own use, and his biographer surmises that he was not pleased with the results, even though Emma Darwin had found Edwards' photographs of her husband to be very 'true to life'.[53] While acknowledging the great popular interest in images of Darwin, Joseph Hooker thought that commercial photographs were entirely inadequate and raised the issue with Darwin in 1864: 'I am very anxious to get [Thomas] Woolner down to take a clay model of your bust, for myself, as you kindly promised I might.... [H]e shall cut it in marble

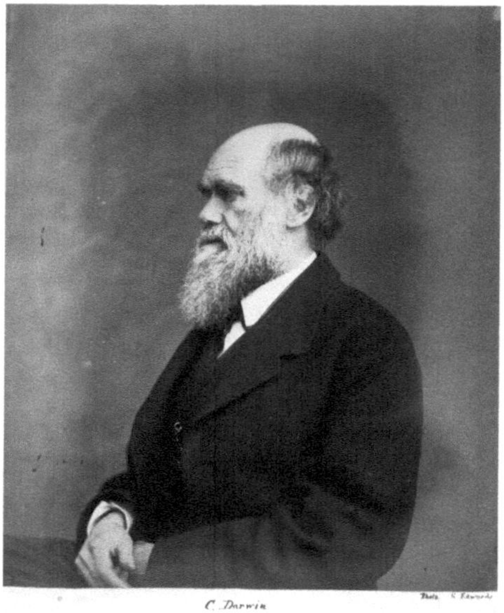

Figure 3.8 Ernest Edwards, *Charles Darwin*, 1865–66, albumen silver photograph. Collection: National Portrait Gallery, London.

at his leisure for me. [S]uch heaps of people want to know what you are like – & the photographs are not pleasing'.[54]

Janet Browne has examined Darwin's specific modes of self-fashioning and his ambivalence about being photographed for the commercial portrait trade.[55] In contrast to the many written tributes to him, which had been the mainstay of earlier publications, 'agreeing to Walford's request [to be photographed] marked the beginning of a different form of commemorative activity', Browne suggests: 'Unused to self-advertisement, and by nature modest, Darwin may have felt that photography for public consumption was somewhat vulgar, unless required by public position'.[56] Contemporary commentators certainly emphasised this distinction and, even though the majority of studio portraits appeared very similar, the purpose and circumstances of their production, for the sitter at least, varied and were not necessarily always consonant with the priorities of the photographer and publisher or indeed the whole concept of celebrity portraiture. For Stephen Thompson, this was part of the uncertainty of portrait photography:

> Some persons have quite a passion for sitting for their portraits – others an invincible repugnance; and these – often the most worthy – submitting to it as a kind of infliction incident to their position.... There are others who are always excusing themselves to *themselves* for sitting. This is really through a proper feeling. No-one would or should sit for a portrait for their own sakes [sic]. The fact of sitting implies something which should have a higher origin – in the region of the affections, or in private or public worth.[57]

Yet, if portraiture was to be reserved as a mark of great honour and 'a spur to posterity to stimulate the same virtues', as Thompson furthermore maintained, popular criticism suggested that many Victorians were still not totally convinced that photography was well-suited to portraiture, given the reputation of the medium for copious, unseemly details.[58] 'There is a certain anomaly in the public estimation of photography which is worth noting', one commentator observed. 'No art is so frequently treated with contempt and contumely', he noted, yet 'none more blindly and unreasoningly admired and trusted'.[59] Even Claudet's early image of Faraday had been inconsistently critiqued, being admired for its conformity to the sitter's appearance in one sentence but condemned for its unflattering fidelity in the next.[60] Equally though, whereas Darwin did note his approval of a large-format, dramatically lit and soft-focus portrait of him taken by Julia Margaret Cameron in 1868 – adding this inscription on the print sent to him by Cameron: 'I like this photograph very much better than any other which has been taken of me' – he also expressed reservations about the same photograph, subsequently describing the portrait to fellow naturalist, Alfred Russel Wallace, as 'heavy and unclear'.[61]

Given the importance of portraiture for many scientists, along with the undeniable popularity of studio photography, it is reasonable to assume that most were active participants in determining the parameters of their portraits and had a clear idea of the image they wished to project. In Faraday's case, a significant factor was his reputation not only as a philosopher of science but also as a figure in the public gaze who was an eloquent and influential professional 'performer' of science. For this reason, his entire approach to portraiture and self-representation stands in stark contrast to Darwin's guarded manner. Faraday's many portrait sittings and his renowned practical demonstrations were aimed at bringing science into the public gaze by way of dramatic, accessible forms of display, albeit through different performative modes. As John Tyndall, his successor at the Royal Institution, described it: 'Except by those well acquainted with his subjects, his Friday evening discourses were sometimes difficult to follow. But he exercised a magic on his hearers which often sent them away persuaded that they knew all about a subject of which they knew but little'.[62] In one of his best-known portraits, taken by Maull and Polyblank and published in October 1857, both performative modes are combined in a re-enactment of a practical demonstration in 1845 of his discoveries of dia-magnetism and the influence of magnetism upon light. Faraday's research was almost entirely pursued in the Royal Institution in Albemarle Street, London, where he delivered his weekly lectures in the great theatre and where he also had his residence.[63] That proximity – that merging, even – between the private spheres of professional research and personal domesticity, on the one hand, and the public, institutionalised realm of activity, on the other, is an important factor in the way he presented his scientific discoveries in his lectures and projected himself in portraits. As Faraday's discoveries 'made their passage upwards from the basement laboratory to the famous lecture theatre', they moved from private space to public domain and underwent a corresponding transformation in their epistemological status, as David Gooding notes, 'from the realm of novel and personal experience to that of publicly witnessed, demonstrable fact'.[64] Commercial portraits of men of science likewise routinely moved between public and private spheres. For Faraday especially, their circulation across those realms replicated the way he demonstrated and disseminated privileged knowledge which, for most of the public, would otherwise have been impenetrable and highly esoteric.

Darwin's approach, on the other hand, was more likely designed to represent himself as a detached and neutral observer. It has been noted of his public musings about his place in Victorian society that he suppressed individual subjectivity on behalf of science and, in so doing, identified himself with the scientific community.[65] Darwin confirmed in his autobiography in later years that he aspired 'to take a fair place among leading scientific men', although whether he was more or less ambitious than a growing number of his colleagues who were formally attached to museums and other institutions, and thus were 'public' figures in every sense of the word, he could not say. Furthermore, while he was anxious to receive the approbation of such distinguished men as Charles Lyell and the Hookers, he 'did not care much about the general public' and was 'sure that [he had] never turned one inch out of [his] course to gain fame'.[66]

In contrast to Darwin, and his reclusive life and status as a gentlemanly amateur, Richard Owen – Britain's foremost public authority in the natural sciences and a long-time opponent of Darwin's natural selection theory – was a consummate publicist. Like Faraday, he was at the very nucleus of a small but growing field of professional scientists. Hence, in 1856, the *Times* anticipated his promotion later that year from the role of conservator at the Hunterian Museum to superintendent of the natural history collections at the British Museum – a position created exclusively for him – by

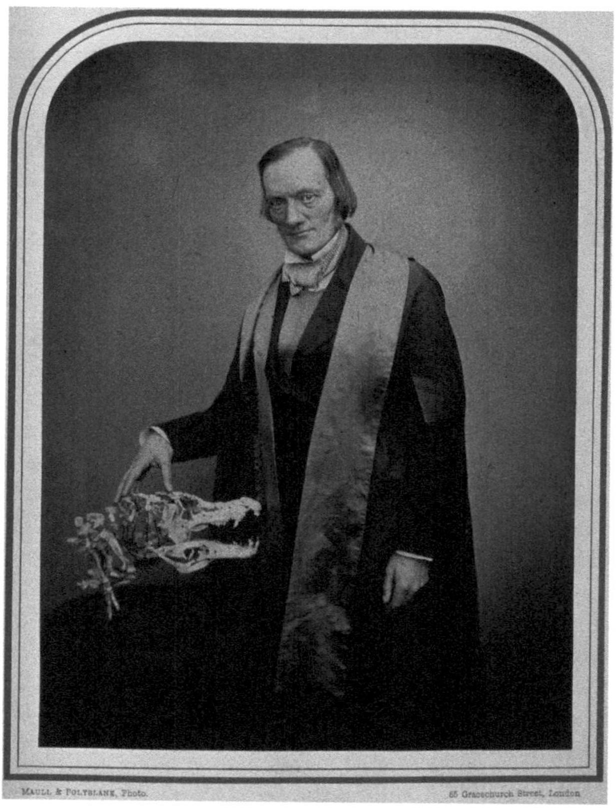

Figure 3.9 Professor Owen, c.1856, plate 1 in Maull and Polyblank, *Photographic Portraits of Living Celebrities*, 1856–59. Collection: Wellcome Library, London.

declaring that there was 'not a more distinguished man of science in the country'.[67] Furthermore, it was Owen who took primary position in Maull and Polyblank's *Photographic Portraits of Living Celebrities*, which commenced that same year (Figure 3.9). The series was promoted as 'a work of national importance and interest', and Owen's appearance in such a lavish and state-of-the-art publication confirmed his considerable professional and social status not only by his inclusion but also by the privilege of being chosen as the opening subject in this serialised publication.[68]

Rather than being defined by national imperatives alone, Owen's work and reputation were global in scope. Few other scientists included in *Living Celebrities* could be unequivocally accorded such imperial influence as Owen was for his promotion of natural history. Yet, unlike most other naturalists – including Darwin in his early years – who built up knowledge and specimen collections during exploratory expeditions, Owen never once travelled to any of Britain's colonies to undertake research. His method was ultimately more strategic, effectively exploiting the pre-eminence of his position at the very centre of imperial science to found a dedicated national museum of natural history, which of course would signify empire as much as nation. Quite pragmatically, Owen's studies centred on the living and fossil fauna of Britain's colonies, especially Australia and New Zealand, and involved a worldwide network of enthusiasts and specialists contributing – either purposefully or indirectly – to that ultimate objective of museum building. Darwin, in contrast, was a private individual and gentlemanly specialist of independent means, even though in many ways he was no less a scientist of empire than Owen – his extensive imperial network was based almost exclusively on written and illustrated observations exchanged between naturalists, although this exchange rarely resulted in the procurement of specimens. The rationale underlying Owen's scientific agenda, Nicolaas Rupke notes, was thus primarily 'institutional, in that his [scientific] papers reflected the policy of specimen acquisition and were part of a planned, systematic enrichment of the collections to establish national hegemony, international parity and imperial grandeur'.[69] A further imperative was to endorse certain orthodoxies of natural history that were by then well established and thus counter the growing support for Darwin's transmutation theory of evolution.

In light of the strategic considerations that went hand in hand with the research aspects of his work, Owen needed to promote himself and his agenda to a far larger audience than his immediate scientific coterie.[70] This involved the projection of natural history to the public in a way that made it accessible yet also presented it as more than merely a recreational pursuit; that is, as a significant yet still emerging profession engaged in credible methodologies and new areas of inquiry and with a tangible social value. The importance of science to Britain's material wealth through industrial and imperial progress was not lost on the Victorians. Indeed, the public was receptive to developments in science. This was evidenced in the substantial popular interest in natural history – and to a lesser extent the physical sciences – which was now beginning to be carefully cultivated by professional scientists and expositors who sought to enthuse laypersons with the exhilarating potential and results of scientific discovery through organisations such as the British Association for the Advancement of Science and the Royal Institution of London, the writings of scientific popularisers, the great international exhibitions and the various avenues of popular culture.[71]

In an article published several months prior to the publication of Owen's portrait in Maull and Polyblank's portrait series, G. H. Lewes likened Owen to Isaac Newton as representing the acme of British scientific achievement. Likewise, he observed, 'The public no more reads Professor Owen than it reads Newton … The number of readers, the number of "copies sold" is only an element in reputation when the object of the writer is to gain a large audience'.[72] Though Newton's place in history as England's greatest natural philosopher remains unparalleled, Lewes' comparison of Owen to the Enlightenment's most iconic figure was not as fanciful as it might seem. Both figures had come to be seen as symbolising the triumph of national endeavour. They were therefore recognisable to a great many in the population who had no acquaintance with the studies of optics and gravity or comparative anatomy but who were nonetheless avid consumers of the celebrity portraits and illustrated biographical publications that couched these cultural achievements in more accessible and popular terms.[73] The contrast between various forms of high culture and the mass-produced objects, such as commercial photographs, demonstrates a broad spectrum of visual and textual media through which both scientific and national identity was constructed and promulgated in the ongoing process of fashioning the individual's public image. By the early nineteenth century, Newton's name had become virtually synonymous with 'genius', and his image had become a symbol of the acquisition of knowledge and the power of an individual's rational thought to fundamentally transform one's perception of the world. For this reason, his reputation was invoked in many contexts outside the realm of natural philosophy.[74] It is not surprising, then, that Newton became the exemplar for commentators such as Lewes when reflecting on their own generation of innovators and intellectuals.

This idea of scientific genius as an exemplary type was embodied in the metaphor of enlightenment – the concept of truth revealed through empirical evidence – which Lewes and others embraced and which derived from Baconian ideals. With fundamental changes in the nineteenth century to both the structure of British society and conceptions of natural knowledge, distinct but complementary notions of exemplarity gave the more influential figures of science a new and broader relevance which, especially in the case of professional scientists, accorded them a more visible role in relation to cultural institutions and activities as well as a range of civic reforms.[75] In the case of Owen, this increased visibility derived from his task as a juror for the Great Exhibition to his recommendations, as a member of the Metropolitan Sanitary Commission, for practical improvements in urban sewerage systems. Notwithstanding Lewes' comparison of him with Newton as the epitome of British achievement, Owen modelled himself on certain prominent and established figures from his own area of the natural sciences. His role models tended to be individuals who, like Faraday, were renowned for demonstrating and popularising their scientific work and who could often be identified in their portraits through motifs representing not only their scientific discoveries but also their public performances.

Most prominent among these figures was Owen's mentor, the Rev. William Buckland, geologist and canon of Christ Church at Oxford. Famous for his important palaeontological discoveries of British fossil fauna, Buckland was invariably portrayed as acting out his excavations or flourishing one of the strange fossils from his spectacular collections. These portraits often refer to the processes of induction and re-vivification, tracking the visual and intellectual methods of the palaeontologist but also taking the viewer back through time. For example, in George Rowe's 1823 portrait of Buckland, there is a visual progression that proceeds from the fossilised jaw bone that Buckland proffers to the articulated skeleton in the top right-hand corner and then to the living animals portrayed just below (Figure 3.10).[76] In most portraits, he wears his

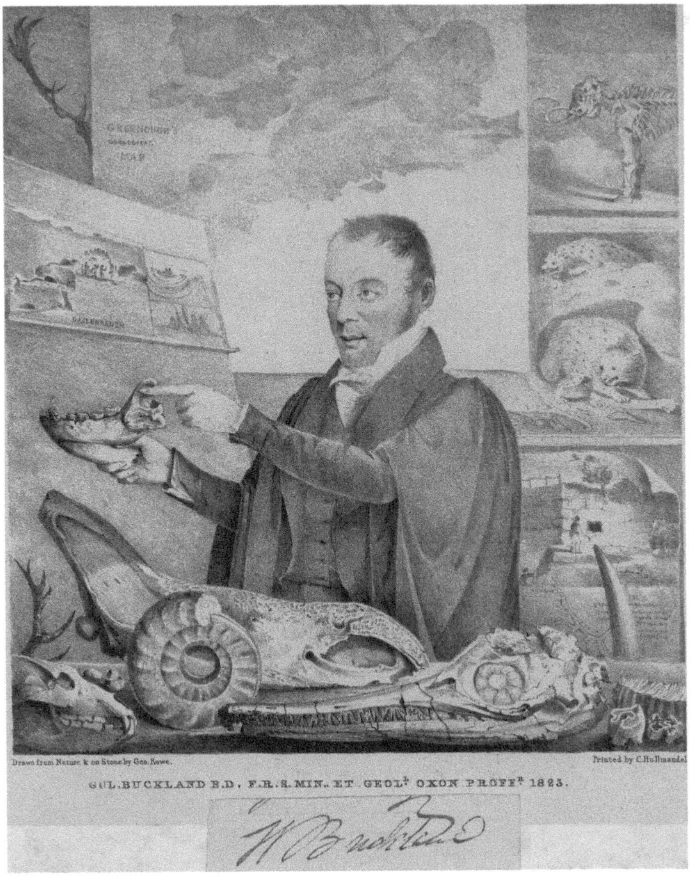

Figure 3.10 George Rowe, *William Buckland*, 1823, lithograph printed by Charles Joseph Hullmandel. Collection: National Portrait Gallery, London.

academic gown and is often shown next to a map of Britain, which together symbolise his academic status and the national focus and significance of his work.

Thomas Phillips' 1832 painting of Buckland proudly holding the fossilised skull of an extinct hyena he had discovered in a Yorkshire cave and accompanied by a pair of handsome ammonites clearly provided the inspiration for his scientific protégé Owen in a number of formal portraits (Figure 3.11). The earliest of these portraits was painted by Henry W. Pickersgill in 1844 and portrays Owen in the act of lecturing, holding the *Dinornis* thigh-bone that would be so crucial to his celebrity (Plate 7). In this portrait, the object of natural history is represented quite conventionally – as identifying motif and prized trophy – compared with the elaborate set-up and extensive detail in some of Owen's later photographic portraits. The virtual absence of fossils depicted in still-life paintings, let alone portraits, prior to this, demonstrates the influential role played first by Buckland, and then Owen, in bringing the new science of palaeontology to popular attention.[77] Another version of this first portrait, painted by Pickersgill

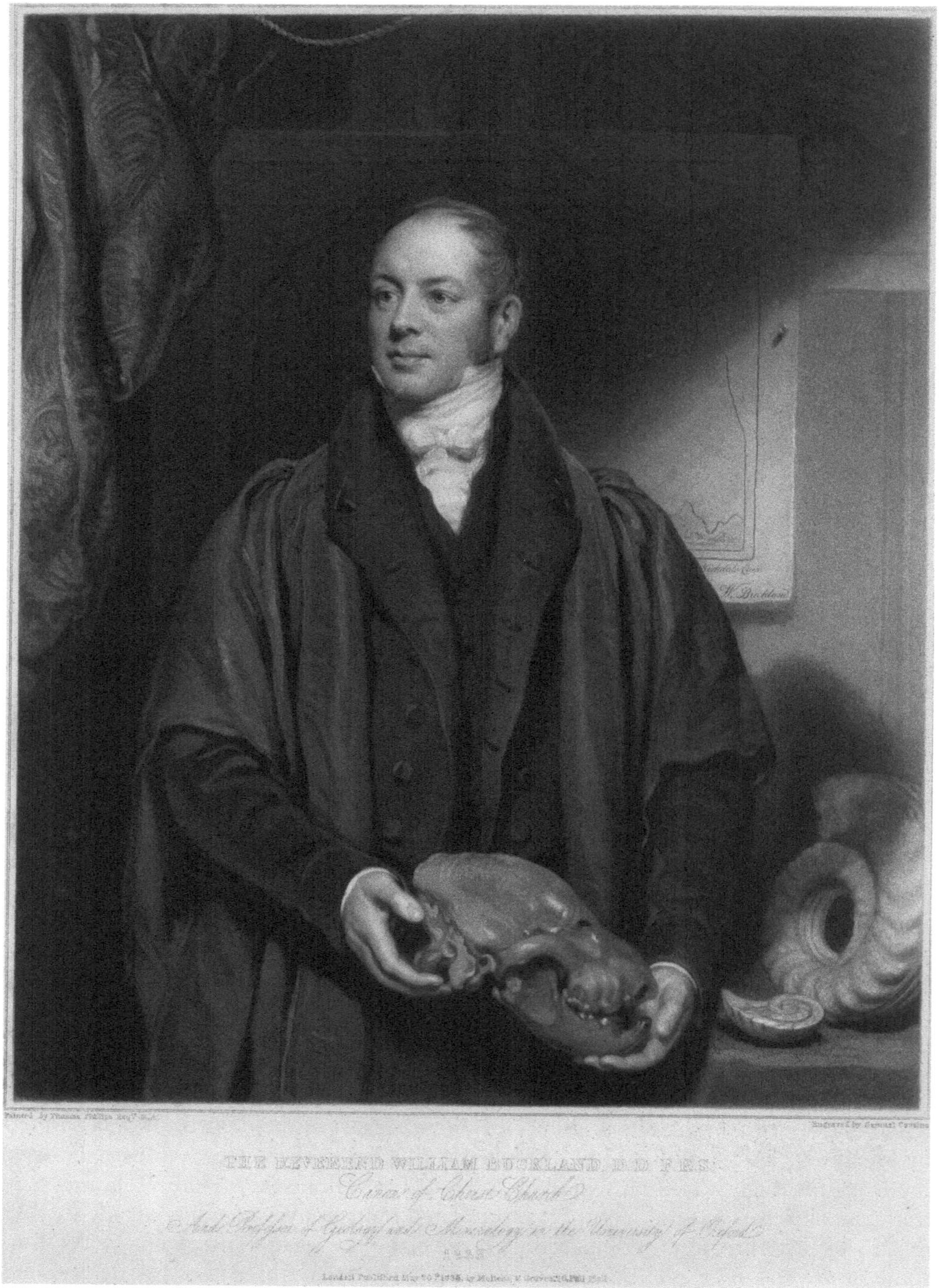

Figure 3.11 Samuel Cousins, after Thomas Phillips, *William Buckland*, 1832, mezzotint, published 1833. Collection: National Portrait Gallery, London.

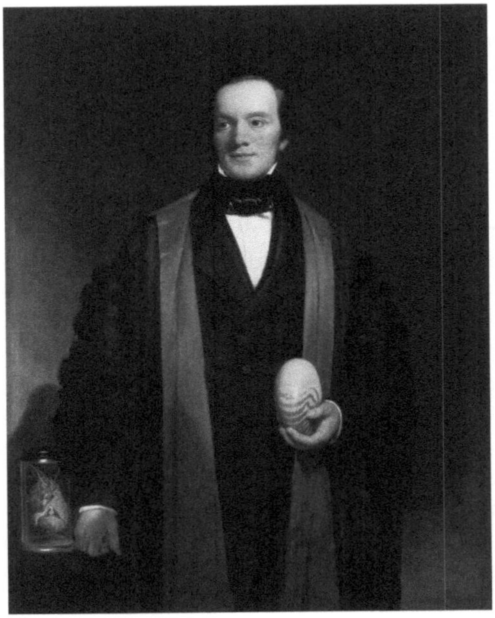

Figure 3.12 H. W. Pickersgill, *Portrait of Richard Owen* (holding the Pearly Nautilus), c.1845, oil on canvas. Collection: National Portrait Gallery, London.

the following year, shows Owen lecturing on the Pearly Nautilus – signifying his first major scientific monograph, *Memoir on the Pearly Nautilus*, 1832 (Figure 3.12).[78] Together, these portraits demonstrate how Owen was clearly influenced by Buckland in the way he portrayed his spectacular discoveries as events of intellectual significance, public interest and national pride. In addition, the subjects chosen for Owen's portraits reflect the increasingly crucial role of empire to his work, with both paintings depicting important specimens that had been sourced from New Zealand, Australia and other locations in the South Pacific.[79] Recent x-rays have revealed that the Pearly Nautilus shell and the jar behind containing the fleshy part of the mollusc are in fact later changes to the painting, with the work originally depicting Owen holding the moa bone as in the earlier painting.[80] By depicting this part of the Pearly Nautilus, the portrait flouts a long-held convention for illustrating shells, which was to show their external form without any indication of the invertebrate's soft interior parts.[81] Similarly, the presentation of the single *Dinornis* bone symbolises Owen's capacity to conceptualise the animal around it. Both portraits allude to the art that Owen had been perfecting in describing an entire animal from parts and, in the process, re-presenting the scientific method as public performance.[82] Although as motifs they were necessarily incomplete, the moa bone and the Pearly Nautilus were nonetheless significant visual and conceptual elements that Owen would exploit even more effectively in later photographic portraits.

Darwin's approach to self-representation in his written narratives, by contrast, illuminates the much more restrained manner he typically adopted for photographic portraits. As he stated in the introduction to his autobiography, produced following a request from a German editor for 'an account of [his] mind and character' and written when his life was 'nearly over': 'I have attempted to write the following account of myself, as if I were a dead man in another world looking back at my own life'.[83] He thus sets himself up as an external, transcendent observer of his own life, a method less reliant on narrating his subjective experiences or offering a psychological self-examination – that is, what his experiences, feelings and thoughts actually meant to him – than on proffering the analytical perspective of a dispassionate, scientific observer.[84] It is said that dead men tell no tales, and, similarly, Darwin promised little in the way of self-examination of his own personality or motives that might compromise or obscure this factual and externalised view of himself. Instead, his intention was simply to lay out the events of his life for his readers to absorb, not to provide a guided tour of the vicissitudes of his mind. In denying his own subjectivity, he effectively presented his life as a series of events, offering himself as a medium through which others were invited to create a persona based on the unembellished facts he recounts and their preconceptions of him derived from his public image. This latter image had taken on a life of its own as a figment of others' imaginations and was a marketable commodity quite separate from his lived experience.

Darwin's approach to autobiography was not dissimilar to the complex role that photographic portraits played in the Victorian era. Although intended to be authoritative and objective, Darwin's autobiography was as much performative as it was an assurance of authenticity, and it allowed him to portray different versions of himself simultaneously. As performance, it recalls Hippolyte Bayard's group of photographs entitled *Le Noyé* (*Self Portrait as a Drowned Man*), 1840.[85] Highlighting the essential difficulties of photographic representation, Bayard's self-portraits played on the paradoxical conflict between the visualisation of death as a mode of authentication and the illusion inherent in the act of representing one's own death.[86] As in Darwin's autobiography, Bayard's desire for impartiality and the effacement of subjectivity through the enactment of his own demise was as much a precondition as the focus of his study. By staging his own death, he presents a self-portrait in which he is simultaneously both absent and present and which questions photography's capacity for truthfulness.[87] Furthermore, in these precociously staged portraits, Bayard places himself amongst plaster figurines and *objets d'art* from his collection, thus picturing himself as just another item of exchange – and simultaneously promoting the medium's capacity to delineate and record the physical characteristics of different types of things.[88] By presenting himself as just another object ripe for visual commodification, Bayard was anticipating the immense role that photographic portraiture would assume in greasing the wheels of the Victorian laissez-faire economy. Moreover, he reveals the photographer and sitter to be not only a mere collector of things, but, as the photographer–director of this mise-en-scène, the creator of his own objectification.[89] As such, he demonstrates two elements that would contribute to the rise of popular photographic portraiture: the role of this photographic genre in portraying collectors of things and delineating the types and range of their collections; and also in producing a further tier of consumers – those who collect the portraits and images of objects to create and supplement their own collections.

With the explosion of commercial photography studios in the 1850s, the enthusiasm for collecting portraits was reflected in a sharp rise in the production of serialised and anthologised portrait publications. In their first publication, *Literary and Scientific Portrait Club*, Henry Maull and George Polyblank, self-described 'artist-projectors and publishers', included a range of figures from vocational scientists such as Darwin to figures from the commercial side of science such as natural history publisher John van Voorst. Their *Photographic Portraits of Living Celebrities* developed from their earlier format and proved to be a highly effective model for distributing images, along with associated biographies, of distinguished figures in British public life. To demonstrate their capacity to produce first-class photographic portraits, Maull and Polyblank pointed to the success of their previous, more exclusive series, copies of which had been distributed to 'Members only' but which could also 'be seen at the Libraries of the Royal Society, the Linnean, Geological, Royal Astronomical, Royal Geographical, and Chemical Societies, and at the Artists' [photographers' studio] [in] Gracechurch Street'.[90] Portraits in the later series were produced in the same size and style as the *Literary and Scientific Portrait Club* but had a wider purview with regard to the public role and professional background of sitters and, ultimately, the intended audience. In its representation of a notionally broad-based scientific fraternity, the *Literary and Scientific Portrait Club* reflected the extended intellectual network of the British scientific community. With many of the same figures also appearing in *Photographic Portraits of Living Celebrities*, scientists could be viewed as contributing to multiple overlapping contexts, a situation that was also highlighted in the biographical texts. In the latter publication, for example, Owen's important recommendations as a member of the Commission of Public Inquiry into the Health of Towns and into Smithfield Market, his assistance to the institution of the Board of Health and his 'invaluable aid in the development of the Great Exhibition scheme' were given almost equal prominence along with his contribution to pure scientific inquiry.[91]

Issued in 40 parts, *Photographic Portraits of Living Celebrities* was launched in May 1856 and continued with instalments more or less monthly until its conclusion in August 1859. Notices for the latest and back issues of this publication appeared quite consistently throughout the three years of its production in the daily press as well as in general interest journals such as the *Athenaeum* and the *Illustrated London News*. Initially, the release of each instalment attracted a review or some other commentary in the press. However, as early as December 1856, after the publication of only eight or so portraits, it was evidently assumed that potential subscribers had by then come to know what to expect, and subsequent notices tended to provide readers with only cursory details of the latest celebrity instalment. In publicising and presenting the photographs with only perfunctory information, the producers of portrait publications presumed that their potential customers already had some familiarity with the reputations, or at the very least the names, of the selected subjects, most of whom would have appeared in the press or popular journals at one time or another.[92]

Serialised and anthologised portrait publications such as these reflected a particular hegemonic order in the medley of faces and names. Even in the illustrious company of figures such as Baron Macaulay, whose profile succeeded Owen's as the second in the series, Owen's portrait nonetheless stands out. His portrait style epitomises his ambitions for the national museum of natural history and symbolises his vocation as a museum naturalist. Wearing robes designating his position as Hunterian professor, he revels in the honorific function of the portrait. While the *Athenaeum* criticised Owen's photograph for this 'unfortunate formality of dress which might', the reviewer suggested, 'with a grain of management have been avoided', such ceremonial attire was a highly visible sign of both his intellectual accomplishments and his social standing.[93]

At the same time, he addresses the viewer directly and gestures purposefully to the mounted skull and upper vertebrae of a Saurian. Other sitters might have been publicly derided for choosing such an audacious object. Referring to a later issue in Maull and Polyblank's series, for instance, *The Athenaeum* ridiculed a prop that had been tucked mischievously between the legs of explorer David Livingstone: he was photographed 'with the inevitable table and damask table cloth, common, very common to photographer's rooms, holding awkwardly and ludicrously enough the twin horns of the greater antelope ... sharp, thick and strong, they form a sort of horny mitre in his hand' (Figure 3.13).[94]

In selecting the Saurian skull for his sitting, Owen was demonstrably aware of not only the commemorative function of the portrait and its cachet in terms of contemporary celebrity but also the visual potency of including such a striking and singular object, as well as its potential to impress a wider audience with an example of his pioneering work in comparative anatomy and to promote his plans for a national museum of natural history. The details revealed by Maull and Polyblank's sharply focussed camera lens and scrupulous processing were critical to the success of this portrait. Its clarity was, however, precisely another aspect of the image to which the *Athenaeum* reviewer took exception, pointing out that 'as mere likeness, it is agreeable; but as a photograph, we miss all attention to anything like artistic resources'.[95] It was argued that the outline of Owen's figure appeared 'hard and cutting against the flat board background'.[96] Thus, the reviewer intimated, the portrait ignored the conventional, domestic-style context of contemporary portraiture and instead concentrated

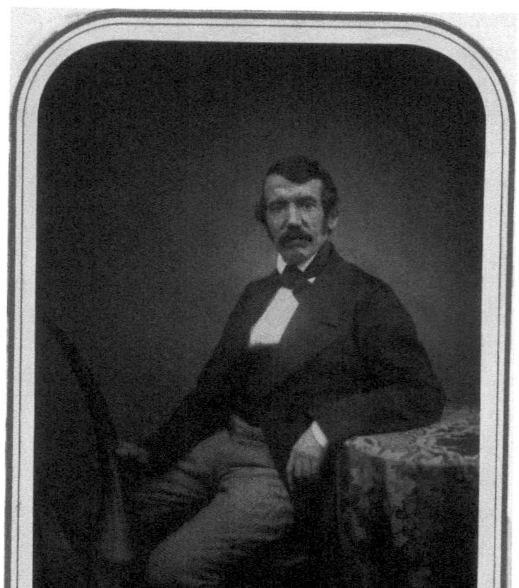

Figure 3.13 David Livingstone, 1857, plate 25 in Maull and Polyblank, *Photographic Portraits of Living Celebrities*, 1856–59. Collection: National Portrait Gallery, London.

the viewer's attention to the detail of the sitter's robe and the unconventional object he proffers. On the other hand, both sitter and publisher were clearly encouraging viewers to attend to these tiny details in the photograph, reinforced in the accompanying biography, which celebrates Owen's role in re-introducing the microscope into scientific practice 'as an essential instrument in anatomical … investigations'.[97]

While celebrity portraits and attendant publications were unquestionably both commercial ventures and vehicles for self-promotion, they were also more complex from the point of view of their conception and presentation of contemporary notions of exemplarity, concerned, as they were, with revealing not only aspects of the visible world but also providing a visual representation of a more abstract socio-political order. Exemplarity functioned in the portraits in two ways: in terms of rhetoric – the identification of a paradigmatic or model individual or group; and as a more tangible cognitive process – the presentation of particular objects or themes, or details thereof, as practical examples or illustrations.[98] Both of these Platonic and Aristotelian ideas of exemplarity were implicit in the rationale and configuration of the published portrait anthologies. However, in these mass-produced studio portraits of the mid-Victorian period, the two kinds of exemplarity tended to be realised in different ways. Apart from being personified in the sitters themselves, some portraits embraced the contemporary preoccupation with popular demonstrations of scientific method and their importance for the relationship between science and the public. In addition to general props such as books, documents and writing materials – which were customary motifs connoting erudition, diligence and status – portraits of naturalists can be seen to incorporate more specific objects, including identifiable specimens and scientific paraphernalia, as demonstrated by Owen's portrait. With the inclusion of such things, sitters not only signified their individual interests and fields of inquiry and their affiliation with others in their field but also revealed something of the nature of their work to a more general audience. Much of the representational power of the *carte de visite* was derived, therefore, from the construction of the self through things – the photographs themselves and the objects with which sitters surrounded themselves.[99] This pertained not only to the allure of the *cartes de visite* as objects to be bought and exchanged but was also drawn from a fascination for the way they depicted everything – whether significant or seemingly trivial – in such inexorable detail.

For the viewer, collecting *cartes de visite* and celebrity photographs, as with other collections of things, involved arranging and systemising them, examining them closely and comparing their finer points. This was nowhere more evident than in portraits of naturalists where the principal methodology being promoted was the Baconian or inductive approach, which, in contemporary debates about the philosophy of knowledge, privileged empiricism above idealism. As Richard Yeo notes, 'a focus on method offered a point of consensus in the midst of increasing specialisation and allowed the scientific community to promote a unified public image'.[100] Indeed, the inductive approach was considered to be the most accessible of scientific methods. Even in the face of increasing specialisation and an emerging generation of professional scientists, empiricism's careful processes of observation encouraged participation by a wide range of contributors, including amateur and even recreational cultivators of science, who remained a vital part of Britain's scientific endeavour.[101] In this respect, contemporary discussion about the philosophy of knowledge and the relative value of theoretical and empiricist approaches to science, as John Stuart Mill asserted, was not merely a 'matter of abstract speculation' but was also an issue 'full of practical consequences'.[102]

For Mill and subsequent commentators, an appreciation of scientific method over and above a superficial knowledge of scientific findings was seen to have broader moral and educational ramifications for the general populace.[103]

As regards the progress of science itself, a consensus about appropriate scientific method bolstered communication within the scientific community and was seen to be crucial in minimising the fragmentation of scientific inquiry brought about by increasing specialisation. Specialisation was considered by many to be detrimental to the future of the broad-based philosophical inquiry epitomised by the Enlightenment and the accomplishments of Newton as symbols of Britain's intellectual achievement. The British Association for the Advancement of Science was in fact founded so as to foster connections and ameliorate the effect of a growing disconnection between scientists from different disciplines, as well as to convey scientific method and discovery to the general public. Because British science was perceived to be a focus for national pride and imperial endeavour, this fragmentation of science was alarming since it alluded to an empire in danger of disintegration. As William Harcourt cautioned, when 'science is detached from the central body, first one fragment falls off, and then another; colony after colony dissevers itself from the declining empire; and by degrees the commonwealth of science is dissolved'.[104] Similarly, William Whewell pondered the spectre of a future science in which intellectual inquiry would be divided into so many specialisations and crumble like 'a great empire falling into pieces'.[105]

It was against this background that portraits of men of science began to be conceived and disseminated for a wider range of purposes, with some specifically concerned with scientific method and the associated issues of professionalisation and specialisation. Because the major part of a naturalist's practice was to be an eyewitness to nature, attending to details such as the specific objects to be included in their own photographic portraits allowed scientific practitioners to not only define themselves as consummate observers of natural phenomena but also to promote their art of scientific 'looking' to the general public, an important aspect of the annual meetings of the British Association for the Advancement of Science. It is unsurprising, therefore, that Maull and Polyblank's portrait of Owen featured in the 'Exhibition of Photographs' at their 1859 meeting in Aberdeen. As the association noted, Aberdeen was 'situated in a district rich in natural phenomena ... and in the midst of a population animated by the desire to share in those pursuits which the Association most desires to cultivate'.[106] It was a significant meeting for a number of reasons. Owen was the outgoing president of the association and gave a keynote address in which he welcomed his successor, His Royal Highness the Prince Consort. With Prince Albert officiating, the events and displays encompassed during the week-long meeting were even more lively and varied than in previous meetings.[107] Reflecting this, the photography exhibition was held in the Music Hall Building, which was also the venue for the evening *conversaziones* and lectures. The venue's whole suite of rooms was thrown open for the events, with the exhibition rooms including a variety of attractions and entertainments, and on tables – extending the whole length of the magnificent hall – were ranged all manner of objects of scientific and artistic interest.[108]

Thousands of visitors attended these events over the week, including members from across the empire and many foreign visitors. For the association, the meeting promoted the practices and discoveries of science to a national and imperial constituency and to an international audience and presented these in ways that were comprehensible and appealing. In the midst of this, Owen's portrait with the Saurian specimen gained enormous exposure and demonstrated not only the exemplarity of the sitter but also the skill required and the virtue to be found in the art of close observation. Owen had become famous for his analytical and deductive work in palaeontology, which resulted in his spectacular functionalist reconstructions from disparate and often tiny fossil remains. His provisional identification and characterisation of the giant, flightless *Dinornis*, or moa, in 1839, which had existed in New Zealand, was one of his most celebrated restorations and earned him the sobriquet of the 'British Cuvier'.

Revelations of extinct species were becoming more frequent at that time, yet the significance of the *Dinornis* announcement was to be found in Owen's facility in describing a previously unknown creature by extrapolating from a minute shard of fossilised bone, which implied that exactly what was being observed was rarely immediately apparent and never entirely unequivocal. There is a clear reference in many of Owen's portraits to the methodologies and practices of comparative anatomy; that is, the study and comparison of the structure of different classes or groups of animals. While Georges Cuvier had developed the science of comparative anatomy in the 1790s using the extensive collections of France's Musée National d'Histoire Naturelle, Owen had Britain's considerable imperial holdings for his source and became a serious rival to Cuvier when he started publishing in the 1830s. It was Cuvier's grasp of the organism as a functioning system of parts that enabled the recreation of extinct specimens from mere fragments of physical remains. 'Using the rhetorical device of synecdoche', Claudine Cohen suggests, 'Cuvier presented with great confidence one of the most powerful and enduring images connected with the palaeontologists' profession: the spectacular practice of reconstruction, which, starting from a small part, produces a gigantic creature'.[109] Like Giovanni Morelli's 'philological' approach to art attribution, or connoisseurship, published in the 1870s, identification through synecdoche required one to concentrate on the minor details rather than the whole in order to compare, extrapolate and verify origins.[110] Relating this to the realm of natural history, we can begin to appreciate that 'more than the *causes* that ruled objects and phenomena, Cuverian science studied the structural laws that determined the *relations* that linked them together in "closed systems"' using what Cuvier called "the correlation of forms in organisms" in conjunction with the functional conception of the organism'; that is, what they look like, how they move, what they eat, and how well adapted they were to their environment.[111] It was this process of restoring and revivifying nature that fuelled the popular fascination for palaeontology, and these issues appeared frequently in the contemporary press; the moa itself – as a significant discovery for Victorian natural history and also the subject of popular interest – became a topic of discussion in newspapers such as the *Illustrated London News* and the *Times*.

Hence, it was the display of conspicuously exotic specimens representing prehistoric fauna, as much as the promotion of his own image, that was most advantageous to Owen in distinguishing his own scientific reputation and pursuing his public campaign for a dedicated museum of natural history. Choosing to pose with the saurian skull in the Maull and Polyblank portrait was undoubtedly a deliberate statement as his work on reptiles mostly related to the skeletons of extinct forms – this work was published in the first important account of Mesozoic land reptiles, later to be termed 'Dinosauria'. His decision to include a natural history specimen thereby serves to distinguish his portrait from most other sitters in *Living Celebrities* and can be seen as reinterpreting certain conventions only recently established for photographic anthologies. Additionally, it is the physical distinctiveness and incongruousness of the specimen itself – a mode of valuation that draws its significance from the very nature of

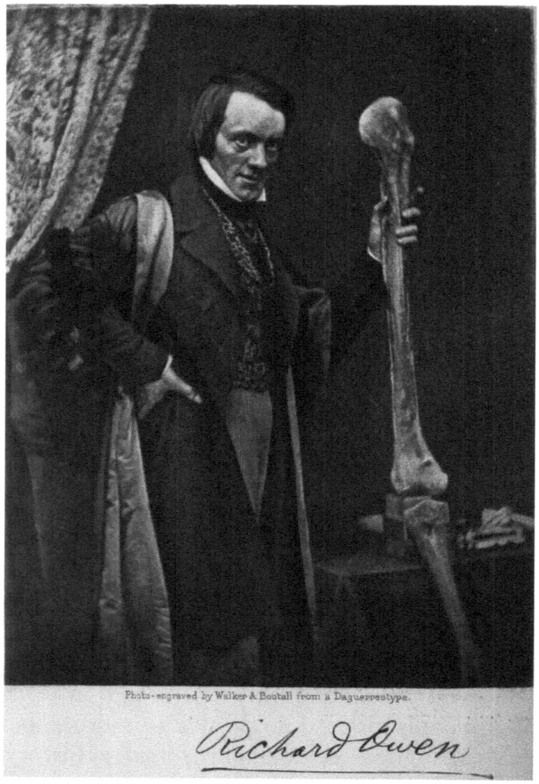

Figure 3.14 Walker and Boutall after unknown photographer, *Portrait of Richard Owen*, c.1845, photo engraving after a daguerreotype in Owen, Rev Richard, *The Life of Richard Owen*, vol. II, John Murray, London, 1894.

the object – that ultimately affects the reception of the overall image. This approach was established in a daguerreotype dated around 1846 – one of the earliest portraits of Owen – which survives as a photo-engraving by Walker and Boutall (Figure 3.14). Clad in academic dress and looking directly into the camera, Owen submits to the viewer evidence of the moa in the form of two enormous bones comprising the upper and lower sections of the animal's massive leg. The sheer size of the objects in relation to the height of their bearer is striking. Judging by the obvious confidence in his pose, its visual impact was clearly not lost on Owen. Any viewer could not help but begin to imaginatively re-construct the giant bird *in toto*, and Owen's probing gaze seemed to invite just that. On another level, the representation of partial or dismembered specimens denoted not only the practice of natural history but also the function of museums in replicating the metonymic structure of the collections and the way in which museum displays demonstrate the rhetorical practice of signifying a totality through parts.[112]

The very presence of the bones and the way they are presented within the composition suggest quite different ways in which this photograph might have been understood by viewers. Most evident is the inclusion of a human figure – Owen himself – to indicate scale, a familiar device used from the earliest years of photography in the documentation of archaeological sites and objects.[113] In this respect, we can see that Owen's photograph straddles various genres and different approaches to photography: it is not only an honorific portrait but also serves other, practical purposes. An even closer examination of the photograph reveals that it has been very carefully planned and perhaps set up by the sitter himself, most likely to ensure that neither his own status is diminished nor the stature of the moa undermined. Seen from slightly below eye level, the viewpoint monumentalises both Owen and the moa bones. He clearly seeks to evoke the spectacle of gigantism. At the same time, however, the bones themselves are truncated by the lower edge of the composition and thus are represented only incompletely. The huge bones are shown to be not only disproportionate to the scale of their discoverer but unable to be displayed within the parameters of the photographic image. Here, the portrait represents the subject of natural history as a measure of man, but by the same token, it also documents a natural history in which man provides the scale for nature in its most gigantic proportions.

The features of this 1846 portrait were reprised in a photograph produced around 1877 and published in Owen's monograph *Memoirs of the Extinct Wingless Birds of New Zealand*, 1879 (Figure 3.15). In this later image, a complete, articulated skeleton of *Dinornis maximus* towers over Owen, who holds the original fossil thigh bone fragment that prompted his speculations that resulted in the spectacular visual

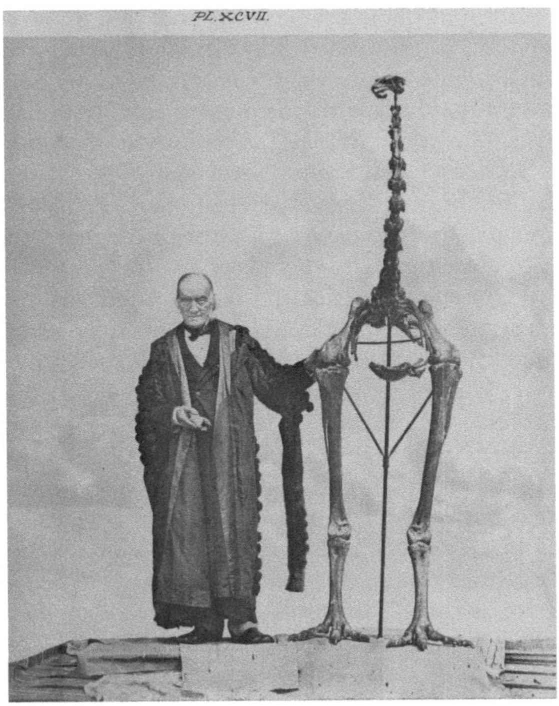

Figure 3.15 Photographer unknown, *Portrait of Richard Owen*, c.1877, albumen silver photograph, plate 97 in Richard Owen, *Memoirs of the Extinct Wingless Birds of New Zealand*, John Van Voorst, London, 1879. Courtesy of the University of Texas Libraries, University of Texas at Austin.

reconstruction of the gigantic moa. The photograph is testament not only to the British Museum's immense resources to gather and reassemble geographically dispersed and long extinct animal species but also to Owen's ambitions for a complete makeover of the museum's public face. Owen's aspirations for the natural history collections called for a massive venue that could accommodate examples of all the largest extinct and living species and which would form a breathtaking centrepiece for the museum. He proclaimed that the largest specimen that could possibly be acquired ought to be exhibited in a new national museum of natural history, which would be the pinnacle of international science and represent the past and present fauna of the empire to its fullest extent.[114] In this way, he clearly articulated the scope of his vision, differentiating the envisaged displays at his proposed museum from the relatively modest displays that he had thus far been supervising at the British Museum. The successive portraits of Owen proffering exotic specimens and gigantic fossil bones thus actually served to prosecute his cause towards a dedicated museum of natural history.

In comparison, Maull and Polyblank's portrait of John Edward Gray, Owen's colleague at the British Museum, shows a more retiring figure and downplays his achievements and public position (Plate 8). While also posed with specimens relating to his natural history studies, these objects seem almost incidental to the composition of the portrait and less significant to the viewer's experience. Gray presents his occupation as museum keeper primarily as 'a collector and organizer of knowledge' rather than as a presenter of public spectacle.[115] On a table beside him in the portrait are three glass spirit jars of specimens preserved in alcohol, and a closed book. Gray, an agent of the process of knowledge formation, is positioned between these two forms of knowledge – raw data in the form of specimens awaiting the naturalist's analysis, and published facts. Although the museum naturalist's role in that procedure is implied through the presence of these objects, Gray does not engage with these specimens or books in any way. The diminutive scale and positioning of the spirit jars within the composition offers only limited information and visual interest since their contents are not obvious to the viewer. Unlike most dry collections with their emphasis on the presentation of the specimens, spirit collections, by their very nature, are not conducive to easy observation. The tall containers restrict and distort one's view of the objects inside, and the alcohol tended to change the colour and toughen the structure of the perishable soft-tissue specimens.[116] Because wet specimens were impractical for most amateur naturalists, they were usually the province of learning institutions or remained in study collections in public museums. These objects therefore denote, rather than explicitly demonstrate, the behind-the-scenes work of the professional naturalist and the process of knowledge production in the museum context. Further, the domestic environment suggested by the standard fit-out of the commercial photographic studio, with Gray seated in an armchair, echoes the transitional status of Victorian natural history. At the British Museum, this space encompassed the highly visible public galleries, to which many of Owen's portraits allude, as well as the more restricted and private areas, including Gray's working space and the residence for his family.

Even when sitting for their own relatively modest *carte-de-visite* portraits, many men of science – whether professional or amateur – were eager to demonstrate their specialisms and competencies by being portrayed with objects of their study as symbols of their practical and scholarly prowess. Becoming customary in the mid-Victorian period, natural history motifs appeared in thematic and celebrity portrait series and persisted well after *cartes de visite* had reached the height of their vogue (Figures 3.16–3.19). Indeed, it was almost *de rigueur* for scientists to pose with specimens and the accoutrements of their work. For instance, when Lovell Reeve visited the studio of Triptree & Sons, he took along a number of objects that would denote both his studies in conchology and his business interests as a successful publisher (Figure 3.20). The resulting photograph illustrates the degree of collaboration between some subjects and their photographers and shows how carefully certain sitters planned their sittings so as to develop the iconography within their own portraits. In the centre of the photograph, Reeve is poised with a sizeable manuscript tucked under his arm. On a table to his right, a pair of shells rest on two substantial books and are framed by the plush fabrics of a table cloth and curtain. At the left side of Reeve, a chart-stand with illustrations showing the forms of various shells is bisected by the edge of the image. It is clearly legible, though, and serves to supplement the two specimens on display. The chart is placed at precisely the same height as the standing figure of Reeve, reflecting the nature and breadth of his knowledge, and thus casts him as no mere *student of science* but as a rather more accomplished *cultivator of science*. Positioned midway between these various artefacts of nature and culture, he gazes out and across to the far left. The style and sheen of his dark dress coat and vest, with matching bow tie and double watch chain; the clarity and specificity of the main group of objects presented; and his purposeful but unceremonious stance are calculated to distinguish him as a

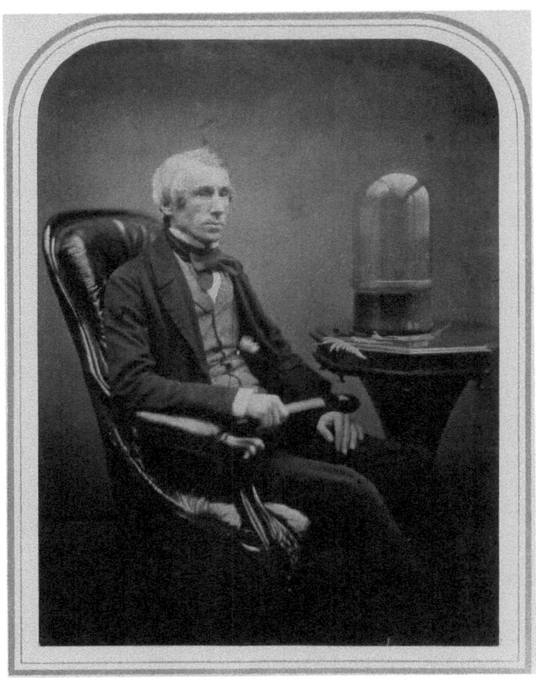

Figure 3.16 Maull and Polyblank, Dr. Thomas Thomson, c.1854, albumen silver photograph in *R. C. A. Prior Album* (assembled from 1854). Collection: Linnean Society, London..

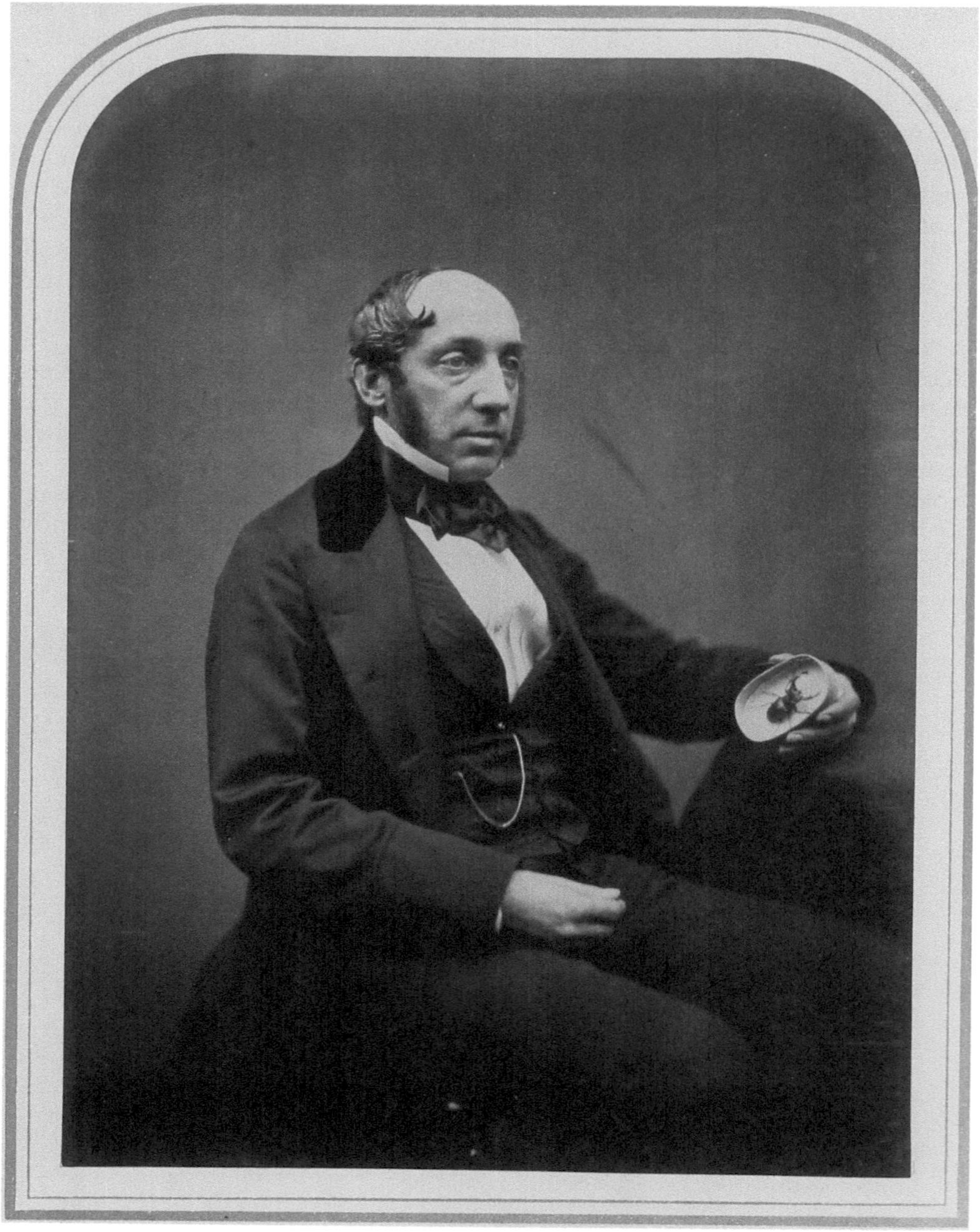

Figure 3.17 Maull and Polyblank, James Scott Bowerbank, c.1854, albumen silver photograph in *R. C. A. Prior Album* (assembled from 1854). Collection: Linnean Society, London.

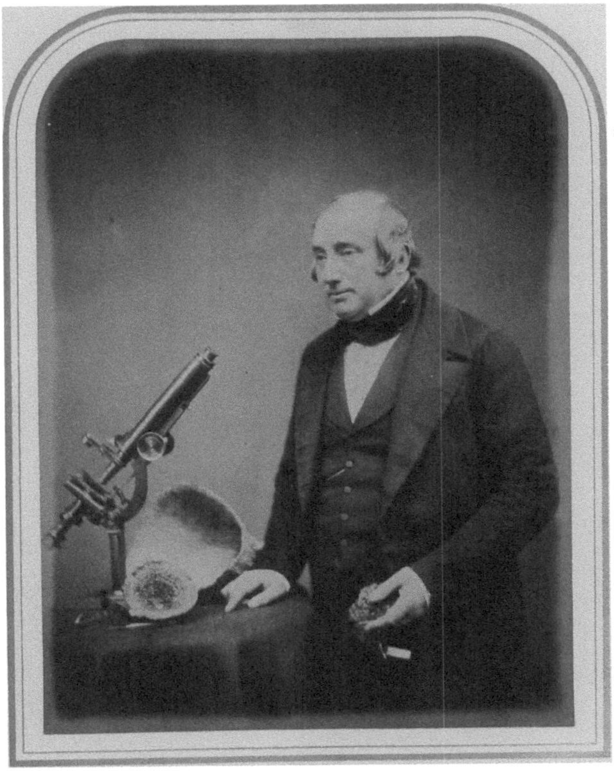

Figure 3.18 Maull and Polyblank, Samuel Stevens, c.1854, albumen silver photograph in *R. C. A. Prior Album* (assembled from 1854). Collection: Linnean Society, London.

person of means: cultured and knowledgeable but also exhibiting the resourcefulness and tenacity of a successful businessman.

While Reeve appears to have done all he could to make these aspects of his identity obvious to the audience, the portrait image does not seem overburdened by these visual cues, as some contemporary commentators might have argued, and indeed the modest dimensions of the *carte de visite* renders much of the detail to a mere notation of its referent. The minute scale of these details does not diminish the interest either; rather, one can imagine a viewer – perhaps a fellow naturalist – pulling out a magnifying loupe to inspect the finer details of each shell. As James Hector, geologist and inaugural director of New Zealand's Colonial Museum, exemplified in his own *carte de visite*, the naturalist's powers of observation and attention to detail were essential skills and characteristics that typically distinguished the most accomplished individuals from others in their field (Figure 3.21). Noticeable disparities between these two studio portraits underscore some typical differences between the self-presentation of naturalists at the centre and the periphery of empire, nevertheless. Reeve enacts his role in Victorian natural history in terms of the culmination of an intellectual enterprise – closely associating himself with weighty publications as the end products of

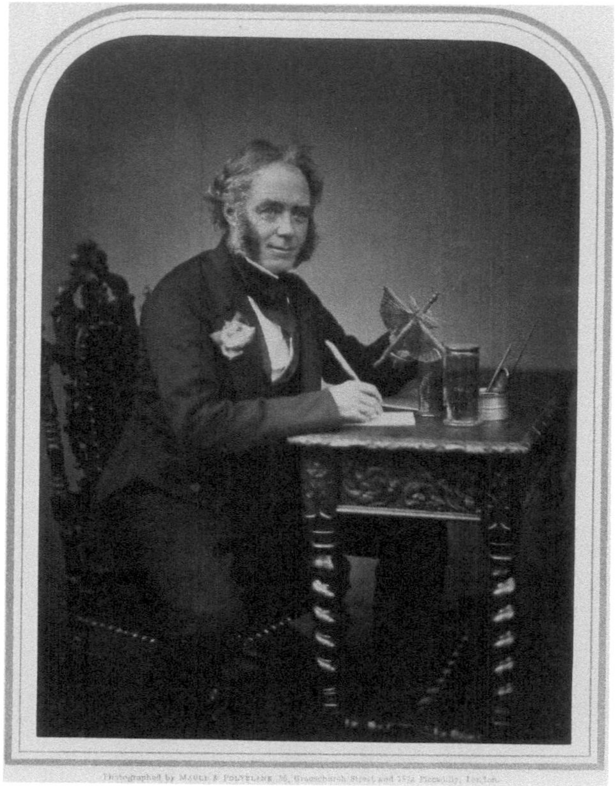

Figure 3.19 Maull and Polyblank, J. O. Westwood, c.1854, albumen silver photograph in *R.C.A. Prior Album* (assembled from 1854). Collection: Linnean Society, London.

the scientific process – whereas Hector is still merely in the process of looking and discovering and hence is engaged in the ongoing behind-the-scenes procedures of an diligent practicing naturalist.

For most other occupations, such distinctive identifying motifs were relatively uncommon. Darwin's portraits diverge from those of other naturalists in the way they rarely deviate from the low-key style that dominated commercial studio portraiture of the day.[117] Like his approach to autobiography, in which, as he explained, he 'had taken no pains about [his] style of writing', for Darwin to do anything more than simply present the unadorned facts of his life as he saw them – for instance, to explicitly affect a conspicuous style or to draw attention to his personality and thus acknowledge his subjectivity – would 'only obscure objective truth'.[118] That is not to say, however, that Darwin did not closely monitor the performative aspects of his career – which ranged from the largely practical work of the naturalist to the imaginative expression of his observations and theories – to convey his authority. The very fact of his public standing and the magnitude of his fame meant that various popular forms of self-display were unavoidable and, while he did not endeavour to make a spectacle of himself, he had nevertheless become visible by virtue of his reputation.[119]

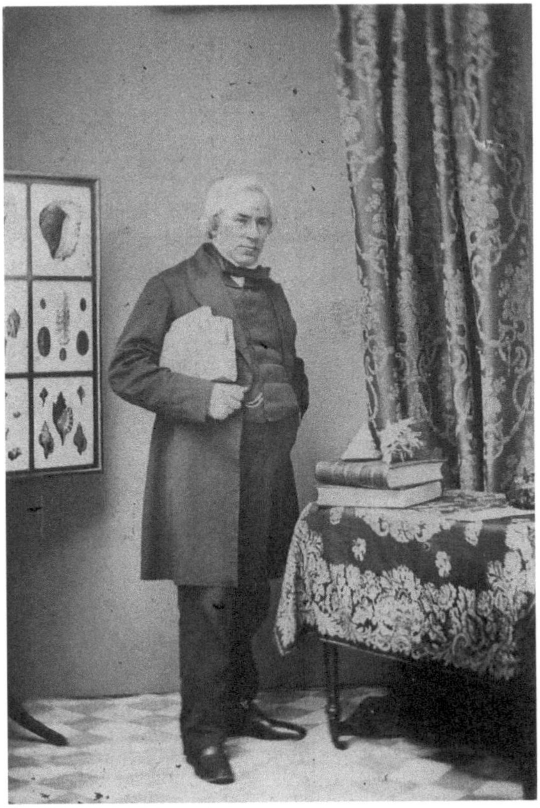

Figure 3.20 Triptree and Sons, *Lovell Reeve*, c.1860, albumen silver *carte-de-visite* portrait photograph. Collection: Linnean Society, London.

When Darwin performed to the camera, he assumed a certain demeanour which, apart from one or two of his earliest and his very last portraits, he used consistently in all of his sittings. Reserved, expressionless and seemingly oblivious to the photographer and the process going on before him, he chose not to engage the viewer and almost invariably appeared introspective as if wholly immersed in his own thoughts.

It might be expected, therefore, that for the general audience, Darwin's portraits were a lesser viewing experience, especially compared to the theatrical approach adopted by his celebrated contemporaries, Owen and Faraday. Or it might also be supposed that his impassive manner would have suggested to some that he was disengaged from the general public or, at the very least, was hesitant to reward their interest, which would almost certainly have promoted more attention. Yet, as Linda Shires observes, in the same way that the literary authority of Victorian writers 'was constructed in the public arena as an occasionally active and an occasionally passive spectacle of power', other prominent figures of the period, including men of science, also inhabited a spectrum of performative self-representation ranging from active display to a more passive demeanour, thereby denoting quite different types of spectacle.[120] In fact, public figures' reticence to animatedly parade themselves, or even openly engage

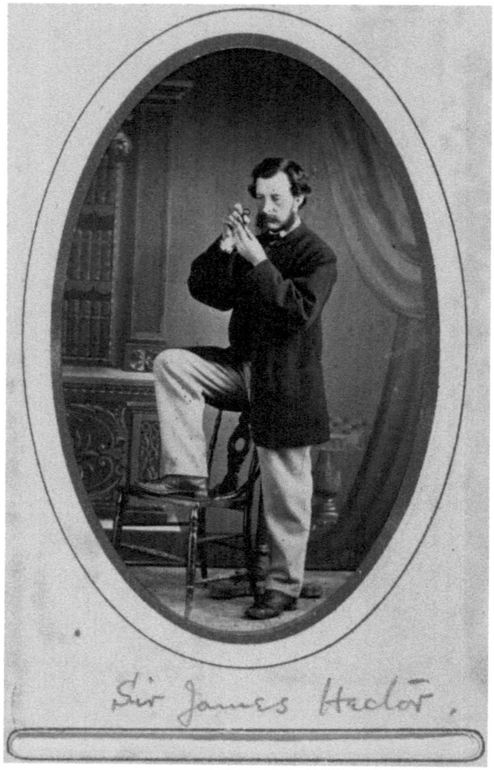

Figure 3.21 John McGregor, *James Hector*, c.1863, albumen silver *carte-de-visite* portrait photograph. Collection: Alexander Turnbull Library, Wellington (The Webster Collection).

with their audience, rarely diminished their allure, for, while viewers were becoming very used to having their adulation acknowledged by the attentive gaze of celebrities in innumerable commercial photographs, the middle-class Victorian audience was also, Leo Braudy argues, 'among the first to become excited by the refusal of its gaze'.[121]

This avoidance of the viewer's gaze became a feature of Darwin's portraits, which he achieved by positioning his head, if not always his body, side-on to the camera so as to be portrayed in almost perfect profile, usually looking out of the left-hand side of the composition. In contrast, the frontal and three-quarter poses more commonly adopted for studio portraits allowed the sitter to open up to the camera and present the full volume of their head and body in a demure yet interesting pose while also looking directly at the viewer. The profile portrait, on the other hand, was less revealing of the sitter's facial and other physical features and hence was also less informative or engaging for the viewer. Yet, other aspects of this portrait format made for a satisfying viewing experience. For instance, the profile pose accentuates Darwin's high forehead – and notionally his high intelligence – thus ostensibly indulging the Victorians'

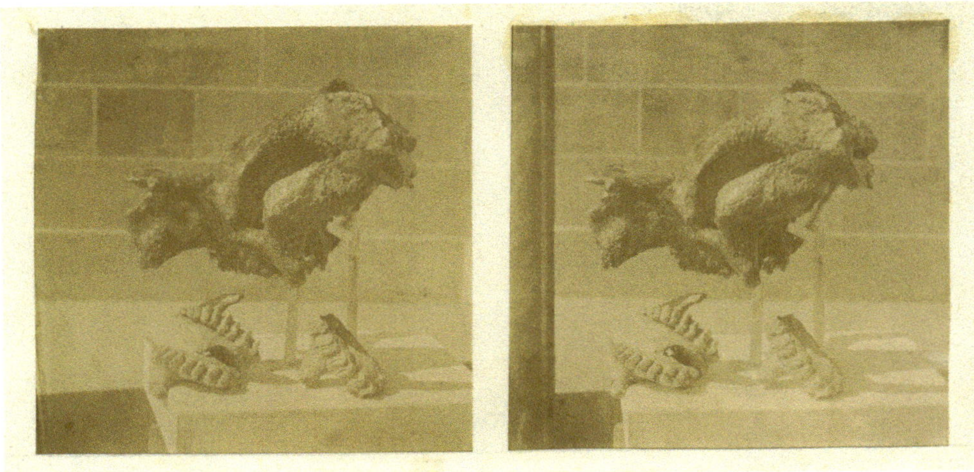

Plate 1 James Glen Wilson, *Skull of Zygomaturus trilobus*, stereoscopic pair of salted paper photographs, 1857. Collection: Geological Society, London (LDGSL 203-3).

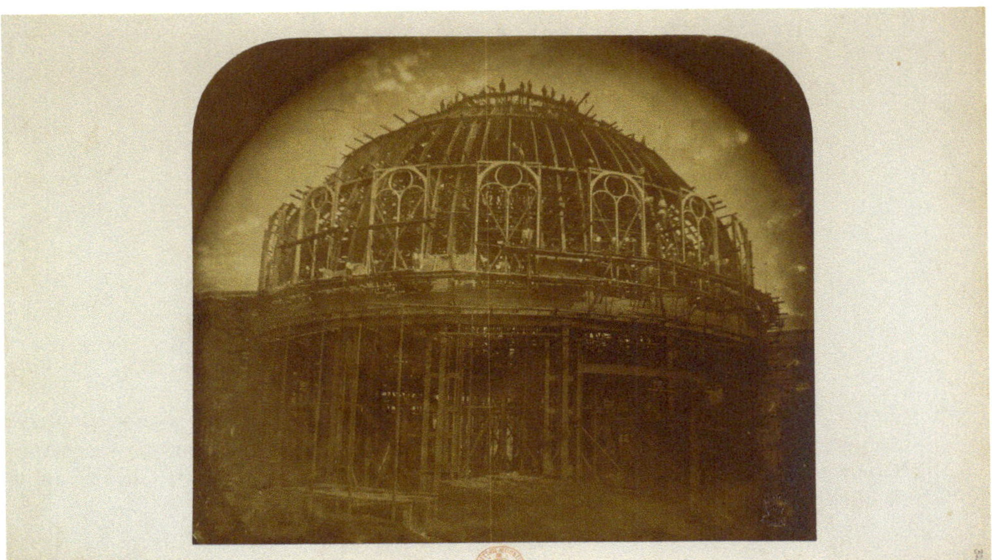

Plate 2 Photographer unknown, Construction of the new reading room cupola, British Museum, p. 32 in *A Collection of Plans and Views of Portions of the Buildings and Contents of Old Montague House and the Present British Museum*, album assembled by John Edward and Maria Emma Gray, 1780–1858. Collection: British Library, London (Shelf mark Maps C.26.f.7).

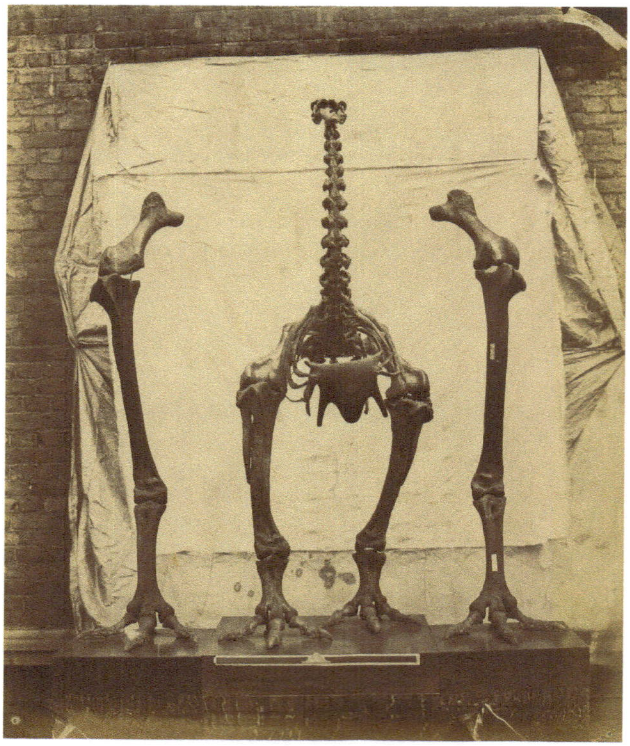

Plate 3 Roger Fenton, *Dinornis elephantopus and leg bones of Dinornis giganteus in the Gallery of Fossils, British Museum*, c.1857, salted paper print from glass negative. Courtesy of the Museum of Modern Art, New York / Scala, Florence. Gift of Paul F. Walter (83.1988).

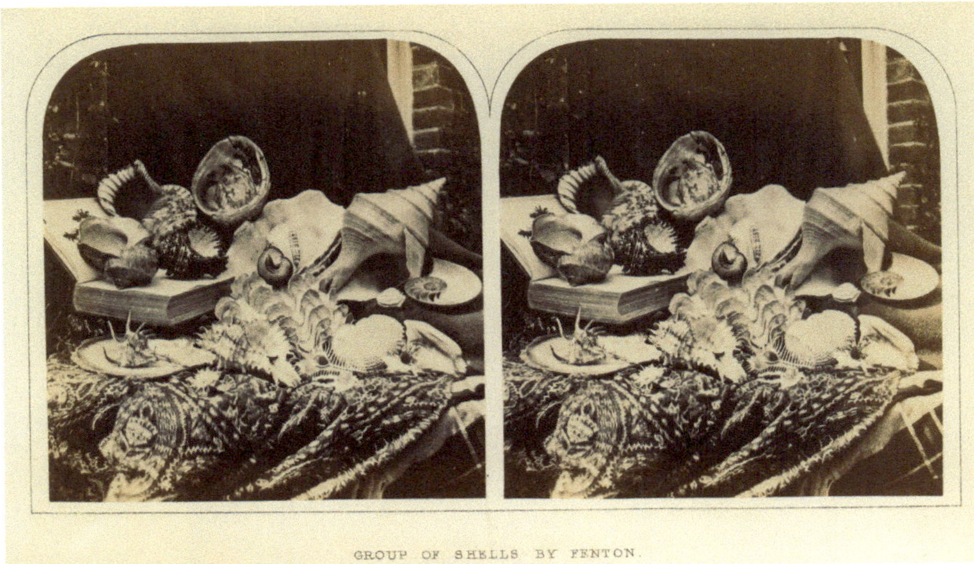

Plate 4 Roger Fenton, *Group of Shells from The Stereoscopic Magazine*, vol.3, no.38, 1861. Rare Books Collection, State Library Victoria.

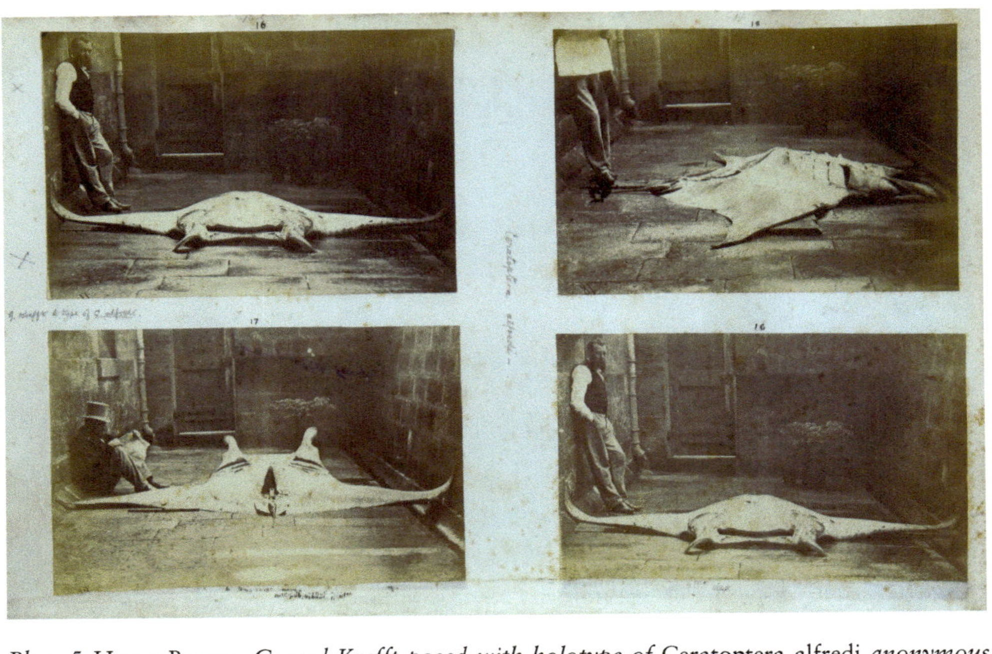

Plate 5 Henry Barnes, *Gerard Krefft posed with holotype of* Ceratoptera alfredi *anonymous (Krefft)*, (manta ray), 1868, albumen silver photographs in *Photograph Album, no. 14, Fishes no.1*, AMS421-14, p.71. Collection: Australian Museum, Sydney.

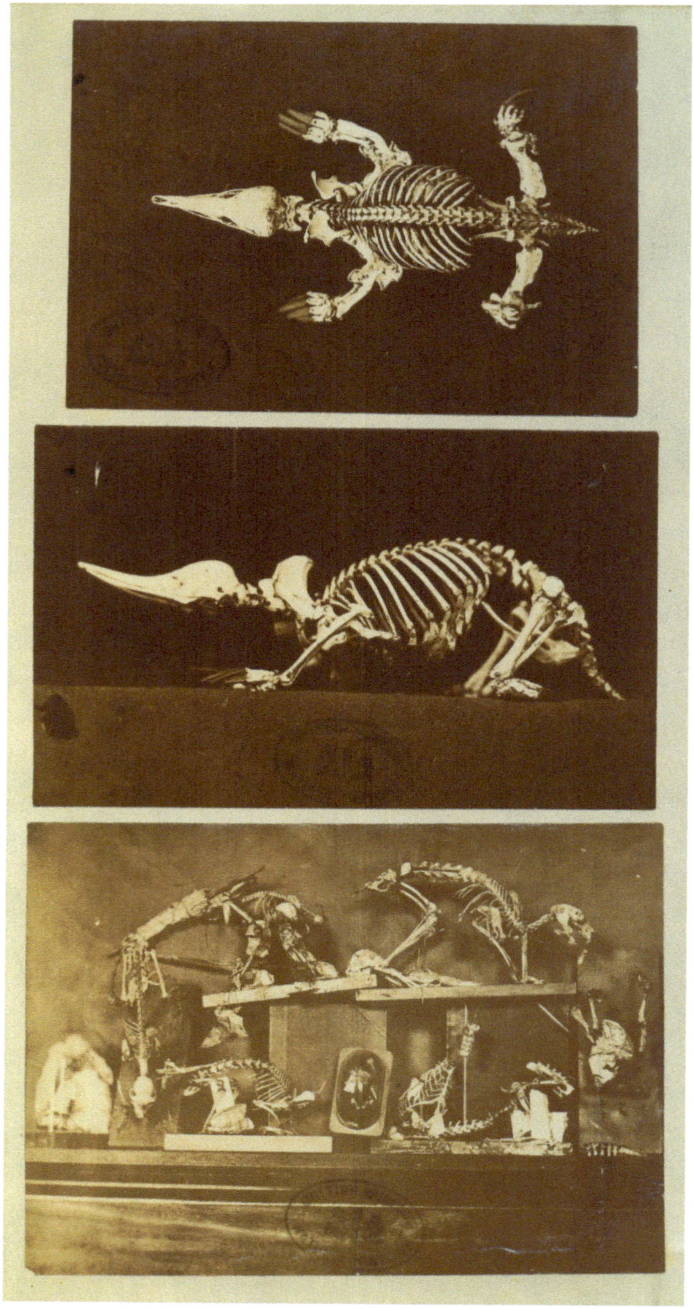

Plate 6 Henry Barnes, *Echidna skeleton and Natural history specimens sent from the British Museum to the Australian Museum*, c.1864, albumen silver photographs, enclosed in a letter from Krefft to Owen, 21 October 1864: Owen correspondence, vol.27, f.3-5. Collection: Natural History Museum, London.

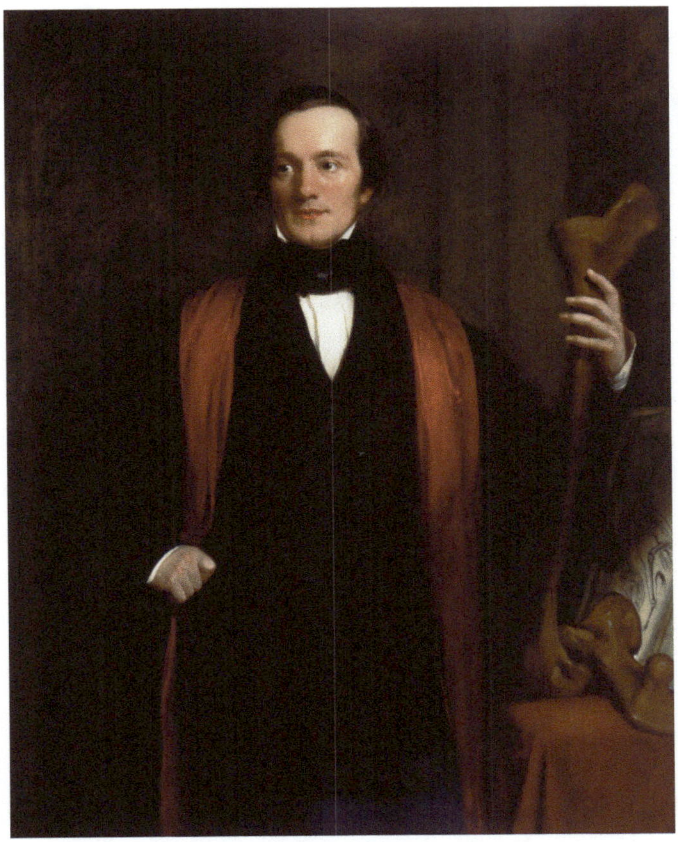

Plate 7 H.W. Pickersgill, *Portrait of Richard Owen* (holding the Dinornis bone), 1844, oil on canvas. Collection: Natural History Museum, London.

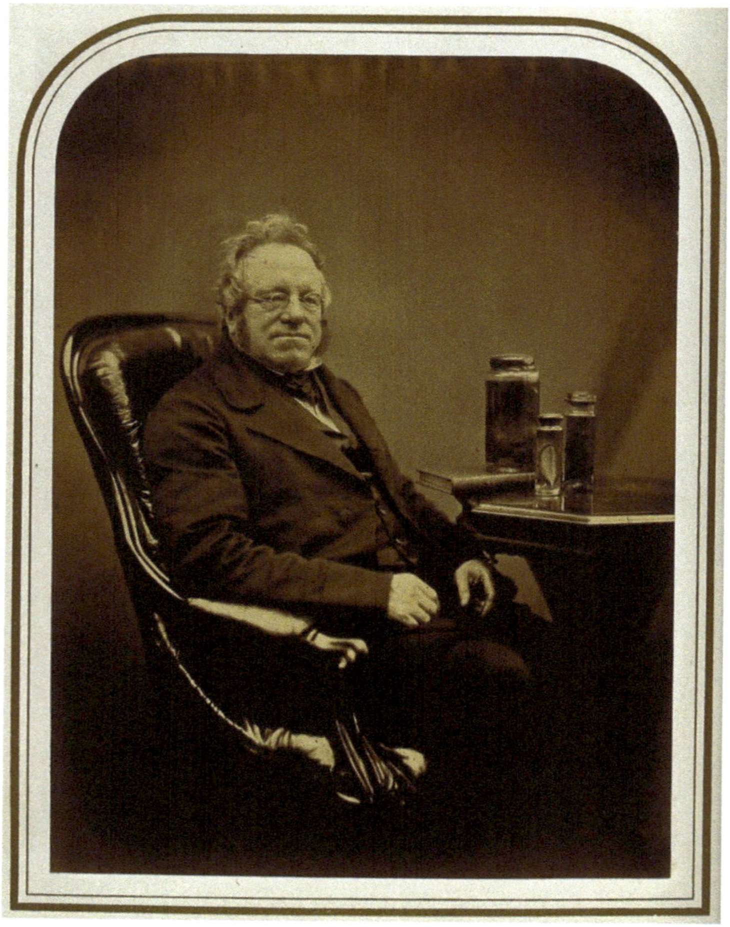

Plate 8 Maull and Polyblank, *John Edward Gray*, albumen silver photograph, 1855. Collection: Wellcome Library, London.

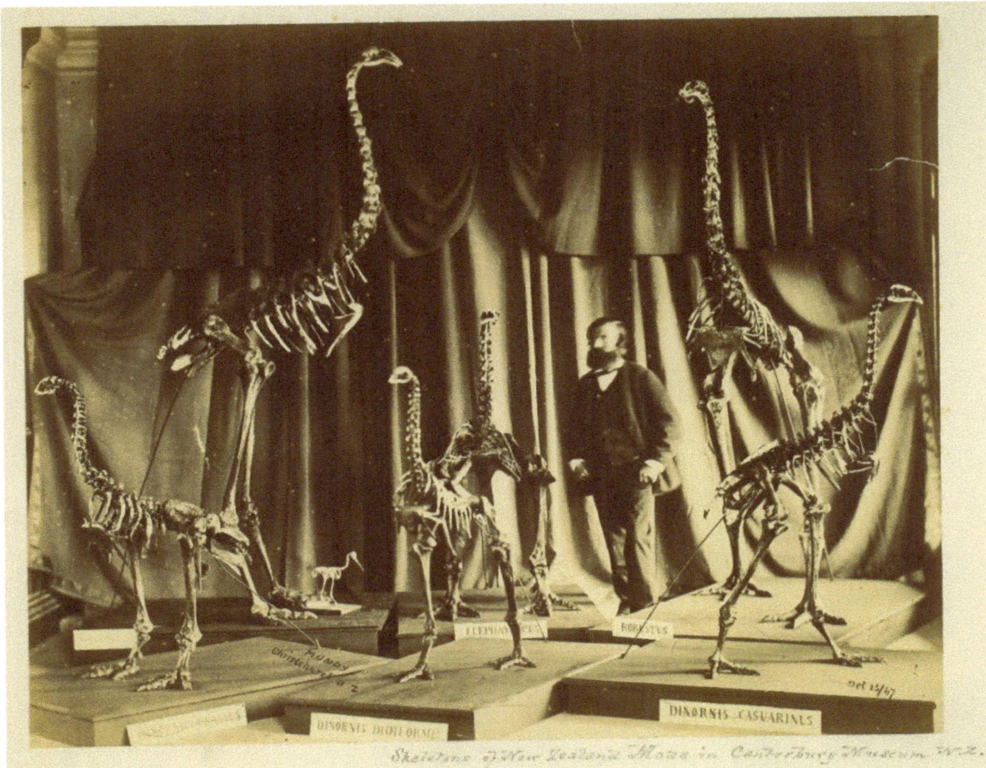

Plate 9 Daniel Mundy, *Skeletons of New Zealand Moa in Canterbury Museum,* N.Z., 25 October 1867, albumen silver photograph. Collection: Sir George Grey Special Collections, Auckland Libraries (589–76).

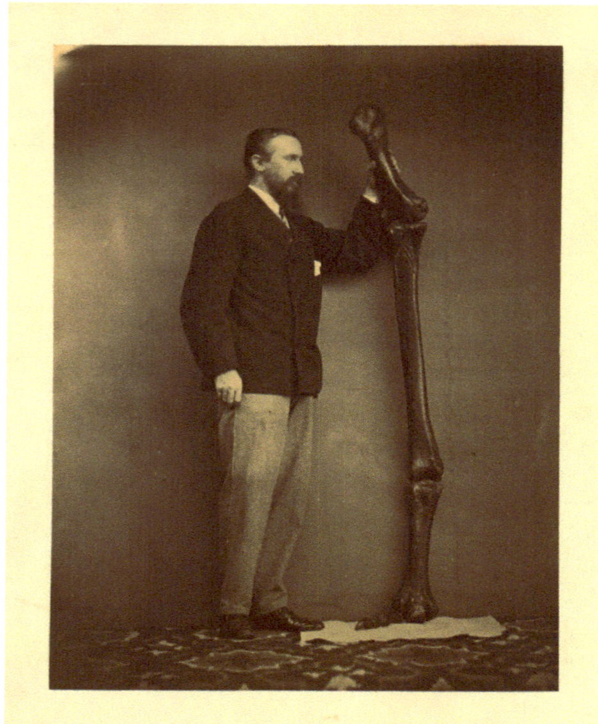

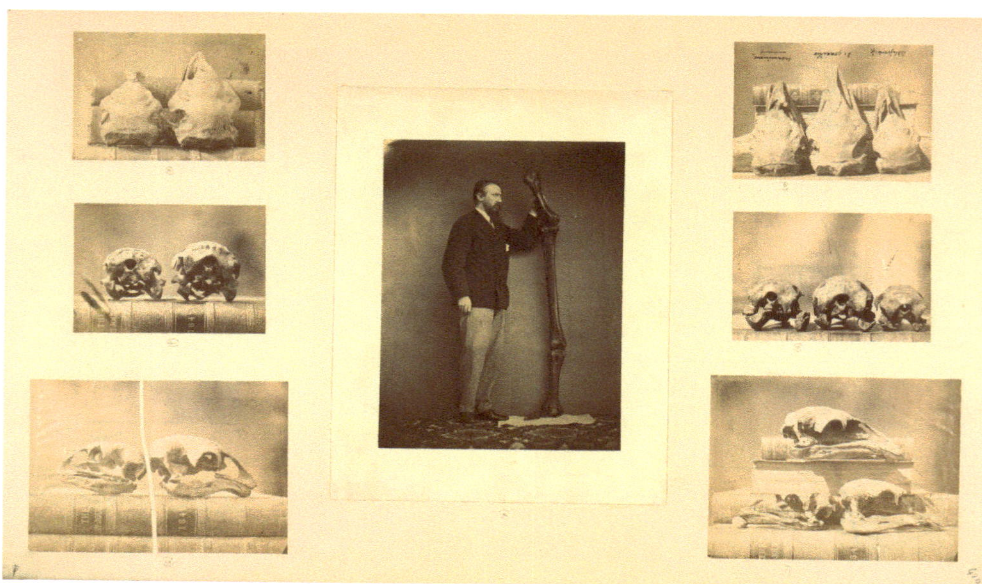

Plate 10 Photographer unknown, *Major J. Michael and bones of Dinornis giganteus*, c.1865, albumen silver photograph, Richard Owen Collection of Palaeontological and Zoological Drawings, f.480. Collection: Natural History Museum, London.

considerable enthusiasm for physiognomy and phrenology.[122] Not least, the subject's averted eyes allowed the viewer's gaze to linger and treat every detail of the image equally. In a portrait by Maull and Fox dated around 1854, Darwin's trousers, waistcoat and necktie, all dazzlingly chequered, activate the image, drawing attention away from the physicality of the man to emphasise his social and professional status as a young, urbane and already relatively accomplished upper-middle-class gentleman.[123] Another portrait, taken by Maull and Polyblank around 1855, similarly demonstrates the sartorial elegance of an amateur naturalist of independent means. However, in this latter portrait, Darwin's frontal pose and the intensity of his stare is far more confrontational, which, as he mused in a letter to Joseph Hooker, showed him with such a bad expression that he eschewed frontal poses for almost all subsequent sittings (Figure 3.7).[124]

The distinction between those public figures who resisted public attention and those who embraced it – contrasted here in the figures of Darwin and Owen – seems relatively uncomplicated in the way it delineates, if only superficially, between scientists of amateur status who were more retiring in the public realm and those who vigorously promoted their work or institution through their professional status. Yet, that division does not adequately address the more complex aspects of self-representation, particularly where those opposite impulses of theatrical parading and passive display were negotiated by a single individual.[125] A successful career from the mid-nineteenth century increasingly meant that one was a public figure, which required the individual to be seen and to contribute to the process of their own commoditisation by the public through self-conscious display. Yet, to many Victorians, Linda Shires agues, 'self-display through acting implied deceit or an inability to be what they thought of as "truly" sincere'.[126] Consequently, there was sometimes an impulse to hold back from this active spectacle and to construct one's identity through self-effacement and detachment as a person of sincerity, integrity and authority.

As we have seen, Darwin epitomised this approach in the way he wrote his autobiography. To use this approach in one's own portraits, however, raised certain complications. Sitting for a photographic portrait was very different to writing an autobiography, with the latter being a fairly rarefied mode of self-representation usually formulated as a narrative unfolding in time and based on the presupposition of abstract individualism.[127] Studio portraiture, by contrast, was a process with which audiences had a considerable familiarity through their own first-hand experience. Viewers of celebrity portraits knew very well the conceits, contrivances and concessions that commercial photographers employed. They were also thoroughly aware of the sorts of choices available to sitters and the amount of control they had over the way they appeared in the final images. Thus, there was a certain acquaintance with the processes and products of celebrity portraiture, although not necessarily a more profound identification with its subjects. For Darwin, or any sitter for that matter, avoiding eye-contact with the camera and presenting himself as oblivious to the activities going on around him – and to have the audience accept that premise – required the complicity of the viewer to disregard the inherently staged nature of the studio portrait. Quite paradoxically, the very qualities that men of science hoped to evoke in their portraits – credibility, integrity, sincerity, authenticity, trustworthiness – were largely dependent on the audience's inclination to suspend their disbelief in the face of the obviously inauthentic staging of the portraits.

Conclusion

The very act of sitting for a photograph involved issues of display – private versus public, passive versus active – that were often quite nuanced. Even when intended for private consumption or a limited circulation, portraiture is very much about performance.[128] For most Victorian naturalists, portraiture was part of the process of self-identification within their social, professional or national milieu. In Britain – unlike Germany, which had a greater level of scientific professionalisation – it invariably brought together both amateur naturalists and a growing number of professionalised practitioners attached to institutions, since it identified and reframed private intellectuals as public figures. In this respect, portraits of scientists were culturally specific, reflecting the achievements of the individuals portrayed but also being steeped in national traditions and policies that went to the heart of public support and promotion of scientific endeavour. Owen's access to almost unlimited imperial collections of specimens was unequalled during this period, and his portraits are symbolic of that acquisitive power and the extraordinary influence he wielded across the British Empire. Although Darwin's status, on the other hand, was that of a gentlemanly amateur naturalist, he nonetheless exchanged observations with – and accrued evidence for his intellectual work from – a wide network of colonial and international correspondents and was as much a scientist of the empire as Owen. Darwin's celebrity, however, was due almost entirely to his controversial theories; it took him years to win over the majority of his colleagues, and during the early years in particular he both captivated and divided public opinion.[129]

The general public encountered the latest scientific theories and discoveries by and large through various forms of popular culture. As one of the more popular of these forms, images of scientists were designed to signify and sometimes demonstrate aspects of their practice and new areas of scientific knowledge. As awareness of his work increased, along with debates about its broader ramifications, Darwin was increasingly regarded as the embodiment of evolution theory.[130] Yet, the very nature of his theory of natural selection, which encompassed vast time scales and gradual changes in nature that were not immediately perceptible, meant that it was almost impossible to represent it by visual means. Given the contentiousness of his work and the reception it initially received, the most crucial issue for Darwin was to address his own credibility. Owen's discoveries, on the other hand, were not only visually spectacular but the result of a painstaking process of visual identification and comparison, which made it possible for him to draw directly on the visual aspects of his work to engage the imagination of contemporary audiences. The different approaches that these and other naturalists used when sitting for commercial portraits reveal the diverse ways in which men of science used photography as a new form of self-representation to make claims about their credibility, authenticity and professional authority. Consequently, studio portrait photographs and associated publications tell us much about the changing structure and perceptions of scientific authority in Victorian England.

The number of *cartes de visite* and celebrity portraits produced in the Victorian era was so immense that, in the past, this material has been viewed as monolithic and, perhaps for this reason, they are often only considered *en masse*. By paying attention to the various sub-genres of portraiture and the specific contexts in which studio portraits were produced and circulated, however, a more nuanced understanding of their purpose and influence can be achieved.[131] In attempting to explicate how certain portraits were formulated and circulated in various personal and public domains, the purpose of this chapter has been to address the importance of this most ubiquitous of photographic modes in the promotion and institutionalisation of natural history, a discipline that – due to the Victorian propensity for accumulation and classification – flourished as part of the work of empire.

Owen was the most visible professional scientist of the empire, as evidenced in the many commissioned and commercial portraits of him, notably those in which he posed with the moa and other examples of prehistoric fauna sent to him from across the empire – thereby presenting his 'discovery' and restoration of giant, extinct species as the defining motif, while also exemplifying his comparative methodology. Commencing in 1844, these portraits were produced at particular defining moments in the escalating trajectory of Owen's career. There was much early conjecture surrounding his reconstruction of the moa until 'ample confirmation came to hand in 1843 and [thereafter] continued to arrive', with naturalists, colonial officials and travellers enlisted to search New Zealand for substantial moa fossils and to forward them to him in London.[132] In the next chapter, I will investigate the role of colonial scientists in this process and describe how naturalists involved in this ongoing work – as well as the moa itself – became celebrities in their own right. By examining the establishment of New Zealand's Canterbury Museum and its early years under the directorship of Julius von Haast, I will demonstrate how colonial naturalists were involved in a process of self-fashioning parallel to that of their counterparts at the centre of the empire, quickly adopting and using portrait photography for related ends. The priorities of museum building and the ongoing process of establishing and maintaining scientific authority were also promoted though portraiture.

For naturalists working at the periphery, the tyranny of distance meant that they had to work even harder to prove their scientific worth and establish their credentials in the imperial realm of science. Museum naturalists in colonial locations were keen to exchange and circulate their portraits as a socially acceptable way of promoting themselves and expanding their professional networks. Unlike the portraits of naturalists at the centre of the empire, which were typically staged in photographers' studios, the portraits of colonial naturalists were also often taken in exhibition galleries and other spaces in the colonial museum, such as the portrait series of Gerard Krefft posing with the manta ray in the Australian Museum courtyard in homage to Prince Alfred (Plate 5). They also occasionally portray the naturalist accompanied by associates such as taxidermists, articulators and others engaged in the behind-the-scenes processes of their vocations. Their portraits appear similar to those of metropolitan savants, nevertheless, in that they often illustrate the particularities and minutiae of scientific work in their representations of the practical work of naturalists in an unfinished state and as just one part of a larger, intellectual process. However, the colonial naturalists' portraits allude not so much to specific methodologies – for instance, the detective work undertaken by comparative anatomists at the centre – but more to colonial naturalists' significant discoveries in the field and the behind-the-scenes processes in colonial museums in an attempt to demonstrate their scientific prowess as well the skills, labour and resourcefulness required on the periphery.

Notes

1 'Portrait of Faraday', *The Art Journal*, 1849, p. 163.
2 Roy Porter, Introduction in W. D. Hackmann, *Apples to Atoms: Portraits of Scientists from Newton to Rutherford*, National Portrait Gallery, London, 1986, p. 11.
3 Stephen Thompson, 'On Portraits and Portraiture', *British Journal of Photography*, 1862, pp. 21–22.
4 Patricia Fara, 'Michael Faraday: Media Man', *Endeavour*, vol. 30, no. 1, 2006, p. 10.
5 Michael Shortland and Richard Yeo (eds.), *Telling Lives in Science: Essays on Scientific Biography*, Cambridge University Press, Cambridge, 1996, p. 31.
6 Jane Davidson, *A History of Palaeontology Illustration*, Indiana University Press, Bloomington and Indianapolis, 2008, pp. 96–106.
7 Bernard Lightman, 'Depicting Nature, Defining Roles: The Gender Politics of Victorian Illustration' in Ann. B. Shteir and Bernard Lightman (eds.), *Figuring It Out: Science, Gender, and Visual Culture*, Dartmouth College Press, Hanover and London, pp. 234–5.
8 Anandi Ramamurthy, 'Spectacles and Illusions: Photography and Commodity Culture' in Liz Wells (ed.), *Photography: A Critical Introduction*, Routledge, London, 2004, p. 197; John Tagg, *The Burden of Representation: Essays on Photographies and Histories*, Macmillan, Basingstoke, 1988, pp. 34–59.
9 Prescott, 'Faraday: Image of the Man and the Collector', pp. 24–5.
10 Gisèle Freund, *Photography and Society*, Fraser, London, 1980, p. 9.
11 Tagg, *The Burden of Representation*; Suren Lalvani, *Photography, Vision, and the Production of Modern Bodies*, State University of New York Press, Albany, 1996.
12 Ludmilla Jordanova, 'Medical Men 1780–1820' in Joanna Woodall, *Portraiture: Facing The Subject*, Manchester University Press, Manchester, 1997, p. 101.
13 Jan Golinksi, *Making Natural Knowledge: Constructivism and the History of Science*, University of Chicago, Chicago and London, 2005, p. 60.
14 Peter Hamilton and Roger Hargreaves, *The Beautiful and the Damned: The Creation of Identity in Nineteenth-century Photography*, Lund Humphries and National Portrait Gallery, London, 2001, p. 54.
15 Adrian Desmond 'Redefining the X Axis: "Professionals", "Amateurs" and the Making of Victorian Biology – A Progress Report', *Journal of the History of Biology*, 34, 2001, p. 3.
16 Morris Berman, '"Hegemony" and the Amateur Tradition in British Science', *Journal of Social History*, vol. 8, no. 2, 1975, pp. 39–40; Jack Morrell, 'Professionalization' in Robert Olby et al. (eds.), *Companion to the History of Modern Science*, Routledge, London and New York, 1990, pp. 980–9.
17 Richard Bellon, 'Joseph Hooker's Ideals for a Professional Man of Science', *Journal of the History of Biology*, vol. 34, 2001, p. 51.
18 Samuel Alberti, 'Amateurs and Professionals in One County: Biology and Natural History in Late-Victorian Yorkshire', *Journal of the History of Biology*, vol. 34, 2001, p. 115.
19 Golinski, *Making Natural Knowledge*, p. 60.
20 Idem.
21 Rebekah Higgitt, *Recreating Newton: Newtonian Biography and the Making of the Nineteenth-century History of Science*, Pickering & Chatto, London, 2007, p. 2.
22 Ruth Barton, '"Men of Science": Language, Identity and Professionalization in the Mid-Victorian Scientific Community', *History of Science*, vol. 41, no. 1, 2003 pp. 73–119; also Jim Endersby, *Imperial Nature: Joseph Hooker and the Practices of Victorian Science*, University of Chicago Press, Chicago and London, 2008, pp. 21–7.
23 Barton, 'Men of Science', p. 81.
24 Ibid., p. 110.
25 James Forbes, *The Danger of Superficial Knowledge*, London, 1849, p. vii.
26 'Portrait of Faraday', *The Art Journal*, 1849, p. 163.
27 Barton, 'Men of Science', p. 90; also Caroline Jordan, *Picturesque Pursuits: Colonial Women Artists and the Amateur Tradition*, Melbourne University Press, Melbourne, 2005.
28 Desmond, 'Redefining the X Axis', pp. 3–50.
29 Shortland and Yeo, *Telling Lives in Science*, p. 37.
30 Idem; Richard Yeo, *Defining Science: William Whewell, Natural Knowledge and Public Debate in Early Victorian Britain*, Cambridge University Press, Cambridge, p. 117.

31 Hamilton and Hargreaves, *The Beautiful and the Damned*, p. 49.
32 Lovell Reeve, *Portraits of Men of Eminence in Literature, Science and Art, with Biographical Memoirs. The Photographs from Life by Ernest Edwards*, Lovell Reeve & Co, London, 1863–67.
33 'On Studio Properties', *British Journal of Photography*, 17 September 1875, pp. 447–8.
34 'Backgrounds – How to Paint and Arrange Them', *Photographic News*, 13 January, 1860, p. 222; 'Backgrounds – How to Paint and Arrange Them', *Photographic News*, 27 January, 1860, p. 246.
35 Roy Porter (Introduction) in W. D. Hackmann, *Apples to Atoms: Portraits of Scientists from Newton to Rutherford*, National Portrait Gallery, London, 1986, pp. 12–13. The portraits of Faraday and Owen are rather conventional compared to many of the other portraits of these two men and quite uncharacteristic of how they usually presented themselves.
36 William Lake Price, 'On Composition and Chiar-oscuro – VIII', *Photographic News*, 5 April 1860, p. 367 (emphases in original).
37 Idem.
38 Idem.
39 Ibid, p. 368 (emphases in original).
40 'John Hunter – from the Painting by Sir Joshua Reynolds', *Illustrated London News*, 2 April 1859, p. 332.
41 William Lake Price, 'On Composition and Chiar-oscuro – VIII', *Photographic News*, 5 April 1860, p. 368.
42 Edward Dunmore. 'On Backgrounds and Portraits Combined', *British Journal of Photography*, 1875, p. 487.
43 McCauley, *A. A. E. Disdéri*, p. 220.
44 Thompson, 'On Portraits and Portraiture', p. 22 (emphasis in original).
45 Idem.
46 A. A. E. Disdéri, *L'Art de Photographie*, 1862, pp. 299–300 – quoted in Michael Fried, *Courbet's Realism*, University of Chicago Press, Chicago and London, 1992, p. 45.
47 Alvar Ellegård, *Darwin and the General Reader: The Reception of Darwin's Theory of Evolution*, University of Chicago Press, Chicago, 1990, p. 25.
48 Ibid, pp. 27–8.
49 Janet Browne, *Charles Darwin: The Power of Place*, Jonathan Cape, London, 2002, pp. 272–4; also Janet Browne, 'I Could Have Retched All Night' in Christopher Lawrence and Steven Shapin (eds.), *Science Incarnate: Historical Embodiments of Natural Knowledge*, University of Chicago Press, Chicago and London, 1998, pp. 240–87.
50 Darwin to Edward Walford, 22 [January–April 1865]: Darwin Correspondence Project Database – http://www.darwinproject.ac.uk/entry-5508 (letter no. 5508; accessed 4 July 2011).
51 Darwin to Edward Walford, 22 [Januray–April 1865]: Burkhardt et al., *Correspondence of Charles Darwin*, vol. 13 (1865), Cambridge University Press, Cambridge, 2005, p. 37; Edward Walford, *Representative Men in Literature, Science and Art*, Alfred William Bennett, London, 1868; Janet Browne, *Charles Darwin: The Power of Place*, p. 272.
52 E. A. Darwin to Emma Darwin, 25 November 1865 in Frederick Burkhardt et al. (eds.), *The Correspondence of Charles Darwin*, vol. 13 (1865), Cambridge University Press, Cambridge, 2005, pp. 314–5.
53 Emma Darwin quoted in Browne, *Charles Darwin: The Power of Place*, p. 272.
54 J. D. Hooker to Darwin, 24 January 1864: Darwin Correspondence Project Database – http://www.darwinproject.ac.uk/entry-4396 (letter no. 4396; accessed 4 July 2011); Darwin did not sit for the bust by Woolner until 1868 – see Darwin to Hooker, 26 November 1868: Darwin Correspondence Project Database – http://www.darwinproject.ac.uk/entry-6476 (letter no. 6476; accessed 4 July 2011).
55 See Janet Browne 'Looking at Darwin: Portraits and the Making of an Icon', *Isis*, no. 100, 2009, pp. 542–70; Janet Browne, *Charles Darwin: The Power of Place*; Janet Browne, 'Charles Darwin as a Celebrity', *Science in Context*, vol. 16, no. 1/2, 2003, pp. 175–94.
56 Browne, *Charles Darwin: The Power of Place*, p. 272; Browne, 'I could have Retched all Night', p.257.
57 Thompson, 'On Portraits and Portraiture', pp. 21–22 (emphasis in original).
58 Ibid, p. 22.

59 'Identification by Photography', *Photographic News*, 26 September 1874, pp. 462–63.
60 'Portrait of Faraday', *Art Journal*, 1849, p. 163.
61 Philip Prodger, *Darwin's Camera: Art and Photography in the Theory of Evolution*, Oxford University Press, 2009, p, 18. Cameron's portraits of Victorian intellectuals were produced as works of art but were also a transparently commercial venture, being autographed by the celebrity sitters at the photographer's behest and marketed at the time by the London dealers, the German Gallery and Colnaghi's – see Colin Ford, *Julia Margaret Cameron: A Critical Biography*, The J. Paul Getty Museum, Los Angeles, 2003; Marta Weiss, *Julia Margaret Cameron: Photographs to Electrify You with Delight and Startle the World*, MACK and V&A Publishing, London, 2015.
62 John Tyndall, 'Michael Faraday' in Leslie Stephen (ed.), *Dictionary of National Biography*, vol. 18, Smith, Elder & Co., London, 1889, p. 201.
63 Edward Walford, 'Professor Faraday, D.C.S., F.R.S.' in Maull and Polyblank, *Photographs of Living Celebrities*.
64 David Gooding and Frank A. J. L. James, *Faraday Rediscovered*, p. 8.
65 Regenia Gagnier, *Subjectivities: A History of Self-Representation in Britain, 1832–1920*, Oxford University Press, New York and Oxford, 1991, pp. 257–65.
66 Francis Darwin (ed.), *The Life and Letters of Charles Darwin, including an Autobiographical Chapter*, John Murray, London, 1887, p. 64.
67 'Of the Three Learned Professions, the Medical has Attained the Highest Character for Disinterestedness', *The Times*, 25 January 1856, p. 6; see also, 'Professor Owen, FRS', *The Times*, 10 July 1856, p. 12.
68 *Photographic Portraits of Living Celebrities*, no. 1, 1856.
69 Nicolaas Rupke, *Richard Owen: Biology Without Darwin*, University of Chicago Press, Chicago and London, 2009, p. 44; Iwan Morus, Simon Shaeffer and Jim Secord, 'Scientific London' in Celina Fox (ed.) *London: World City 1800–40*, Yale University Press and Museum of London, New Haven, 1992, pp. 135–6.
70 Rupke, *Richard Owen: Biology Without Darwin*, p. 26.
71 George Basalla, William Coleman and Robert H. Kargon (eds.), *Victorian Science: A Self-Portrait for the Presidential Addresses to the British Association for the Advancement of Science*, Anchor Books, New York, p. 3.
72 G. H. Lewes, 'Professor Owen and the Science of Life', *Fraser's Magazine for Town and Country*, vol. 53, 1856, p. 79.
73 Lewes, 'Professor Owen', p. 81.
74 *The Historic Gallery of Portraits and Paintings, and Biographical Review*, vol. 1, Verner, Hood & Sharpe, London, 1807, unpaginated – quoted in Fara, 'Faces of Genius', p. 61.
75 S. S. Shweber, 'Scientists as Intellectuals: The Early Victorians' in James Paradis and James Postlewait (eds.), *Victorian Science and Victorian Values*, Annals of the New York Academy of Science, vol. 360, 1981, p. 2.
76 Similarly, in a silhouette portraying Buckland with his family produced by Auguste Edouart, it is naturalist's practice and the performative nature of his work, rather than the sitters' appearances, which governs the composition – Auguste Amant Constant Fidèle Edouart, *William Buckland and his wife and son Frank, examining Buckland's natural history collection*, c.1828–9, cut-out full-figure silhouettes from matte black coated white wove paper mounted to beige wove paper, some elaborations in black pen and wash. Collection: Museum of Fine Arts, Boston (Mary L. Smith Fund): MFA 66.964.
77 W. D. Ian Rolfe, 'Fossils' in J. M. Chalmers-Hunt (ed.), *Natural History Auctions 1700-1972: A Register of Sales in the British Isles*, Sotheby Parke Bernet, London, 1976, p. 32.
78 Richard Owen, *Memoir on the Pearly Nautilus (Nautilus Pumilus Linn.): With Illustrations of its External Form and Internal Structure Drawn Up by Richard Owen*, W. Wood, London, 1832.
79 George Bennett, *Gatherings of a Naturalist in Australasia*, John Van Voorst, London, 1860.
80 National Portrait Gallery, London – correspondence with the author, 2006.
81 S. Peter Dance, 'Ruskin the Reluctant Conchologist', *Journal of the History of Collections*, vol. 16, no. 1, 2004; E. C. Spary, 'Rococco Readings of the Book of Nature' in Marina Frasca-Spada and Nick Jardine (eds.), *Books and the Sciences in History*, Cambridge University Press, Cambridge, 2000, pp. 255–75.

82 A further portrait of Owen was commissioned by the prime minister, Sir Robert Peel, in 1844 as a pendant to a Pickersgill portrait of Cuvier that hung in the gallery at Peel's residence – 'Portrait Gallery at Drayton Manor' *The Gentleman's Magazine* , vol. 27, 1847, pp. 289–91; 'The Walhalla at Tamworth', *The Living Age*, vol. 36, 1853, pp. 444–5; George Peel (ed.), *The Private Letters of Sir Robert Peel*, John Murray, London, 1920, p. 214.

83 Darwin (ed.), *The Life and Letters of Charles Darwin*, p. 27.

84 Alexis Harley, 'Autobiography, Authority, Atheism: The Evolving Charles Darwin', unpublished ms., Fisher Rare Books, University of Sydney: PrEs, 2001, p. 9; see also Paul White, 'Darwin's Emotions: The Scientific Self and the Sentiment of Objectivity', *Isis*, vol. 100, no. 4, 2009, pp. 811–26; Paul White, 'Darwin Wept: Science and the Sentimental Subject', *Journal of Victorian Culture*, vol. 16, no. 2, 2011, pp. 195–213.

85 Hippolyte Bayard, *Le Noyé (Self portrait as a Drowned Man)*, 18 October 1840, direct positive prints. Collection: Societé Française de Photographie, Paris.

86 Michael Sapir, 'The Impossible Photograph: Hippolyte Bayard's "Self portrait as a Drowned Man"', *Modern Fiction Studies*, vol. 40, no. 3, 1994, pp. 620–3; Geoffrey Batchen, *Burning with Desire: The Conception of Photography*, MIT Press, Cambridge (Mass.), 1997; Linda Nochlin, *Realism*, Penguin, Harmondsworth, 1971, pp. 61–2.

87 Michael Sapir, 'The Impossible Photograph', p. 624.

88 Julia Ballerini, 'Recasting Ancestry: Statuettes as Imaged by Three Inventors of Photography' in Anne W. Lowenthal (ed.), *The Object as Subject: Studies in the Interpretation of Still Life*, Princeton University Press, Princeton, 1996, p. 45.

89 Idem.

90 'Photographic Portraits of Living Celebrities', *The Athenaeum*, 5 April 1856, p. 410; Maull and Polyblank published the *Literary and Scientific Portrait Club* in parts around 1854; the set of photographs was issued on subscription and presented loose in a folder.

91 'Richard Owen', *Photographic Portraits of Living Celebrities*, Maull and Polyblank, London, 1856.

92 Prescott, *Fame and Photography*, pp. 51–2.

93 'Fine Arts: Photographic Portraits. Maull & Polyblank', *The Athenaeum*, 21 June 1856, p. 782.

94 'Fine Arts: Fine-Art Gossip', *The Athenaeum*, 2 October 1858, p. 434.

95 'Fine Arts: Photographic Portraits. Maull & Polyblank', *The Athenaeum*, 21 June 1856, p. 782.

96 Idem.

97 'Richard Owen', *Photographic Portraits of Living Celebrities*, Maull and Polyblank, London, 1856.

98 Fay, 'Portrait Galleries', p. 151.

99 John Plunkett, 'Celebrity and Community: The Poetics of the Carte-de-Visite', *Journal of Victorian Culture*, vol. 8, no. 1, 2003, p. 55.

100 Richard Yeo, 'Scientific Method and the Image of Science, 1831–1890' in Roy Macleod and Peter Collins (eds.), *The Parliament of Science*, Science Reviews, Northwood, 1981, p. 67.

101 Idem.

102 J. S. Mill, *Autobiography* – quoted in Yeo, 'Scientific Method and the Image of Science', p. 66.

103 Karl Pearson, *The Grammar of Science*, A & C Black, London, (1892) 1911, p. 9; Yeo, 'Scientific Method and the Image of Science', p. 81.

104 William Vernon Harcourt, 'The Founding of the British Association for the Advancement of Science' in Roy Macleod *et al.* (eds.), *Victorian Science: A Self-Portrait from the Presidential Addresses to the British Association for the Advancement of Science*, Doubleday, New York, 1971, p. 32.

105 William Whewell, 'On the Connexion of Physical Sciences', *Quarterly Review*, vol. 51, 1834, p. 59.

106 General Committee in the Town Hall, Leeds, 27 September 1858, Mtg Item 5, f.132–3 – Minutes of meetings of the General Committee, signed by the President, 1848–68, 1:

Papers of the British Association for the Advancement of Science (BAAS), Bodleian Library, University of Oxford.

107 'The Past Week', Wednesday, 21 September 1859 & 'Conversazione', f.1, 405, Press cuttings and photographs: Papers of the BAAS, Bodleian Library, University of Oxford.

108 Idem.

109 Claudine Cohen, *The Fate of the Mammoth: Fossils, Myth and History*, University of Chicago Press, Chicago and London, 2002, pp. 116–7.

110 Edgar Wind, *Art and Anarchy*, Duckworth, London, 1985, pp. 40–1; Carlo Ginzburg, 'Clues: Morelli, Freud and Sherlock Holmes' in Umberto Eco and Thomas A. Sebeok (eds.), *The Sign of Three*, Indiana University Press, Bloomington & Indiana, 1988, pp. 81–118.

111 Cohen, *The Fate of the Mammoth*, pp. 116–7 (emphases in original).

112 Amy Boesky, '"Outlandish-Fruits": Commissioning Nature for the Museum of Man', *ELH*, vol. 58, no. 2, 1991, p. 322.

113 Julia Ballerini, 'The In-visibility of Hadji-Ishmael: Maxime Du Camp's 1850 Photographs of Egypt' in Kathleen Adler and Marcia Pointon (eds.), *The Body Imaged: The Human Form and Visual Culture Since the Renaissance*, Cambridge University Press, Cambridge, 1993, pp. 147–60.

114 Richard Owen, *On the Extent and Aims of a National Museum of National History*, Saunders & Oatley, London, 1862, p. 13.

115 R. J. Cleevely, 'Gray, John Edward (1800–75)', *Oxford Dictionary of National Biography* – http://www.oxforddnb.com/view/article/11344 (viewed 22 August 2007); Albert E. Gunther, *A Century of Zoology at the British Museum: Through the Lives of Two Keepers, 1815–1914*, Dawsons, London, 1975, p. 86.

116 Anne Larsen, 'Equipment for the Field' in N. Jardine, J. A. Secord and E. C. Spary (eds.), *Cultures of Natural History*, Cambridge University Press, Cambridge, 1996, p. 359.

117 Browne 'Looking at Darwin', p. 548.

118 Darwin (ed.), *The Life and Letters of Charles Darwin*, p.27; Gagnier, *Subjectivities*, p. 259.

119 Linda M. Shires, 'The Author as Spectacle and Commodity: Elizabeth Barrett Browning and Thomas Hardy' in Carol T. Christ and John O. Jordan (eds.), *Victorian Literature and the Victorian Visual Imagination*, University of California Press, Berkeley, 1995, p. 202

120 Ibid., p. 201.

121 Leo Braudy, *The Frenzy of Renown: Fame and its History*, Oxford University Press, New York, 1986, pp. 398 and 402; Shires, 'The Author as Spectacle and Commodity', p. 201.

122 I am indebted to Barbara Larson for this observation and regarding the dual meaning in the Victorian period of the terms *highbrow*, *middlebrow* and *lowbrow*. On the significance of these aspects of portraiture for the emerging eugenics movement, see Fae Brauer, 'Framing Darwin: A Portrait of Eugenics' in Barbara Larson and Fay Brauer (eds.), *The Art of Evolution: Darwin, Darwinisms, and Visual Culture*, Dartmouth College Press, Hanover, 2009, pp. 124–54. Although the study of physiognomy and phenology continued to be influential in Victorian wider culture and for the progenitors of eugenics (most notably Darwin's cousin, Francis Galton), Jonathan Smith argues that Darwin was at pains to disassociate his own scientific work (particularly his *Expression of the Emotions in Man and Animals*, published in 1872) from these popular areas of study, which by then had become largely discredited in elite scientific circles – Jonathan Smith, *Charles Darwin and Victorian Visual Culture*, Cambridge University Press, Cambridge, 2006, p. 199.

123 See Janet Browne, 'I Could Have Retched All Night'.

124 Darwin to J. D. Hooker, 27 May 1855: Burkhardt et al., *Correspondence of Charles Darwin*, vol. 5 (1855), p. 339.

125 Shires, 'The Author as Spectacle and Commodity', p. 202.

126 Ibid., p. 199.

127 Gagnier, *Subjectivities*, p. 39.

128 Prescott, *Fame and Photography*, p.66; Lalvani, *Photography, Vision, and the Production of Modern Bodies*, p. 59.

129 Ellegård, *Darwin and the General Reader*, pp. 56–61.

130 Janet Browne, 'Commemorating Darwin', *British Journal for the History of Science*, vol. 38, no. 3, 2005, p. 252.

131 *Cartes-de-visite* culture in Victorian science has become the subject of increased scholarly treatment – see Janet Browne, Geoffrey Belknap, Paul White; also Jennifer Tucker, *Nature Exposed: Photography as Eyewitness to Science*, John Hopkins University Press, Baltimore, 2005.

132 Reeve, *Men of Eminence,* 1863–67, p. 52.

4 Nature as spectacle

Encountering the moa from Christchurch to Madras via London and Paris

> In the Old World, nations and the distinctions of their civilization form the principal points in the picture; in the New World, man and his productions almost disappear amidst the stupendous display of wild and gigantic nature.[1]

Introduction

In 1880, Alexander McKay, a New Zealand colonist and prominent geologist, published *The Canterbury Gilpin; or, The Capture and Flight of the Moa* under the *nom de plume* of Dinornis Sumnerensis. In this epic poem, he conjures a thrilling escapade involving fellow scientist and director of the Canterbury Museum in Christchurch, Julius von Haast; and a living specimen of the extinct giant bird of New Zealand, the *Dinornis* or moa. This encounter with the moa is as significant for its improbable occurrence as it is for the specificity of its setting.[2] McKay's literary allusion is to the comical story of John Gilpin in William Cowper's verse of 1782; yet, the inspiration for his own saga was an episode in natural history much closer to his own circumstances, both in location and time:

> Of late years frequent reports have been circulated – the wish being father to the thought – that live Moas have been seen in various parts of the South Island … Fortunately for the Moas perhaps … these never reached their destination, which was said to be Christchurch. I have imagined that one of the Moas did actually reach town, and that the various incidents connected with its arrival and subsequent departure did take place. Having once given the rein to imagination, it was but natural to suppose that such an acquisition to science … should be displayed in public.[3]

McKay's poem was just one of a number of revivifications of the moa to appear in popular and scientific literature and imagery. The existence of this long-extinct bird, and its characteristics, had been famously identified and extrapolated in 1839 from a single shard of fossilised femur. Richard Owen's process of retrodiction was re-enacted and celebrated in successive honorific portraits, as examined in Chapter 3. However, from the 1860s onwards, other, more speculative and promiscuous images had begun to appear. These featured in scholarly and popular science publications and in exploration journals – as well as in museums – and ranged from romantic vignettes by the great populariser of natural history, P. H. Gosse (Figure 4.1), to the

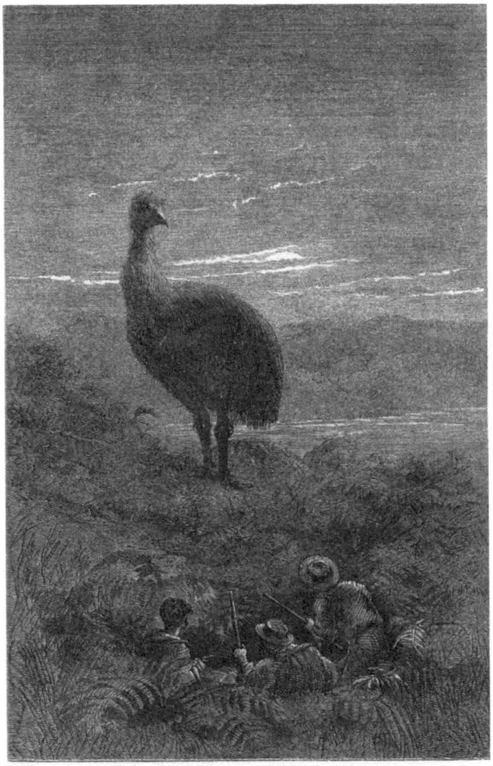

Figure 4.1 'Encounter with a Moa' in P. H. Gosse, *The Romance of Natural History*, second series, James Nisbet & Co, London, 1861.

detailed, systematic observations in Ferdinand Hochstetter's *New Zealand: Its Physical Geography, Geology and Natural History*, 1867.[4] In addition to these well-publicised and widely distributed images, a number of more modestly produced photographs were strategically circulated with the purpose of presenting an alternative to Owen's imperial portraits. In this chapter, I analyse the origins of photographs and other visual representations of the moa, their significance in promoting colonial natural history to metropolitan and local audiences and the ways in which they helped build the reputations of museums as well as individuals within the global scientific community.

In these images, as in McKay's poem, the moa has been relocated from the prehistoric past of the New World into the modern, colonial era, thereby creating a palimpsest of time. Such a temporal conflation was not entirely new or unique during this period as other images demonstrate a similar displacement of the present into the historical past or, conversely, from the past into the present, such as in Pierre Boitard's *Paris before Men*, 1861; Eduard Riou's illustrations for Louis Figuier's *Earth before the Deluge,* 1863; and the complex historicism represented in paintings by the Pre-Raphaelites and their associates, who saw no contradiction in embracing history and modernity simultaneously.[5] In various images and displays of the moa, aspects of the distant past entered into a dialogic relation with the nineteenth-century present, suggesting a correspondence and immediacy rather than difference and distance both in spatial and temporal terms. The Reverend Richard Taylor's *New Zealand Fauna and Flora of Past and Present Times*, 1870, for example, portrays pre-settlement New Zealand with a Maori man witnessing a procession of extinct and living birds and plants – including moa, kiwis, weka, cabbage trees, flax, toitoi, nikau and lancewood – some of which would never be observed by European eyes (Figure 4.2). Indeed, much of the Victorian age's fascination with time was concerned with progress – not so much as an inexorable or straightforward procession from the present to the future but as manifest in an ever-accumulating past. Such a preoccupation was largely reliant on imagination and conjecture; that is, in making visible that which was unseen.[6] A wide range of Victorian cultural forms expressed this fascination with unearthing and exposing – but also reanimating – the unseen and the unknown.[7] The Victorian perspective on modernity, therefore, did not involve an unambiguous break with the past or a new beginning but instead, as Lynda Nead contends, 'a set of processes and representations that were engaged in an urgent and inventive dialogue with their own historical conditions of existence'.[8] It was no coincidence that the period during which the moa came to light and attained its greatest prominence comprised those same decades that witnessed the rise of the modern museum and during which the emergent historical sciences of geology, palaeontology and archaeology were coming under the public gaze in an unprecedented way.[9] As Edward Bulwer Lytton observed: 'Every age may be called an age of transition – the passing-on, as it were, from one state to another never ceases; but in our age the transition is *visible*'.[10] The challenge for Victorians was to locate that sense of change within a broader panorama. Museums became a priority in the nineteenth century in order to secure the future by reconstructing the past, attempting to fill an ontological void even as the breadth and abundance of their collections threatened to expand out of control.[11]

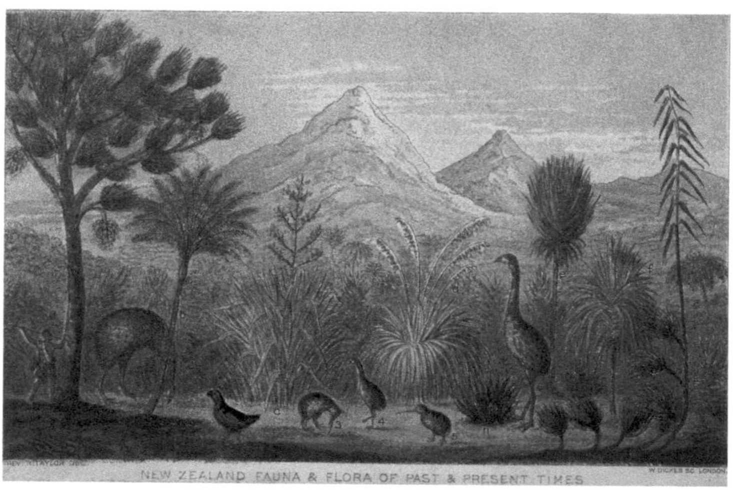

Figure 4.2 Rev. Richard Taylor, del. W. Dickes, *New Zealand Fauna and Flora of Past and Present Times*, 1870, chromolithograph. Collection: Alexander Turnbull Library, Wellington, New Zealand (A-247-037).

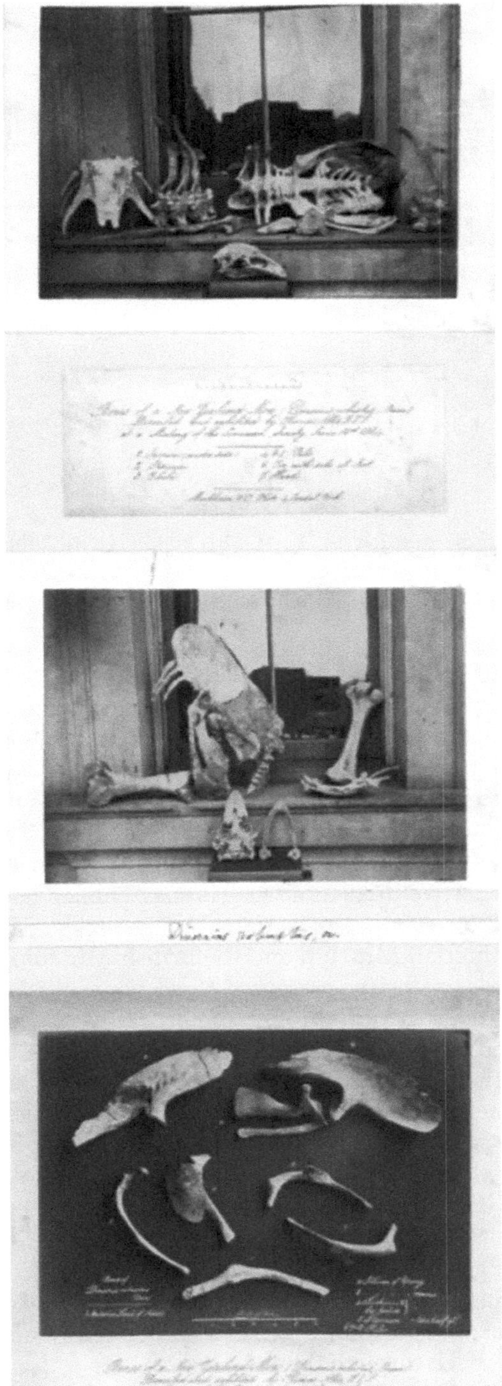

Figure 4.3 Monkhouse and Co., *Bones of a New Zealand Moa (Dinornis robustus* Owen*). Described and exhibited by Thomas Allis F.L.S. at a Meeting of the Linnean Society, June 16th 1864*, albumen silver photographs, Richard Owen Collection of Palaeontological and Zoological Drawings, f. 487. Collection: Natural History Museum, London.

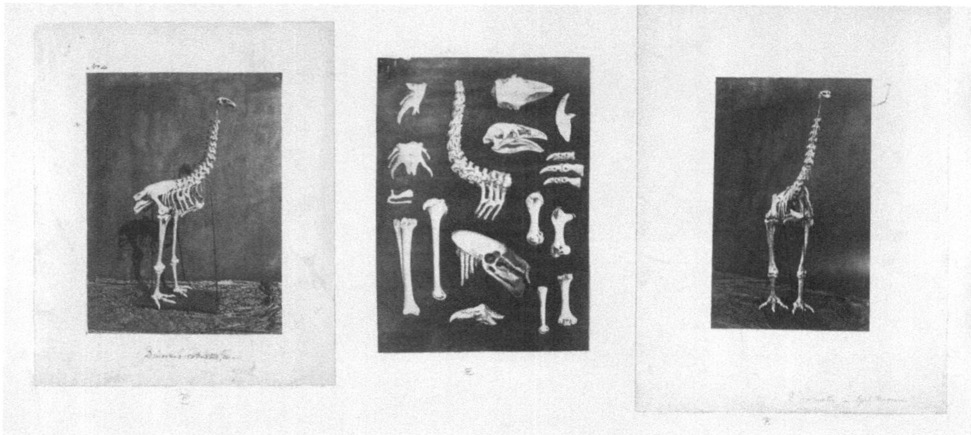

Figure 4.4 Photographer unknown, *Dinornis robustus in York Museum*, c.1876, albumen silver photographs, Richard Owen Collection of Palaeontological and Zoological Drawings, f.488. Collection: Natural History Museum, London.

Nineteenth-century images of the moa and other prehistoric fauna can be seen as a variant of the genre of natural history illustration but based on scenes from deep time. The emergence of these images is associated with the rise of the museum due to the large-scale collecting and cataloguing that facilitated the piecing together of fossil evidence so as to create ever more credible reconstructions of a hitherto unseen past.[12] Accounts of the methodological processes used by France's Georges Cuvier and his English disciple Richard Owen describe the practice of comparative anatomy as a sequence of deduction – or 'retrospective prediction' as Thomas Huxley termed it – that became increasingly more precise as relics of extinct animals in the Musée Nationale d'Histoire Naturelle, the Hunterian Museum and the British Museum collections were augmented with discoveries of better preserved fossil bones.[13] Indeed, it is the Baconian notion of knowledge construction through the processes of accrual and inductive reasoning that is inferred in individual specimens as recurring motifs in honorific portraits of Owen and other Victorian scientists, as emphasised in the previous chapter. Nonetheless, this assumes that visual reconstructions of prehistoric nature portrayed merely the processes of scientific discovery; that is, as if the prehistoric world was revealed in ways that were provisional only for the time it took for better specimens to be found, which would gradually permit more authentic representations of deep time.[14] Just as crucially, it disregards the complex circumstances surrounding how and where these images emerged and were circulated. Sometimes described as keyholes to the past, but perhaps better understood as a form of virtual witnessing, the conceptual and constructed nature of these images belies any notion that they represent an unmediated view back through time.[15]

Images of prehistoric fauna, especially the moa, have much to tell us much about their own time.[16] Their inherent multivocality meant that they were used in asserting the scientific reputations of regional British museums within a national context and also in advocating the independence of colonial science. In provincial England, for example, the museum of the Yorkshire Philosophical Society, one of the earliest purpose-built museums in the country, was among the first to acquire specimens of moa bones from excavations in the 1860s. Like most images of the moa circulated by museums during this period, the fossils were also incomplete but nevertheless of great interest to a scientific coterie that was dispersed throughout the empire. 'Magnificent Moa bones were exhibited at the Linnean Society the other night', Joseph Hooker wrote from London to the New Zealand geologist James Hector, and they were 'apparently as fresh as if they had just left the butchers, with a long piece of ligament attached to the head of one femur! The appearance was most extraordinary'.[17] Publishing his observations from that presentation in *The Linnean* in 1864, Thomas Allis, museum curator and secretary of the society at York, took advantage of growing interest in the moa to assert the scientific reputation of the York Museum on the national stage. He subsequently ensured the acquisition of a complete articulated moa skeleton – described at the time it was acquired for the museum in 1876 as the largest and most perfect specimen known to science. Richard Owen received letters, publications and elaborate photographs detailing each of these events and acquisitions, the latter of which he retained as part of his paper museum of images sent from correspondents based in provincial and colonial museums around the world (Figures 4.3–4.4).[18]

In the colony of New South Wales also, Australian Museum trustee and renowned geologist Rev. W. B. Clarke, together with curator Gerard Krefft, scoured fossil fragments received from expeditions to the Wellington Caves and Queensland's Peak Downs and disseminated photographs of specimens they believed to be related to New Zealand's giant flightless bird, *Dinornis* (Figure 1.9).[19] These specimens formed the basis of a number of scholarly papers and articles in the press, which were used by Clarke and Krefft to advocate the independence of colonial science. Clarke wrote to the *Sydney Morning Herald* in 1869, stating he was glad to read in that newspaper Krefft's announcement of one of their findings 'as too many of our discoveries are first made in England'.[20] In a letter written to Richard Owen in 1876, Clarke mentioned his having sent Owen photographs of these discoveries. Clarke had intended to send Owen a plaster cast of a section of the extinct bird's pelvis, but in attempting to make it, Henry Barnes, the Australian Museum taxidermist, 'failed to preserve the bone from fracture' and 'it was resolved therefore to photograph it'.[21] The numerous moa specimens that circulated between the various colonial, provincial and metropolitan locations were often put to a range of uses. Taking the Canterbury Museum in New Zealand as my primary focus, in this chapter I address the key role played by photographs, and other images, of vestiges of extinct fauna in the institutionalisation of natural history. These images promoted natural history as a public spectacle and offered a potent symbol for emergent colonial aspirations during the 1860s, which would become a counterpoint to London's domination of imperial natural history.

The origin of the natural history collections at the Canterbury Museum

In June 1881, a year after McKay's poem had appeared and just two months after Richard Owen opened the doors of the new British Museum (Natural History) in South Kensington, Julius von Haast, as founding director of the Canterbury Museum, delivered his annual address before the local Philosophical Institute.[22] In this lecture, he reflected on the origin and early progress of the museum, which at that point had been established for nearly 20 years and open to the public since 1867.[23] While the role of the public museum in cultivating cultural and civic values provided a starting point for Haast's address, his focus promptly turned from the rationale for the museum to the process of collecting. He devoted the major portion of his speech to enumerating the variety of associations and transactions that had been initiated during that period to develop the museum from one that contained just a handful of unexceptional objects to an institution that was known around the world and courted for its access to unique and spectacular specimens.[24] The mainstay of the museum's early exchanges with local and distant naturalists and institutions was New Zealand's living avifauna, which was extremely distinctive and thus highly desirable to collectors and other consumers of natural history.[25] Yet, Haast's greatest feat proved to be the unearthing of large quantities of moa bones at a crucial time in the museum's development. Realising that the discovery of unprecedented numbers of this giant, extinct bird was extremely valuable and would allow him to carefully and assiduously cultivate his connections, Haast began an extensive program of exchanging and selling moa bones to prominent museums and other natural history consumers across the world. As a result, he saw his own collections 'become so extensive, that it was utterly impossible to find space for them'.[26]

The authority of London's scientific elite exerted considerable control over these transactions from the start.[27] However, this influence was increasingly supplemented, and sometimes even displaced, by an expanding web of intercolonial and transnational connections.[28] The change took place in the latter decades during which Owen's cycle of portraits with the moa skeleton appeared. It also coincided with the period in which Haast had commissioned several series of photographs of himself posing with spectacular moa displays, ensuring the circulation of these images across his wide professional network and into the public arena. Well before 1877, the year Owen sat for one of his most influential published portraits with the articulated moa skeleton (Figure 3.15), Haast had already been very purposefully and publicly advertising himself and the Canterbury Museum along with what Jacob Gruber describes as New Zealand's 'most important contribution to the rapidly expanding body of natural history in the middle decades of the nineteenth century'.[29] While Owen used fossil remains of the moa as the dominant visual motif in his honorific portraits, thereby representing his significant discoveries in the field of palaeontology and defining his status in imperial science, Haast's objective for his photographic commissions was to expand his natural history networks. By using a similar pictorial device, his primary aim was to ensure the viability of his colonial museum by promoting its collections as being both spectacular and unique.

As Haast spelled out in his annual address to the Philosophical Institute, the Canterbury Museum had begun with exceedingly modest resources and prospects, consisting originally of 470 rock samples and a herbarium that Haast had collected initially as a member of the 1859 overland expedition of the North Island led by the distinguished Austrian geologist, Ferdinand Hochstetter; and subsequently in his role as New Zealand's first provincial geologist.[30] Between 1860 and 1862, when the museum was established in two rooms of the Provincial Government Building, Haast drew on local and German–Austrian contacts to broaden his collection, which included a cast of a moa skeleton purchased from Dr. Gustav Jaeger, director of the Zoological Gardens of Vienna. Haast himself had found those 'perfect skull and bones' in the limestone caves of the Aorere Valley in Nelson and sent them on to the Royal Geological Society, which 'commissioned the restoration by Jaeger [who] made thirty copies in plaster-of-Paris [that] were being sold to museums at 100 florins each'.[31] Haast advised the Provincial Council that this 'cast of a skeleton of *Palapteryx ingens* [moa] was ordered by [him] in Vienna and presented to the Museum in 1862'.[32] 'We were then greatly rejoiced at obtaining this cast', Haast recounted, as 'we little dreamt that a few years later the Canterbury Museum would be able to boast of possessing a collection of moa skeletons unrivalled by any other museum'.[33] Jaeger's rather rigid and evidently inaccurate reconstruction of the skeleton was documented – and probably intended to be marketed to other museums – by way of two large original albumen silver photographs showing the assembled and disassembled skeleton, which were tipped into the published account of his restoration of the moa.[34] As a copy of a specimen widely acknowledged for its importance to the natural history of New Zealand, the circuitousness of that particular acquisition of the cast by the Canterbury Museum suggests that, initially at least, Haast's attitude was one of deference to the metropolitan scientists and societies. He was later to change his position, putting his energy into developing and promoting the distinctiveness of his own institution.

Like Owen's campaign for a national museum of natural history in England, Haast's longer-term strategy for a museum in Christchurch was indicative of his own tenacity and demonstrated, without doubt, that the public and professional profile of a museum director was a principal factor in determining the success of natural history museums in the colonies – just as it was in London, at the at the very centre of imperial science.[35] Crucially, Haast recognised that the museum's early progress depended on two key elements, each as indispensable as the other in ensuring its ongoing viability. The first of these was the identification of and undisputed claim to local New Zealand specimens that were acknowledged by other naturalists to be both distinctive and impressive. For that recognition to be possible in the first place and to capitalise on the interest it generated, the second consideration was a need for a well-directed strategy to ensure maximum publicity to a broad-based public as well as exposure to important institutions and colleagues. In 1861, Haast had been commissioned by the provincial government, following a recommendation by Hochstetter, to undertake a major survey of Nelson and to produce a report on the topography and geology of Nelson. He had assiduously sent an account of his explorations to a long list of scientific luminaries, societies, university libraries and newspapers in New South Wales, Europe, London and New York.[36] Consequently, he had already made some headway in gaining access to the global network of scientists that had developed as part of the exchange of information and materials by learned societies, museums and official expeditions.

Haast's report ran to 150 pages, bolstering his status as a professional scientist, one of the few naturalists in the colonies of Australia and New Zealand at the time – or, indeed, in England and across the British Empire – who could make that claim. Among the most noticeable features of this and subsequent publications is the obvious forethought he had given, even while undertaking the expedition itself, to cultivating prospective supporters of his work. For example, in his Nelson report, Haast had begun naming the South Island's great natural landmarks in honour of contemporary English and European scientific luminaries, a process that he would continue on virtually all of his subsequent surveys. This had the effect of literally transplanting the achievements of metropolitan science into the heart of colonial scientific investigation and created an enduring association between Haast in Christchurch and the most eminent scientists of science of the day. More practically, this process created an ideal opportunity to initiate correspondence with these influential figures. As he wrote to Owen in August 1861, having not had the pleasure of a response to an earlier letter, he had since 'been engaged in extensive explorations in the interior of [the South] Island':

> The result of the first of these journeys is embodied in the [enclosed] ... *Report of a topographical and geological exploration of the western districts of the Nelson Province, New Zealand, undertaken for the Provincial Government by Julius Haast Esq....* As your name is closely connected with the Natural History of New Zealand, I did myself the pleasure of naming a mountain range and a river after you, as a small and just tribute to one, whose name is not only known to every scientific man in the civilized world, but is also on the mouth of every Colonist here, even the most uneducated, associating Prof. Owen with Moa bones.[37]

While such towering figures were accustomed to these sorts of tributes, it was nonetheless an honour that would have been impossible to pass over. Consequently, even if Haast's intention was possibly more aimed at furthering his own ambitions, scientific celebrities from London to Vienna found themselves corresponding with him, initially on matters not exactly scientific, because they were indebted to his admiration; these included Charles Lyell, who wrote to Haast in 1861 after having some mountains and a river named after him.[38]

With each subsequent survey report, the list of New Zealand landmarks named after a roll-call of contemporary English, German and Austrian scientific luminaries grew, gradually creating a dense grid of British and European imperial names all over New Zealand's most impressive alpine landmarks. These included Mount Tyndall, the Hochstetter and Murchison glaciers, Mount Forbes, Mount Darwin, Mount Haidinger, Mount Petermann and, winding its way through much of this distinguished company, the River Haast, which he named after himself 'by order of the Provincial Government of Canterbury'.[39] On explaining his rationale for this process to Sir William Hooker, he recounted how, when beginning his survey of the Southern Alps, he had planned to 'create a kind of Pantheon or Walhalla for [his] illustrious contemporaries amongst those never-trodden peaks and glaciers'.

> I had ... the pleasure to call a magnificent glacier ... descending from the south-western side of Mount Cook, the Hooker glacier, and the river issuing from its icy vault, the River Hooker ... I would have chosen a mountain had I not already named in the Nelson Province ... the Hooker Range, in honour of your son, Dr. J. Hooker, as the author of our admirable flora.[40]

Haast also informed Hooker of his intention to send him a map of the wilderness he described, accompanied by watercolour drawings for his son Joseph Hooker. As he assured his correspondent, 'You will become better acquainted with the general features of your Godchild, and you will confess that our Alps can not only rival but

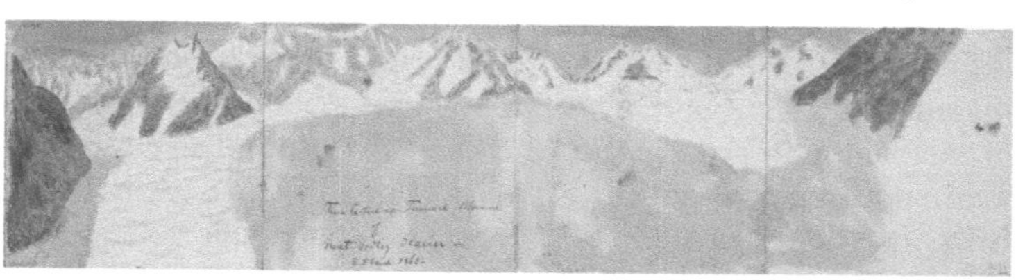

Figure 4.5 Julius Haast, *From Central Terminal Moraine of Great Godley Glacier, 5 March 1862*, pencil, ink and watercolour. Collection: Alexander Turnbull Library, Wellington, New Zealand (C-097-060).

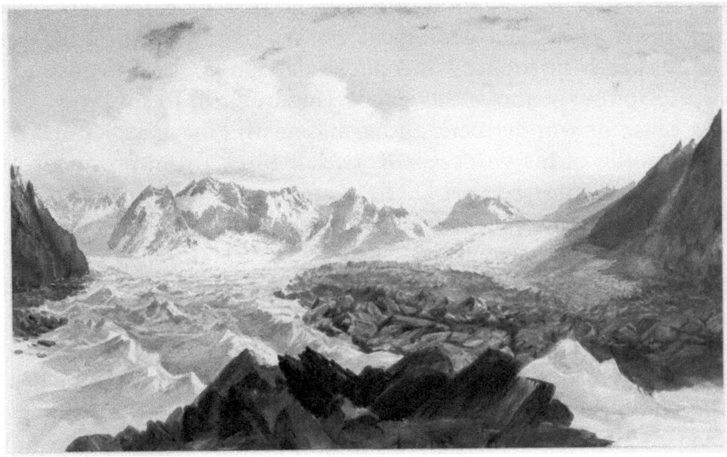

Figure 4.6 John Gully, *From Central Terminal Moraine of Great Godley Glacier, 5 March 1862*, 1863, watercolour. Collection: Alexander Turnbull Library, Wellington, New Zealand (D-037-001).

surpass the European Alps in many respects'.[41] This emphasis on the dramatic and gigantic – sublime – aspects of New Zealand's nature became a recurring motif in Haast's scientific work and his promotion of the Canterbury Museum.

As his first scientific publication, Haast's 1861 report on the western districts of Nelson was entirely textual, with more than half of its length given over to a descriptive narrative of his journey and the latter part detailing the physical geography, geology, roads and usable land in that part of the province, along with such general observations on zoology and botany as the difficulties of the terrain permitted him to make. Many of the geological surveys he undertook during the 1860s and 1870s resulted in fastidiously drawn maps and sections of geological formations that appeared as appendices to his detailed reports. They also inspired him to produce landscape views – often panoramic – in pencil and as subtly tinted watercolour studies. More generalised accounts of some of his reports appeared in London journals.[42] To illustrate a paper on the mountains and glaciers of the Canterbury Province, which he presented before the Royal Geographical Society in 1864 and subsequently published in their journal, Haast sent to the society 12 'well-executed drawings representing the alpine scenery of the Canterbury Province'. These were sketched by Haast and then reproduced in watercolours by the artist and surveyor John Gully.[43] As Haast earlier commented to Sir William Hooker, his first-hand experience of New Zealand's mountains revealed that they were not only a match for the European Alps but also, in many respects, exceeded them.

Although his systematic but unelaborated renderings of geological features recorded all essential information, his commissioning of Gully to produce the final series was clearly intended to amplify and capitalise on the visual impact of New Zealand's most spectacular landscapes. As scientific documents, Gully's watercolours were possibly no less accurate than Haast's own visual records, but they were clearly aiming for a more exhilarating experience for the viewer.[44] For instance, Haast's use of colour in his landscape *From the Central Moraine of Great Godley Glacier, 5 March 1862* is very restrained; moreover, he writes the location and date across the glacier and records the distant mountain forms as distinct geological masses (Figure 4.5). Gully's more detailed version of this sketch introduces and accentuates a rocky foreground that situates the observer precipitously above the glacier (Figure 4.6). In this way, Gully's watercolours added a sense of wonderment and grandeur to Haast's scientific observations and raised the profile of his work. Sir Roderick Murchison, president of the society, for instance, was so impressed that he drew attention to this 'most important account of the highly interesting journeys of the provincial geologist Dr. Haast'.

> In the course of these explorations he has [taken] the opportunity of remarking and commenting upon the features of glacial action … and illustrated [them] by a series of very beautiful, coloured sketches upon a large scale. The sketches are now deposited in the archives of our Society, and, together, with the admirable description which accompanies them, will enable the physical geographer to compare the results of Dr. Haast's observations with those which have been made amid the glaciers of the Alps and of the Himalaya Mountains.[45]

In a footnote to his address, Murchison added that it was his sincerest 'hope that chromo-lithographic copies of those very remarkable coloured sketches of the New Zealand Glaciers will soon be published [as] Glacialists and Alpine travellers should possess them'.[46] However, these promised colour prints were never reproduced and, as the most influential scientific periodicals still included very few illustrations, it is not surprising that neither these nor any other images appeared with the published paper. Indeed, Haast's most complete and visually compelling reports were published in New Zealand by the provincial Government Printer and other local imprints.[47] As with his first report on the Nelson Province, he distributed these later publications to all corners of the globe in the hope of gaining international recognition for his work in geology and glaciology.

It is clear from Haast's written reports and published addresses – even without illustrations – that he sought to not only observe aspects of nature and disseminate his findings as widely as possible but that he was also attempting to capture readers' attention more profoundly by stirring their imagination through his dramatic descriptions of New Zealand's evolving landscape and the sense of time and change invoked in his representations of glacial phenomena. He anticipated that the metropolitan savants would not see New Zealand simply as a collection of facts and figures or an accumulation of curios representing just another corner of the empire but would instead recognise the country as grand, spectacular and vivid. Haast's objective was to make his own natural history realm impressive and uniquely memorable by drawing on the more Romantic aspects of Victorian culture as much as on scientific empiricism. Together with the founding of the local Philosophical Society and his plans for the Canterbury Museum, Haast was clearly attempting to transform colonial natural history from an under-acknowledged, dependent culture beholden to metropolitan science into an autonomous and internationally regarded community of endeavour. As Charles Darwin wrote after he received a copy from Haast of the Nelson survey report along with his inaugural speech to the Canterbury Philosophical Society: 'I have seldom in my life read anything more spirited & interesting than your address. The progress of your colony makes one proud, & it is really admirable to see a Scientific

Institution founded in so young a nation ... You will easily believe how much I have been interested by your striking facts on the old Glacial period. And I should suppose the world might be searched in vain for so grand a display of Terraces. You have indeed a noble field for Scientific research & discovery'.[48]

Haast's diligence in circulating and publicising his discoveries characterised his attitude to colonial endeavour from the start. He was keen to ensure that even his early work in science had a visual impact that would arouse the imagination of a range of audiences. It also distinguished him from other colonial figures who had sought to carve out a career in natural history. For example, Walter Mantell – son of Gideon Mantell, the renowned British geologist and discoverer of the *Iguanodon* – arrived in New Zealand from England having already established impeccable connections, through his father, to Owen, Lyell, Darwin and others from among the metropolitan scientific elite.[49] Mantell's ready-made network was in stark contrast to Haast's pressing need to create professional contacts in the metropolis through his fieldwork, a situation that was familiar to most other naturalists active in the colonies. Although he made significant discoveries of moa remains during that crucial period when Owen and others were attempting to interpret this strange vestige from the past, Mantell's work nevertheless typified the accepted role for colonial science during the 1840s and 1850s. His obligation was to consign his discoveries to the British Museum for others to describe and publish, with little public acknowledgement of his own critical role, a situation that made him especially vulnerable to exploitation. This perpetuated and reinforced the status quo whereby naturalists active in the colonies continued to be perceived as mere collectors rather than a vital source of knowledge about the local environment. The written and visual accounts of Mantell's collecting expeditions into New Zealand's interior are suffused not only with a sense of the physical exertions of overland exploration but also the burden of external expectations:

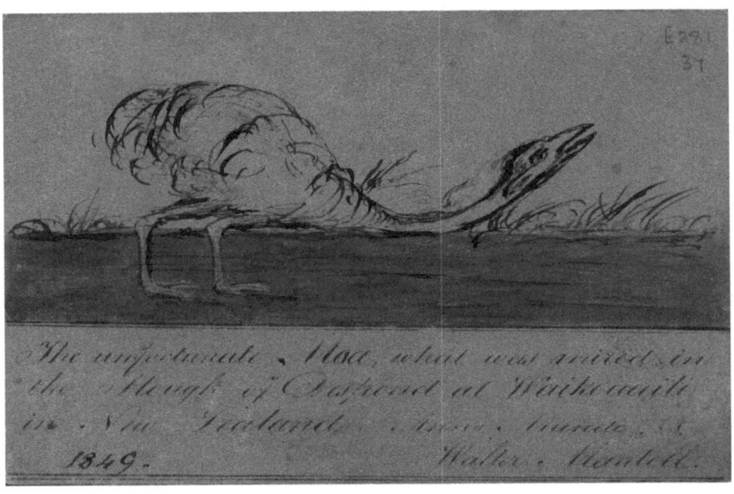

Figure 4.7 Walter Mantell, *The Unfortunate Moa, What Was Mired in the Slough of Despond at Waikouaiti in New Zealand. Anno Mundi???*, 1849, pen, ink and wash. Collection: Alexander Turnbull Library, Wellington, New Zealand (E-281-q-037).

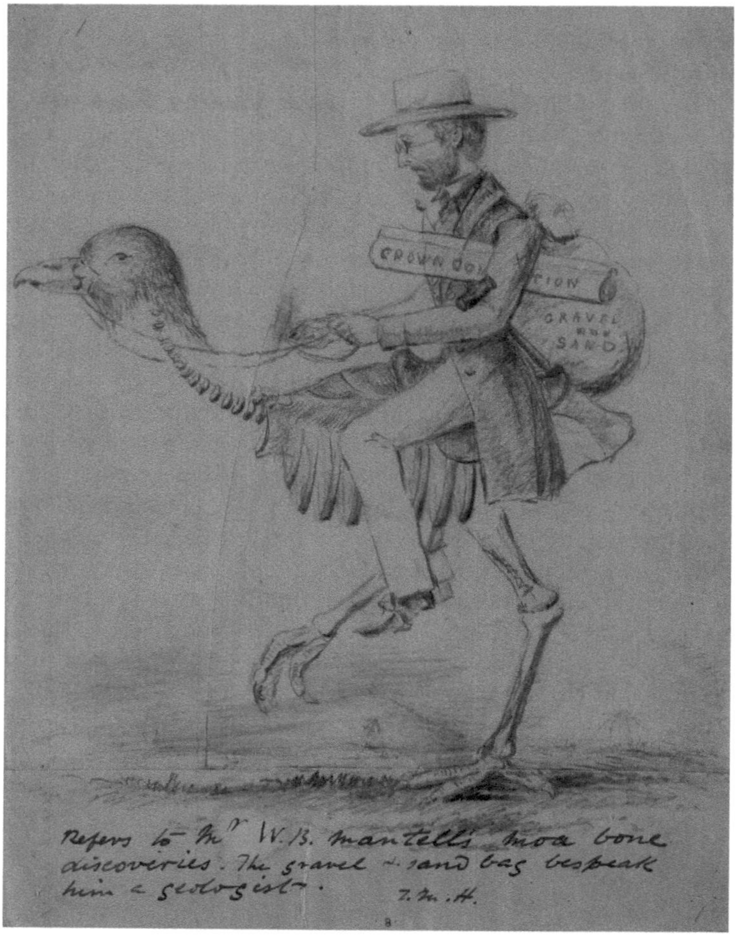

Figure 4.8 James Brown, *Refers to Mr. W. B. Mantell's Moa bone discoveries. The gravel and sand bag bespeak him a Geologist*, c.1851, pencil on paper. Hocken Collections, Uare Taoka o Hakena, University of Otago (11.594).

> I have just returned from my Moa bone hunting expedition and ... I am extremely tired having had since daybreak this morning a long, solitary and fatiguing walk through the forest at the back of the Mountain ... I only found a few toe bones & claws and a few more fragments of egg shells. My travelling companion Charles Nairn whom the Colonel has engaged as interpreter had been directed by him to collect on the way as he, the Colonel, had promised some bones to Prof. Owen but has as yet been unable to fulfil his promise.[50]

Mantell's moa collections, while haphazard, were more sizeable – and thus more significant to metropolitan science – than the occasional finds that preceded his concerted explorations.

Mantell's drawings from that time are neither heroic nor celebratory. Reflecting back on an unknown time in the prehistoric era of the moa, he conjectures on the last of the species, visualising it in its dying days up to its neck in mud and mournfully calling as the sun sets into the foothills of the mountains behind it. While the circumstances and time period of the moa's extinction was a matter of ongoing debate, Mantell's drawings seemed to be a reflection not only of the moa's demise but also an allusion to the difficult terrain he himself had to traverse to unearth its remains and his own hardships as a naturalist in the colonies. Mantell's grey and gloomy images express not only melancholy for the moa's fate but also foreshadow an emerging dissatisfaction and a growing cynicism amongst colony-based researchers regarding metropolitan savants and the long delays – sometimes decades – before their findings were evaluated and published, due in part to the sheer volume and variety of specimens being accrued in London and other centres of authority (Figure 4.7).

At the same time, the colossi in London were still reluctant for naturalists in the field to undertake any significant part of that work or to get any of the credit. Becoming increasingly frustrated and impatient with Owen's intractability, Mantell wrote to his father: 'I am amused with the Professor ... but you cannot think how small he appears at the end of a cool perspective of fourteen thousand miles'.[51] Settling into his colonial life, Mantell held a number of public positions before embarking on his moa expeditions and gradually came to value the importance of local expertise and autonomy.[52] James Brown encapsulated the issue in drawings produced at the time of Mantell's discoveries. In one image, he portrayed Mantell arriving back from an expedition victoriously carrying a brimming basket full of live, though somewhat diminutive, moa above his head.[53] However, the spurs, thigh-high boots and fitted breeches that he sports make him look like a nomadic gaucho or cowboy. Imagining Mantell somewhere between the sort of itinerant travellers who collected specimens purely for profit and the more dedicated men of science, Brown portrays him as something of a transitional figure: engaged initially as a mere casual gatherer for imperial science but now increasingly desirous of having his work acknowledged. In a companion piece, Brown imagines the naturalist riding an articulated and obviously domesticated moa skeleton (Figure 4.8). With a sack clearly marked as containing sand and gravel, the image alludes to Mantell's profession as a geologist; however, the official document secured firmly under his arm reads 'Crown Domination', thereby spelling out the limitations imposed on colonists by imperial science. A mountainous landscape appears faintly in the distance – unmistakably identifying the location and situating Mantell, his important discoveries and the dilemma of colonial science unambiguously in New Zealand. The image of the moa conveying colonists into New Zealand's outlying regions would become a familiar symbol of the vagaries of scientific explorations and other speculative ventures, such as in a cartoon published in *Dunedin Punch* in 1865, which showed a rag-tag expedition of prospectors riding a harnessed moa towards the West Coast in search of gold; and Alexander McKay's provocative poem of 1880, *The Canterbury Gilpin*. James Brown's drawing of Walter Mantell riding the articulated moa skeleton also prefigured a succession of photographs portraying Julius Haast proudly posing with the Canterbury Museum's early displays of impressive moa skeletons, which were assembled from his own moa bone discoveries in the 1860s.

In the early days of the Canterbury Museum, good moa specimens were scarce in New Zealand collections as they were mostly being despatched to museums in London and Europe. To build up the Canterbury Museum collection, Haast initially relied on skins and eggs from the smaller living avifauna as his currency to exchange for other specimens. By employing collectors to catch birds and process skins from such prized species as the kiwi and kakapo, the museum registered several thousand native bird skins in its early exchange books as having been sent to museums in Europe, North and South America, Australia, India, South Africa and England.[54] The range of specimens and objects received in return was indicative of the eclecticism and profusion of specimens that characterised Victorian museum collecting and display.[55] This display of assorted exotic items, Sharon Macdonald suggests, 'were claims of the capacity to know and to govern; signs too for a visitor that theirs was a nation, or a locality, that also played on the global stage'.[56] But the Canterbury Museum's early transactions seemed insignificant compared with the volume and quality of specimens exchanged by Haast after December 1866, when large quantities of moa bones were unearthed in swampy ground on the local property of Glenmark near Christchurch. As he characterised it, 'a new era of the Museum began, and to which, after that time, its rapid growth has principally to be attributed'.[57] The Glenmark moa discovery brought with it an unparalleled opportunity. Haast describes how 'the bones excavated under [his] direction, filled a large American four-horse wagon'.[58] So substantial was his cache of specimens from indigenous fauna, including the moa, that the trustees recorded in their report of 1873 that the museum's achievements thus far were mostly due to the 'the system of exchange carried on by the Director' and that the 'actual amount of money expended in securing the large and valuable collection as it now stands is altogether insignificant'.[59] As well as supplying much large-scale and highly desirable material that would complement other institutions' spectacular palaeontological displays, the excavation provided 'the first seven moa skeletons in the Museum [to be] articulated', an immediate and visually dramatic showpiece that was of local, national and global significance.[60]

As 'boundary objects', the value and meaning of which was constantly being translated as they moved between individuals with different agendas and across various sites of interpretation, the moa were emblematic of the nexus of diverse interests, institutional imperatives and the broad, increasingly shifting terrain of popular natural history.[61] These complex interconnections revealed the elaborate networks that gave rise to many exchanges between centre and periphery. In New Zealand at this time, the number of men who were employed as professional scientists – predominantly on geological surveys and in museums – totalled no more than a dozen.[62] This was typical of Victorian museums of natural history, which included people from diverse but intersecting social worlds, only some of whom saw themselves as men of science. These included collectors; commercial dealers; curators; taxidermists and articulators; carpenters, artists and photographers; trustees; government officers; politicians; and publishers. Like the philosophical societies that provided the impetus for many Victorian museums, the establishment and ongoing development of museums required the involvement of a spectrum of participants and supporters whose expertise and interests were often in fields unrelated to the museum's scientific activities. While some of these stakeholders were trying to pursue and promote a particular field of natural history, either as professionals or amateur naturalists, others were focussed on the civic ideal or practical management of the public museum, while still others were more concerned with social advancement or pursuing their own business interests. Barbara Black argues that, as an emerging site of activity for the new middle class and the industrious working classes, 'the museum solicited not only attendance but

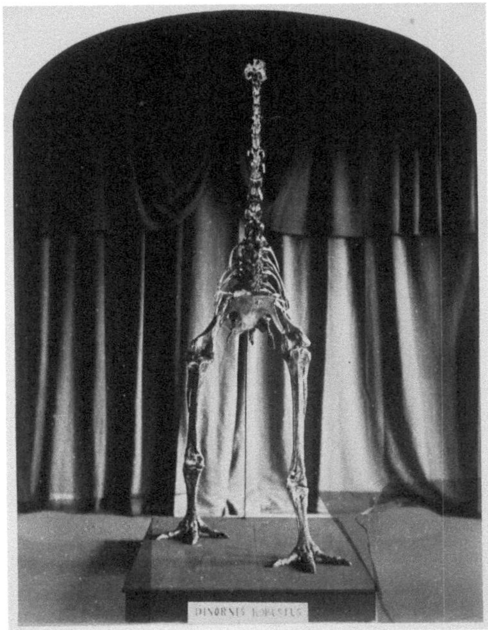

Figure 4.9 Daniel Mundy, *Dinornis robustus* (frontal view), 25 October 1867, albumen silver photograph. Collection: Sir George Grey Special Collections, Auckland Libraries (589–77).

employment as well. Such an opening up of opportunity made the museum the realm of new possibilities for self improvement and upward mobility'.[63] Many stakeholders were also motivated by political interests and patriotic sentiments, variously imperialist, nationalist or colonialist. These separate but intersecting interests enabled a range of relationships and collaborations.

In addition to the organisational complexity of the nineteenth-century public museum, the natural sciences – including palaeontology – were becoming progressively more complex. Fossils and other specimens were no longer regarded as finite or static objects, contained in 'grand, fixed schemes' and waiting to be merely discovered and collected.[64] The public's appreciation of natural history was also changing with the development of more sophisticated anatomical restorations and the production and circulation of new and more elaborate pictorial representations. Thus, while natural history discoveries increasingly harked back to pre-human times, it was the modern visual and cultural technologies, including photography and museum display, and mass culture that revealed them and 'revivified' them for a modern audience, bringing the 'spectacular otherness of "the ancient world" vividly to the public imagination'.[65] In view of that, it is not surprising that, as soon as the first group of Glenmark moa skeletons was ready for display, Haast began to commission photographs of them; mostly, these photographs were only of the moa, but occasionally they incorporated other exhibits as well. Haast would routinely position himself amongst or overlooking the museum's impressive display of giant articulated moa,

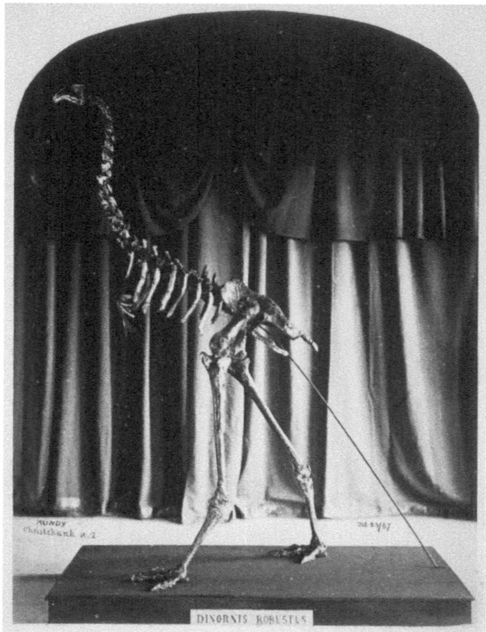

Figure 4.10 Daniel Mundy, *Skeletons of New Zealand Moa in Canterbury Museum, N.Z.*, 25 October 1867, albumen silver photograph. Collection: Sir George Grey Special Collections, Auckland Libraries (589–76).

thereby promoting his own role in their discovery, exhibition and dissemination in the photographs, which were distributed to various museum stakeholders and other interested parties.

A new era of the museum begins: the debate over order versus spectacle

Haast's first series of photographs comprised 13 views. Taken by Daniel Mundy, each glass plate negative was carefully inscribed with the photographer's name and the place and time – Christchurch, N.Z., 25 October 1867. The photographs captured the first six, fully articulated skeletons from the Glenmark cache, with 12 of the studies showing individual skeletons set up to present both frontal and side views (Figures 4.9–4.10). Each skeleton was mounted on an individual plinth, clearly marked with a large label to identify the species. Arranged as vertical compositions, with the camera lens aimed slightly downwards towards the subject, each view has been carefully organised to follow a strict visual code and to maintain the same legibility across the entire series. At the same time, however, Mundy has also followed the Victorian fashion of printing larger photographs with arched tops, thus emulating the elegant presentation of society portraits and grand landscape views. In the most striking image from this commissioned series, Haast poses with the full group of Glenmark skeletons (Plate 9). Here, he introduces a kiwi as a comparison of size, and he has pushed the six plinths together to emphasise the number, range and different sizes of the skeletons.

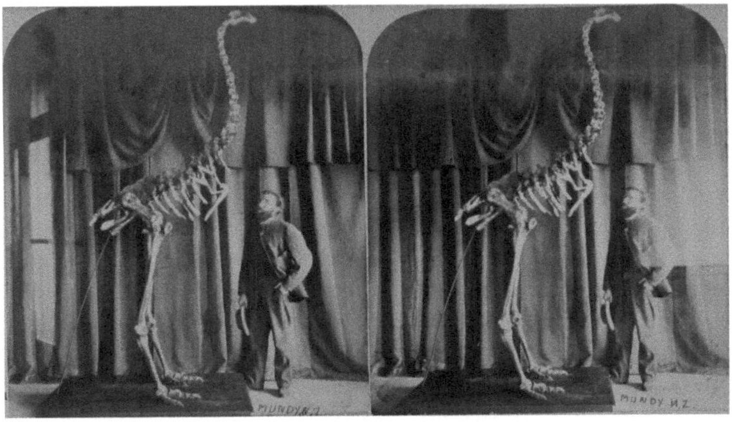

Figure 4.11 Daniel Mundy, *Julius Haast standing beside a moa skeleton on display at Canterbury Museum*, 1867, stereo pair of albumen silver photographs in John Buchanan scrapbook, 1856-c1900. Collection: Alexander Turnbull Library, Wellington, New Zealand (E-207-q-043-1).

Haast is positioned near the centre of the display and is flanked by the two largest moa skeletons facing each other, their sinuous necks extending in a curvilinear form far up above his head toward the ceiling like two architectural features. With the phenomenal height of these two moa dwarfing him, this group portrait is clearly as much about the impressive size of the skeletons as the peculiarities of the species. The sheer contrast in scale between Haast and the specimens, and the theatrical backdrop of double-height curtaining, ensured that the scene was not only instructive but spectacular. Similarly, a stereograph, also produced by Mundy, shows a single moa skeleton, the largest of the group, towering well above the figure of Haast (Figure 4.11). With his head reaching only as high as the creature's hipbone, he plants his feet firmly, tilts his neck back and gazes straight up to observe the full height of the enormous creature. These photographs were almost certainly intended to pique the interest of naturalists from an entrepreneurial, as well as a scientific, point of view and to receive wider circulation beyond the scientific community.

Back in the metropolitan centres of science, in particular, directors and other museum staff acknowledged that they were now operating in an expanding cultural marketplace in which visitors encountered many competing attractions. The idea of nature as spectacle was crucial to the Victorian museum enterprise and its capacity to maintain and attract existing and new audiences. Increasingly, museums displayed their natural history collections so as to achieve maximum effect. For those who hoped to remain untarnished by the commercial imperatives of low-brow attractions – such as the confected curiosities of fairgrounds and circuses – the desire to present natural history in a way that was both meaningful and appealing required a careful balance between information and entertainment. 'A museum of Nature does not aim, like one of Art, merely to charm the eye and gratify the sense of beauty and grace', Owen posited, when putting forward his case for a national museum of natural history.

> But there are forms of animals which excite wonder by their bulk and power ... For purely scientific purposes, size needs only to be accurately measured and recorded; but a Museum destined to gratify the curiosity of the people, and afford them subjects of contemplation, ought to exhibit the maximum of the characters afforded by the dimensions of certain species of a class which is peculiar for the vast bulk of the species.[66]

According to David Murray in his 1904 study *Museums: Their History and their Use*, however, such a recourse to spectacle was a mark of the non-scientific character associated with earlier, private museums and did not reflect the serious purpose of modern public institutions, which were 'arranged and displayed according to scientific method'.[67] In these earlier collections, 'the object in view was to create surprise rather than afford instruction', Murray maintained, and therefore they 'had a tendency to represent the abnormal rather than the normal, what was rare rather than what was common'.[68]

The modern approach to museum display was to convey the underlying taxonomies of typical specimens, thus representing, as Edward Forbes explained, 'a general index of science, and, as it were, Encyclopaedia of reference'.[69] Alternatively, as the economist W. Stanley Jevons put it, one should be able to comprehend a museum in 'a distinct and appropriate *coup d'oeil*'.[70] However, if natural history was transformed from the fetishisation and exposition of singular curiosities in private cabinet displays into the presentation of orderly, representative collections in the modern museum, the desire to display grand, iconic objects never really abated due to their visual impact and the kudos they attracted.[71] The grand central halls and entry displays of nineteenth-century museums were in fact designed precisely for this purpose.[72] In this respect, much of the contemporary discourse about the ideal modern museum was somewhat at odds with the actual approach of many museums towards the display and promotion of their collections, and this conflict remained a continual challenge for museum directors who courted support from the broader public. For Owen, the case for museums displaying extinct species from past eras and the recourse to spectacle in this regard was compelling. Significantly, though, the impressive reconstructions of extinct fauna that dominated museum displays could convey different narratives depending on their location. In Britain's metropolitan and provincial museums, their novelty and imposing presence were symbols of the nation's enduring imperial power and its pre-eminence in the natural sciences. In Britain's colonies in Australia and New Zealand, on the other hand, almost identical museum displays were signs of the colonies' own aspirations and progress, based on their capacity to develop their own museums with valuable collections and to actively participate in imperial science.

Rather like naturalists' paper museums, the fossils that were collected and subsequently articulated suggested a range of possibilities and configurations that ultimately created entirely new entities and stories which superseded the individual narratives lying behind each of their parts.[73] As Owen observed, 'No triumph of science has appeared more marvellous to the intelligent mind than the reconstruction of a form of life that has passed away long ages ago, and the representation to the visual sense of such [an] animal by its framework, so as to leave little to the imagination in realising a complete idea of the once living figure of the extinct beast'.[74] At the same time, extinct animals held an enormous imaginative potential for general audiences. The common experience of museum officers, Owen found, was that no other natural history specimens 'so much excite the interest and wonder of the public, so sensibly gratify their curiosity [and] are the subjects of such prolonged and profound contemplation, as these reconstructed skeletons of large extinct animals'.[75] Even Carl Gustav Carus, the experienced scientific observer of nature and Romantic artist, was no less prone to the sense of wonder invoked by such spectacular displays, as demonstrated in his description of a visit to London's Hunterian Museum in the company of the King of Saxony in 1844, where he is particularly awed by the New Zealand moa:

> On the very entrance the attention is immediately arrested by the rarest fossil animals; on the right the great armadillo from Buenos Ayres (*clyptodon clavipes*) with its massive bony scales, almost like an immense egg, of the size of the largest drum. Opposite to it, on the left, is a gigantic creature of the sloth tribe (*mylodon robustus*).... In all directions are presented to the eye of the connoisseur things of the rarest description, in particular the remains of that immense New Zealand bird of the primitive world (the *dinornis*), which was more than one-half as large again as the ostrich.[76]

In their desire to 'show to the people the proportion of the creatures of former worlds that science has so restored', as Owen put it, the task of museums was twofold: to demonstrate the massive scale of many of the prehistoric species and the sheer quantity of extinct animals unearthed and pieced together.[77]

Exploiting these aspects of natural history contributed significantly to its commoditisation, a process that was as important for securing broad acceptance of the burgeoning science of palaeontology as it was for creating a popular audience to ensure the progress of new public museums. More pragmatically, Mundy's photographs of the moa exhibits in the Canterbury Museum offered practical guidance for those museums considering the acquisition, preparation and display of these giant creatures. In this regard, such photographs – exchanged and circulated among naturalists or appearing in the press – performed a similar function to the contemporary catalogues published by natural history dealers. Most prominent among these was Henry A. Ward's *Catalogue of Fossils*, 1866, which offered casts of fossils held in the principal museums of Europe and America. Comprising numerous woodcuts, many of which are after photographs, along with brief descriptions of the listed fossils, Ward's catalogue remains a remarkable record of the state of palaeontology at that time. While some of the images, as Ward acknowledges, are 'of very mediocre execution' – being, for the sake of economy, '*transfers* of figures found in various books' and intended as 'simple illustrations of a price list rather than for study' – they are nevertheless full of information about contemporary museum display and collectively represent the desiderata of any natural history museum of the period, regardless of size or location.[78]

Most conspicuous among the illustrations are those showing fossils as part of elaborate museum displays, often with figures standing alongside to indicate the scale of the original specimen and thus that of each same-size replica being offered for sale. As Ward noted, and as Haast argued in his submissions to Canterbury's Provincial Council, an extensive, well-arranged and accessible collection was a necessity for a modern museum.[79] 'But ... in order to serve its end, [it] must possess a certain completeness ... [and] there are many difficulties in achieving this completeness in a Museum of Palaeontology'.[80] For practicing naturalists, access to a full set of specimens was the key to developing their expertise. For the general visitor, on the other hand, it was the spectacular qualities of fossil specimens – particularly the unusual appearance and size of the prehistoric vertebrates compared to living species – that were most captivating. To meet both these needs, Ward's recommendation was to complement collections with plaster casts from the best available fossil specimens from the principal museums of Europe and America.[81]

Ward's catalogue is quite prescriptive in the way the casts have been adapted according to their size and form for various positions in a collection and in the information provided to the viewer. Furthermore, while the provenance of the cast objects – the museums and other collections where the original fossils were currently held – is described in detail, the prominent numbering of the casts identifies them permanently as part of Ward's commercial enterprise.[82] While Ward sought to offer an exhaustive inventory of known fossils, the specimens themselves, however, were often only partial representations. For example, the class of birds is covered by only a few 'portions of some gigantic individuals from New Zealand and Madagascar', comprising casts from the 'wrecks of the dodo' in the British Museum and its display-case companion, the right foot and metatarsal of one of New Zealand's moa species, *Palapteryx ingens*.[83] In the context of this trade in fossil casts, it is not surprising that Haast also saw an ideal opportunity to allow the Canterbury Museum to fill a gap in the market by becoming entrepreneurial himself. The challenge for Haast, however, was how to publicise his spectacular moa specimens in such a way as to leverage his museum to maximum effect while upholding and promoting his own good reputation as a disinterested scientist. His networks proved invaluable. Various museum trustees, businessmen and other Canterbury settlers contributed to the publicity by way of their own contacts and skills. Mundy participated in this process by delivering lectures about his photographs to various scientific societies and presenting an illustrated talk at the Photographic Society of London.[84]

Portraits and the press: the colonial naturalist as publicist

Within three weeks of Mundy photographing the moa display at Christchurch, Haast sought permission from the Provincial Government to dispatch eight sets of the views to various stakeholders.[85] His correspondence indicates that, while the local secretary for public works had commissioned the photographs, the decision to distribute them and the choice of recipients was entirely Haast's. One set was to be retained by the Canterbury Museum and another was to be sent to Mr. G. H. Moore of Glenmark Station, where the moa fossils had been excavated. There was also a set for the Royal Society of New Zealand in Wellington, which had held its inaugural meeting in August of that year. For the most part, however, Haast's list comprised influential museum figures in Britain, Europe and America, such as Richard Owen at the British Museum; William Flower at the Royal College of Surgeons' Hunterian Museum; Ferdinand Hochstetter at the Imperial-Royal Polytechnic Institut in Vienna; prominent geologist and palaeontologist the Vicomte d'Archiac at the Museum National d'Histoire Naturelle in Paris; and Professor Agassiz at Harvard's Museum of Comparative Zoology. Each of these men were to receive an identical series of 13 views by which means, Haast envisaged, 'not only the aims of Science [would be] furthered, but by which valuable exchanges for the collections will be obtained'.[86]

Their response was encouraging for the fledgling museum and its director. Without a doubt, Alexander Agassiz at Harvard most accurately encapsulated their value in global terms when he remarked: 'What a [great sight] your room of extinct birds must be to judge from the photographs you sent us. A single room like this is enough to make all the museums in the world jealous of you'.[87] Such an affirmation from Alexander Agassiz and his father, Professor Louis Agassiz, both from the prestigious and well-endowed Museum of Comparative Zoology, was highly valued by contemporary museum professionals.[88] As Owen had recently noted as part of his own petition for a national museum of natural history, 'Prof. Agassiz, throughout his brilliant and productive career, has associated the practical labours of the museum curator with those of the public teacher, and of the original scientific investigator. Of all my contemporaries and fellow-labourers, he is the one in whose opinion as the scope and aims of the Public Museum of Natural History I have the greatest confidence'.[89] Due partly to Agassiz's earlier significant achievements in science – including pioneering work on glaciers and both fossil and living fishes – the museum at Harvard had a higher profile and was more generously endowed than most colonial museums. On the other hand, the movement of photographs between sites of observation and interpretation – at the periphery and centre of metropolitan science – created a common ground of visual experience that seemed to not only lessen the distance between these remote places but also promised to reduce the disparity between the individuals and institutions in those different locations.[90] As an ostensibly incontestable form of visual data, it is tempting to regard the photographs that Haast circulated as stable and unimpeachable documents, or 'immutable mobiles', to use Bruno Latour's term. However, it is probable that Haast never intended that these images should have a single, fixed purpose that would remain forever harnessed to his initial rudimentary interpretation of the material depicted. Indeed, Haast's distribution of these photographs would prove to be more extensive and strategic than he had initially indicated to the Provincial Government.

For another eminent recipient, Charles Darwin, Haast's photographs provided another perspective on Owen's restorations of the moa, to which Darwin had referred in his *Origin of Species*, in his discussion of the succession of faunal types within the same geographical areas during the late Tertiary Period.[91] Darwin's characterisation of New Zealand's fauna as similar to those of an oceanic island was a significant part of his argument about the distribution and modification of species. As he claimed: 'Oceanic islands are sometimes deficient in animals of certain whole classes, and their places are occupied by other classes; thus in the Galapagos Islands reptiles, and in New Zealand gigantic wingless birds, take, or recently took, the place of mammals'.[92] While the species that inhabit oceanic islands are few in number compared with those on equivalent continental areas, the proportion of endemic kinds is often extremely large, which pointed to the possibility that many moa species had once existed. Upon receiving Mundy's photographs of the various distinct species represented in the Canterbury Museum's collection, Darwin replied immediately to thank Haast 'for the very curious photograph of the skeletons of the gigantic birds' and to let him know that everyone to whom he had shown it had been astonished.[93] Compared to Owen's portraits with the moa and other images and descriptions of Owen's various and sometimes disputed reconstructions that had been undertaken over the years using piecemeal skeletal remains obtained from numerous correspondents in different locations, the immense volume of whole skeletons recovered from Haast's excavations at the one site near Christchurch provided a far more accurate indication of the range of species in the *Dinornis* family and the differences between them. As Haast's photographs clearly showed, the museum had already identified and articulated six handsome and complete specimens representing quite different species, all obtained exclusively from the peat fields at Glenmark.

Haast and Darwin had, in fact, been correspondents since 1862, at which time Haast had made it clear to Darwin that he would be glad to undertake any observations on his behalf on the subject of natural history.[94] Over the following years, these observations proved to be wide ranging, including, for instance, notes on geology; glaciology; fossil fauna; the flightless *Kakapo,* or 'marvellous ground parrot' as Darwin described it; and the naturalisation of European animals in New Zealand, such as the change in nesting habits of the New Zealand Duck. As Darwin exclaimed to his correspondent, there was 'hardly a point in the world so interesting with respect to geographical distribution as New Zealand'.[95] During those years, Haast's geological work had been especially useful to his correspondent, with mention of his 'excellent researches' into New Zealand's glaciers being included in the *Origin of Species* when Darwin undertook the fourth edition in 1866.[96] From 1867, Darwin was engaged in the research that would culminate in the publication of *The Expression of Emotions in Man and Animals*, 1872. It was at this time and in response to a list of questions from Darwin about the expression of emotion in the Maori population – which Haast had carefully duplicated and forwarded on to scientific gentlemen and missionaries across New Zealand's North and South Islands – that Haast had taken the opportunity to enclose the 'small photograph' showing the museum's moa display.[97] In his letter, Haast identified the different species represented and provided precise dimensions – ranging in size from the *Dinornis giganteus* at 9 feet and 10 inches [3 metres] down to the *Dinornis didiformis* at 4 feet and 4 inches [132 centimetres] – and, as he spelled out to Darwin, he had 'introduced the skeletons of a Kiwi (Apteryx 'Owenii') & a human figure for comparison'.[98] Having introduced the subject by way of a remark that he had been very busy lately articulating the six Dinornis skeletons for their museum – which formed a really curious and, he believed, unique group – Haast cannily conveyed the importance of the moa for the museum as well as for science in general.[99]

As his correspondence with Darwin shows, Haast's alertness to opportunities for promoting the museum was equally assiduous when corresponding with figures outside the museum profession as well as within it, and was usually expressed in terms of the extraordinary size and uniqueness of the moa, and, increasingly, its antiquity. Accordingly, the photographs and lists he sent to museums and naturalists around the world in late 1867 represented not the culmination but merely the first stage of Haast's promotion of New Zealand's extinct avifauna as the principal attraction of Canterbury's new museum. Of course, this needed to be formalised by being published in a scholarly format for Haast to secure a stake in that area of research. To this end, he described the measurements of various moa species from the Glenmark excavations before a meeting of the Wellington Philosophical Society on 28 July 1868, which was subsequently published in the first issue of transactions and proceedings of the Royal Society of New Zealand.[100] In this publication, he refers to the information 'sent previously to England' – specifically to W. H. Flower at the Hunterian Museum – and the results from more recent excavations 'undertaken for the Provincial Government of Canterbury', which he hoped to include in a future paper describing 'the ground in which the bones were buried, the probable causes through which the numerous specimens were destroyed, and to which they owe the preservation of their osseous remains'.[101] Numerous references to Professor Owen's

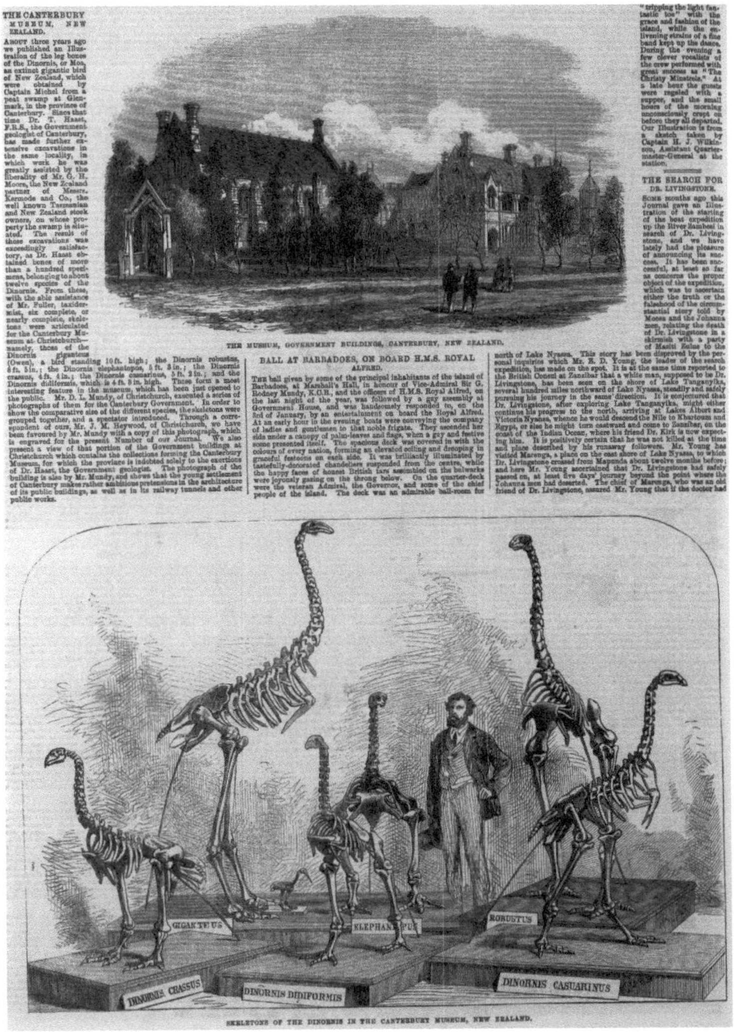

THE CANTERBURY MUSEUM, NEW ZEALAND.

About three years ago we published an Illustration of the leg bones of the Dinornis, or Moa, an extinct gigantic bird of New Zealand, which were obtained by Captain Michel from a peat swamp at Glenmark, in the province of Canterbury. Since that time Dr. T. Haast, F.R.S., the Government geologist of Canterbury, has made further extensive excavations in the same locality, in which work he was greatly assisted by the liberality of Mr. G. H. Moore, the New Zealand partner of Messrs. Kermode and Co., the well known Tasmanian and New Zealand stock owners, on whose property the swamp is situated. The result of those excavations was exceedingly satisfactory, as Dr. Haast obtained bones of more than a hundred specimens, belonging to about twelve species of the Dinornis. From these, with the able assistance of Mr. Fuller, taxidermist, six complete, or nearly complete, skeletons were articulated for the Canterbury Museum at Christchurch—namely, those of the Dinornis giganteus (Owen), a bird standing 10 ft. high; the Dinornis robustus, 8 ft. 5 in.; the Dinornis elephantopus, 5 ft. 3 in.; the Dinornis crassus, 4 ft. 4 in.; the Dinornis casuarinus, 5 ft. 2 in.; and the Dinornis didiformis, which is 4 ft. 8 in. high. These form a most interesting feature in the museum, which has been just opened to the public. Mr. D. L. Mundy, of Christchurch, executed a series of photographs of them for the Canterbury Government. In order to show the comparative size of the different species, the skeletons were grouped together, and a spectator introduced. Through a correspondent of ours, Mr. J. M. Heywood, of Christchurch, we have been favoured by Mr. Mundy with a copy of this photograph, which is engraved for the present Number of our Journal. We also present a view of that portion of the Government buildings at Christchurch which contains the collections forming the Canterbury Museum, for which the province is indebted solely to the exertions of Dr. Haast, the Government geologist. The photograph of the building is also by Mr. Mundy, and shows that the young settlement of Canterbury makes rather ambitious pretensions in the architecture of its public buildings, as well as in its railway tunnels and other public works.

THE MUSEUM, GOVERNMENT BUILDINGS, CANTERBURY, NEW ZEALAND.

BALL AT BARBADOES, ON BOARD H.M.S. ROYAL ALFRED.

The ball given by some of the principal inhabitants of the island of Barbadoes, at Marshall's Hall, in honour of Vice-Admiral Sir G. Rodney Mundy, K.C.B., and the officers of H.M.S. Royal Alfred, on the last night of the year, was followed by a gay assembly at Government House, and was handsomely responded to, on the 3rd of January, by an entertainment on board the Royal Alfred. At an early hour in the evening boats were conveying the company of ladies and gentlemen to that noble frigate. They ascended her side under a canopy of palm-leaves and flags, when a gay and festive scene presented itself. The spacious deck was covered in with the colours of every nation, forming an elevated ceiling and drooping in graceful festoons on each side. It was brilliantly illuminated by tastefully-decorated chandeliers suspended from the centre, while the happy faces of honest British tars assembled on the bulwarks were joyously gazing on the throng below. On the quarter-deck were the veteran Admiral, the Governor, and some of the chief people of the island. The deck was an admirable ball-room for "tripping the light fantastic toe" with the grace and fashion of the island, while the enlivening strains of a fine band kept up the dance. During the evening a few clever vocalists of the crew performed with great success as "The Christy Minstrels." At a late hour the guests were regaled with a supper, and the small hours of the morning unconsciously crept on before they all departed. Our Illustration is from a sketch taken by Captain H. J. Wilkinson, Assistant Quartermaster-General at the station.

THE SEARCH FOR DR. LIVINGSTONE.

Some months ago this Journal gave an Illustration of the starting of the boat expedition up the River Zambesi in search of Dr. Livingstone, and we have lately had the pleasure of announcing its success. It has been successful, at least so far as concerns the proper object of the expedition, which was to ascertain either the truth or the falsehood of the circumstantial story told by Moosa and the Johanna men, relating the death of Dr. Livingstone in a skirmish with a party of Mazitu Zulus to the north of Lake Nyassa. This story has been disproved by the personal inquiries which Mr. E. D. Young, the leader of the search expedition, has made on the spot. It is at the same time reported to the British Consul at Zanzibar that a white man, supposed to be Dr. Livingstone, has been seen on the shore of Lake Tanganyika, several hundred miles northward of Lake Nyassa, steadily and safely pursuing his journey in the same direction. It is conjectured that Dr. Livingstone, after exploring Lake Tanganyika, might either continue his progress to the north, arriving at Lakes Albert and Victoria Nyanza, whence he would descend the Nile to Khartoum and Egypt, or else he might turn eastward and come to Zanzibar, on the coast of the Indian Ocean, where his friend Dr. Kirk is now expecting him. It is positively certain that he was not killed at the time and place described by his runaway followers. Mr. Young has visited Marenga, a place on the east shore of Lake Nyassa, to which Dr. Livingstone crossed from Mapunda about twelve months before; and here Mr. Young ascertained that Dr. Livingstone had safely passed on, at least five days' journey beyond the point where the Johanna men had deserted. The chief of Marenga, who was an old friend of Dr. Livingstone, assured Mr. Young that if the doctor had

SKELETONS OF THE DINORNIS IN THE CANTERBURY MUSEUM, NEW ZEALAND.

Figure 4.12 'The Canterbury Museum, New Zealand', *Illustrated London News*, 6 February 1868, p. 144.

publications on the *Dinornis* provide a point of comparison throughout and give legitimacy to his own research paper.

More prominently, however, the following year, Mundy's photograph of Haast posing with the group of six moa skeletons was reproduced as an engraving in the *Illustrated London News*, along with another image by Mundy showing the part of the Provincial Government buildings containing the Canterbury Museum collections (Figure 4.12).[102] Just as in the description accompanying the photographs Haast had sent to various scientific colleagues and supporters around the world, the newspaper carefully named each species and their different heights, ranging from the *Dinornis giganteus* at 10 feet down to the *Dinornis crassus* at 4 feet, 4 inches, and it also drew attention to the spectator in the photographs to 'show the comparative size of the different species'.[103] As the newspaper's correspondent, J. M. Heywood, a Christchurch shipping agent, noted, 'These form a most interesting feature in the museum, which has been just opened to the public' and for which, he noted, the province was 'indebted solely to the exertions of Dr. Haast, the government geologist'.[104] In promoting these images of the museum and its moa display, Haast and Mundy, with the help of the popular press, managed to deftly steer the British public's imagination away from the idea that exotic specimens were solely brought back to England for study and exhibition towards an awareness of the importance of endemic species, such as the moa, for colonial museums. Hence, if social and professional kudos could accrue from being included in celebrity portrait anthologies in the company of powerful, high-ranking politicians and other worthy persons – which, as Owen had demonstrated, could leverage institutional prestige – from Haast's point of view, there was much to be gained by being photographed in the company of the moa. If palaeontology – along with geology and glaciology – represented natural history on a grand scale, it seemed fitting that the moa was set to become New Zealand's most identifiable and influential celebrity *du jour*.

The reputation of the moa as a popular and grand attraction was becoming globally recognised, and it was valued by museums throughout the empire. For instance, the Madras Museum had not only established a *Dinornis* display but, more conspicuously,

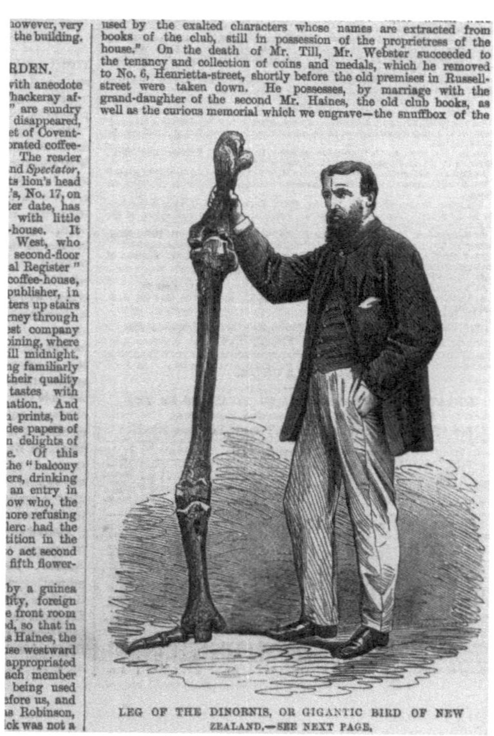

Figure 4.13 'Leg of the Dinornis, or Gigantic Bird of New Zealand', *Illustrated London News*, 10 June 1865, p. 54.

had devised a spectacular image of the moa that emphasised its gigantism in order to promote the museum in the British press. Although there is no ostensible connection between the moa and India's indigenous fauna, it was the principal point of interest in circulated photographs and in illustrated press articles promoting that colonial museum and its superintendent, Captain Jesse Mitchell, during the 1860s. A handful of moa bones, which had been discovered at Glenmark previously, had found their way to the Madras Museum and were brought to the attention of the *Illustrated London News*. In May 1865, the newspaper noted 'a unique specimen of the egg of the extinct moa and the leg bone of that gigantic bird' was on show in the New Zealand Industrial Exhibition at Dunedin.[105] The following month, an entire article was dedicated to the moa as a curiosity, noting that 'a photograph of a much more remarkable specimen, which has been obtained and placed in the Madras Government Museum ... consists of the leg bones – namely the femur, tibia, metatarsus and outer toe, considerably larger than those in Dunedin. They were found at Glenmark, about forty miles north of Christ Church [*sic*] in the Canterbury province of New Zealand'.[106] As with the paper's later report on the moa collection at the Canterbury Museum, the article concluded with statistical information noting the length and circumference of each of the bones retrieved from the enormous bird, and it reproduced an engraving from a photograph of Major J. Michael of the Madras Staff Corps holding the gigantic leg bones so described.

Major Michael's pose appears to be a restaging of Owen's imperial portraits (Figure 4.13, Plate 10). Superficially at least, it seems clearly conceived as a tribute to the original classifier of the *Dinornis* and the principal figure of imperial science. It is also explicitly designed, however, to raise the profile and status of the Madras Museum in the British public's eye by presenting it not as a parochial institution with a focus only on local exhibits but as an all-embracing model of the Victorian museum, in which India could be viewed in relation to the rest of the empire. Like many other museums of its time, the Madras Central Museum began in the eighteenth century as a storehouse of curiosities and, although it became a public museum in 1851, it did not begin to collect systematically until the appointment in 1885 of Edgar Thurston, who was its first full-time superintendent.[107] Thurston acknowledged the visitors' desire for spectacle in the museum and conceded that the appearance of objects, more than their meaning or arrangement, was the main drawcard for the general public:

> For the great mass of visitors to museums in India, who come under the heading of sight-seers and who regard the museums as *tamásha* [show] houses, it matters but little what exhibits are displayed or how they are displayed, provided only that they are attractive. I am repeatedly amused by seeing visitors to the Madras museum pass hurriedly and silently through arranged galleries, and linger long and noisily over a heterogeneous collection of native figures, toys, painted models of fruit, etc.[108]

While the subcontinent had much in the way of spectacular living fauna, it was the museums in England that usually benefited from the discoveries of impressive prehistoric species from that region. For the Madras Museum, where displays of fossils were scarce, the moa bones were unique as well as physically and visually imposing, and consequently they were given quite some prominence in the museum's galleries as an attraction and curiosity for visitors.[109]

For Major Michael, who had acquired the Madras moa bones, and Captain Jesse Mitchell, superintendent of the museum, both of whom were officers of the British military in India, their interest in the *Dinornis* also had an intellectual basis.[110] Military officers performed a crucial role in the collection and dissemination of knowledge in and about colonial India.[111] As British army surgeon and amateur photographer John McCosh advised, officers in India 'should make such a faithful collection of representations of man and animals, of architecture and landscape, as would be a welcome contribution to any museum'.[112] Mitchell's own pursuits were necessarily eclectic and focussed not only on the museum collections but also on the economic productions of the region, ranging from a *Catalogue of the Mollusca in the Collection of the Government Central Museum, Madras* to the publication of his microscopical studies of Indian and other cotton fibres undertaken with the purpose of improving the local industry.[113] Mitchell's interest in the moa and his decision to promote it in the context of his own museum arose out of similar cultural and economic impulses. At this time, Thomas Oldham, superintendent of the East India Company's Geological Survey of India, was eager that his scientific staff should spread 'the knowledge of geology and its allied pursuits' to improve scientific competency.[114] An extensive series of lectures on geology was organised at various institutions, such as the College of Engineering in Madras, from 1862.[115] Additionally, palaeontological and geological specimens were collected from the survey, which facilitated exchanges with British and European collections – including casts of the renowned Indian Siwalik fossils from the British Museum, a series of English fossils sent by Murchison on the British Geological Survey and a group of moa bones from the Royal College of Surgeons – and formed the basis of a company museum in Calcutta.[116] As a consequence, the study of geology and palaeontology, including a familiarity with ongoing research into the moa, was flourishing in India through the 1860s and 1870s.

Whereas Michael's discovery of moa bones during his travels in New Zealand might previously have been considered in the Madras context as merely an interesting souvenir and simply another curiosity to add to the museum collections, he was determined instead to use the publicity he had gained from the *Illustrated London News* as an entrée to contact Owen at the British Museum and inform him of his suspicions that he was in possession of the vestiges of an entirely new moa species:

> I have a leg of a *Dinornis* complete with all the Bones of one toe, two of the Vertebrae and part of the Pelvis – The leg stands over 6 feet and the bones are in excellent condition. I am under the impression that they are larger than any hitherto found and *may* be a new species. I should indeed be pleased to acquire Prof Owen's opinion as to whether my bones belonged to a larger species than the *Dinornis ingens* or *D. giganteus* if so I would, of course send them to England, in the meantime they are *lent* to the Madras Museum ... I beg to enclose a Memo of the measurements of the Bones and if, Sir, you would do me the great favour of informing me whether you are of [the] opinion they are of a new species or not.[117]

It is clear from his letter that, if he were correct, the bones would be of more value to imperial science if they were offered to the British Museum. Until they could be formally identified and their status decided, their display in the Madras Museum was only a temporary arrangement. As it transpired, the bones were found to belong to the *Dinornis maximus*, a species that Owen had previously described. Nevertheless, Michael's discovery and details of the three giant leg bones were described and illustrated in Owen's *On Dinornis* (Figure 4.14).[118] This was immediately noticed by

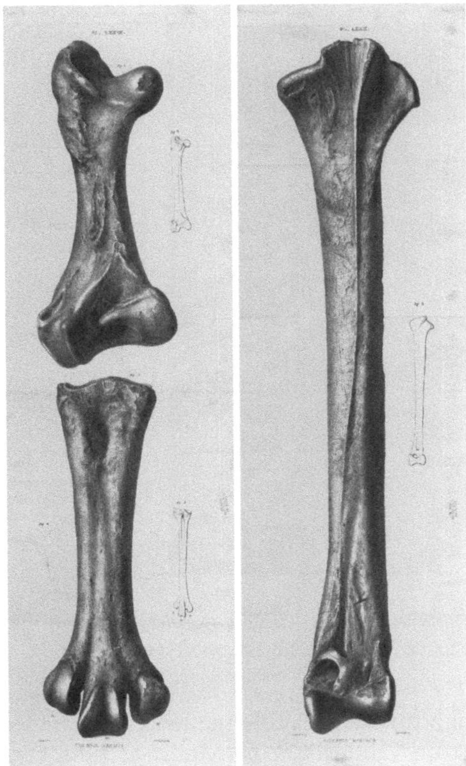

Figure 4.14 Plates 79 and 80 (including Major Michael's specimen) in Richard Owen, *Memoirs of the Extinct Wingless Birds of New Zealand*, John Van Voorst, London, 1879.

Haast, who, on reading Owen's paper, recognised them as belonging to an incomplete skeleton held in the Canterbury Museum. As he wrote to Owen:

> The XII Memoir containing the description of bones of *Din.*[ornis] *maximus* was particularly useful to me because I observed ... that the 3 leg bones of Major Michael belong to the identical specimen of which we have the pelvis, right femur, tibia & fibula & the first dorsal vertebra in our Museum. It is thus evident that we could not succeed in finding other bones as they ... disappeared mysteriously from Glenmark. The men – in excavating the drain – got 3 leg bones out, which they consider to have been 7 feet to 7 feet 6 inches together in length & which Mr. Moore inspected & confirmed their statement. They placed them carefully in the grass but when they returned after dinner to their work, the bones were gone. I hope they will turn up sometime.[119]

After the Madras moa bones had been forwarded to the British Museum for Owen's inspection and finally identified in 1867 as an existing species, they were then sent on to the Exposition Universelle in Paris, where arrangements had been made to show

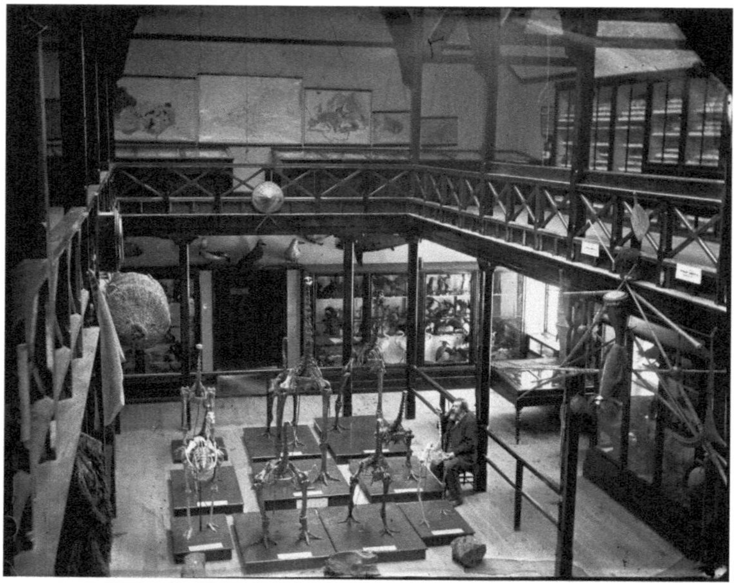

Figure 4.15 Dr. A. C. Barker, *Johann Franz Julius von Haast at the Canterbury Museum Skeleton Hall*, c.1870. Collection: Canterbury Museum, Christchurch (1944.78.66).

them in the New Zealand Court.[120] Through this remarkable sequence of transfers to diverse locations, including the journey from Canterbury to Madras, these vestiges of *Dinornis maximus* illustrate a complex process of world making in which natural history was not simply the product of a local or colonial vision of nature or even an imperial perspective but, instead, a global panorama encompassing a variety of intercolonial and transnational exchanges.[121]

With the substantial leverage gained from its moa discoveries and associated publicity, the Canterbury Museum was re-established in a dedicated building in 1870. While the moa exhibits themselves were impressive, their display had always been compromised during the years the museum was housed in the Provincial Government Buildings, since only parts of the collection could be displayed, and these were often disorganized and crowded.[122] Haast had been anxious to obtain more space and additional showcases to more effectively arrange the collections on hand as they had been installed in two opposite sections of the government buildings.[123] As Joseph Hooker's eldest son, William Henslow Hooker, wrote to his father during a visit to Christchurch in 1869, 'We arrived in Lyttelton Harbour ... on Monday 8th February, where I was met by Dr. Haast ... On the next day ... I saw the Christchurch Museum – but I do not think [it] is nearly as good as the Wellington Museum, though it beats it in one thing and that is the specimen of the Moa, or *Dinornis*'.[124] With the move to the new, purpose-built museum building the following year, the Glenmark moa were given a prominence matched only by their growing international reputation. Haast again commissioned photographs of the museum's displays, this time taken by Dr. Alfred Barker. Conspicuously, it is the grand 'moa hall' – the central and most imposing space in the new museum – that is the subject of nearly all of these images.

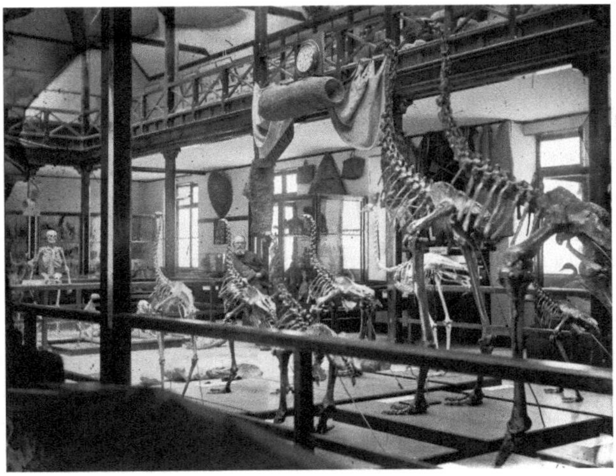

Figure 4.16 Dr A.C. Barker, *Alfred Charles Barker at the Canterbury Museum Skeleton Hall*, c.1870. Collection: Canterbury Museum, Christchurch (1944.78.33).

Many of the photographs include Haast, but Barker also produced several self-portraits among the moa displays (Figures 4.15–4.16). While Barker was not a naturalist himself, his involvement with the museum had introduced a more informal level of exchange of natural history specimens and photographs between New Zealand and England, undertaken between Barker's acquaintances and members of his family. In the case of Owen and Haast, on the other hand, given the public and political dimensions of their respective plans for their museums, both men needed to get their message across to a much larger audience than their immediate circle of colleagues. It was not enough to acquire specimens and to catalogue and display them. It is for this reason that photography was perceived as being so integral to the formation and progress of the nineteenth-century museum.

Barker's exchange of images and objects, however, suggests other concerns. His correspondence with his brother Matthias in England describes in detail the difficulties associated with photographing in the colonies, ranging from problems in acquiring or fashioning reliable equipment, plates and chemical supplies to the unpredictability of the terrain and climate. Like the early collectors and dealers of colonial natural history, Barker's photographic 'specimens' were not easily obtained, and his choice of subjects was often serendipitous, particularly early on. Nevertheless, they became a valuable resource and a form of currency used in exchange for items of photographic equipment and a wide range of photographs taken in England.[125] The images received by Barker, usually from his brother, were mostly commercially produced photographs representing popular views of metropolitan trends and culture, ranging from celebrity portraits to iconic museum masterpieces.[126]

The accrual and dissemination of knowledge between Barker and his family and acquaintances mirrored the more official exchanges between the periphery and centre of the empire, negotiated by the museum and Haast himself. Similarly, Barker did not hesitate to emphasise the uniqueness and gigantism of the moa in his photographs of them and thus adopt the rhetoric of spectacle. As he wrote to Matthias in 1867, prior to his official commission at the new museum, 'I send you with this [letter] some photos I have taken about [two-thirds] the natural size of [five] different species of Moa – I shall hope by next month to have taken photos of a group of eight different species which we have mounted in our Museum – *a sight unequalled in the world*'.[127] This was followed up with actual specimens, which were more than likely from the Glenmark cache that was in the process of being excavated: 'I enclose with [a few] "bones", a photo of the Moas – taken by myself – which was too large to [put] in a letter'.[128]

Compared to Haast's strategies and interventions, Barker's exchanges played a much more limited role in the way natural history and other aspects of colonial experience came to be known and represented, both imaginatively and empirically. Yet, as an individual strategy, this more informal type of knowledge formation was no less important for the dissemination and recognition of colonial expertise. Barker's approach to photography was extremely systematic from the start. Concerned with recording the particularities of the places and objects he photographed, he usually inscribed specific details of the location or person portrayed and the date, sometimes even the time of day, on the glass photographic plates.[129] His work offers an extensive chronicle of the cultural and civic life of Canterbury at that time. We are shown glimpses of how the museum's moa display was assembled and are provided with clues as to the wider

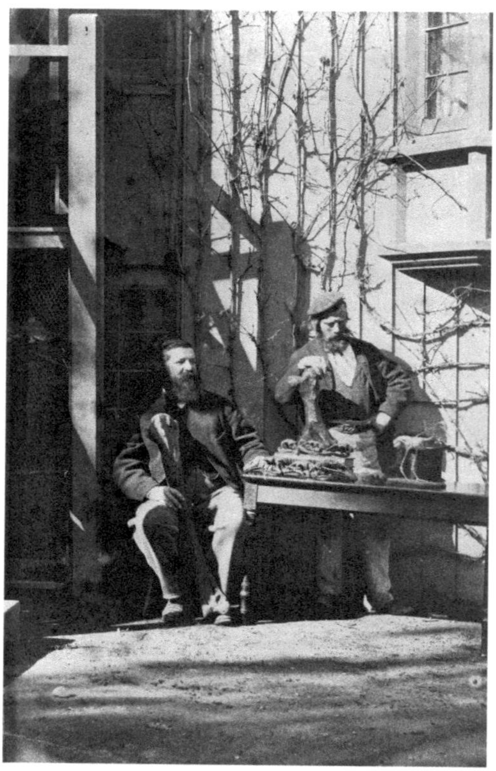

Figure 4.17 A. C. Barker, *Julius von Haast and taxidermist Frederick Fuller*, c.1866, albumen silver photograph. Collection: Alexander Turnbull Library, Wellington, New Zealand (PAColl-5381-01).

range of individuals involved in bringing them to public view. For instance, Frederick Fuller, the museum's taxidermist, who constructed the entire moa display from the mound of bones excavated at Glenmark, had also accompanied Haast on arduous expeditions and prepared and set up the whole of the museum's ornithological and mammal collections.[130] In some ways, Fuller was as pivotal to the workings and success of the museum as Haast himself. In Barker's c.1866 portrait of the two men, however, the contrast between their different roles in creating the public face of the museum is apparent. Having selected the largest femur bone with which to pose, Haast – shown gazing into the distance – has rested it against his own lower leg to demonstrate the femur's massive scale and alluding to its potential for spectacle. Fuller, on the other hand, supports a much smaller tibia bone and, with hand on hip and brow furrowed, he contemplates a kiwi skeleton and some bird skulls and is depicted more as an artisan than as a showman (Figure 4.17). Although the two men were certainly not peers in either a social or professional sense, their work behind the scenes of the museum was certainly collaborative. Barker's double portrait of Fuller and Haast is perhaps the only visual record of the close professional relationship of these two men.

As a trustee and supporter of the museum but not a naturalist himself, Barker well understood that the workings of a colonial museum relied upon a varied group of people from diverse intersecting social worlds, only a number of whom would have seen themselves as men of science. While some stakeholders were focussed on the civic ideal or practical management of the public museum, others, like Barker, were in fact questioning the philosophical basis of natural history as it was being recast in the light of Darwin's publications. For Daniel Mundy, photographing the moa exhibits at the original Canterbury Museum had not been so very different from other photographic assignments he had undertaken. For instance, in letters to Haast, he recounted the accolades he had received from scientific societies and prominent figures in London and Europe for his book of photographs of the thermal region Rotomohana, and in the same letters he expressed his desire to produce larger prints from his negatives of the moa group: 'not framed, but simply mounted on boards; you have no idea the beauty and value of these beautiful permanent prints until you see them'.[131] Mundy had considerable experience photographing on expeditions, and his concern was always to create an accurate but visually compelling and marketable record of the geology and other natural features he encountered.[132] Barker's involvement with the moa project had a perceptibly different impact. Issues of vision, reliability and, ultimately, truth were provoked by his reading of Charles Lyell's *Geological Evidences of the Antiquity of Man*, Charles Darwin's publications and George Campbell's *The Reign of Law* during the same period that he was involved with photographing and promoting Haast's moa project.[133] When he had first become acquainted with Haast, Barker was immersed in the study of Darwin's theories and, unsurprisingly, the two men struck up a close association through their mutual interests in science and culture.[134]

Photographs of the Christchurch moa displays taken by Mundy and Barker were part of the continual flow of exchanges that went on between individuals located at the centre and periphery of Victorian science. Haast had sent his portrait in a letter to Darwin in 1863 after initiating correspondence with him by way of Joseph Hooker the previous year.[135] This, Haast suggested, was 'to enable [Darwin] to know [his] correspondent by sight', and, also, he declared, he would be grateful if Darwin would favour him with one in return.[136] For the majority of Darwin's correspondents, even many of those within his social milieu, *cartes de visite* and more informal portraits offered the only opportunity to put a face to one of the most talked about figures of the period.[137] As Darwin noted to the German naturalist Fritz Müller, 'one likes to have a picture in one's mind of any one about whom one is interested'.[138] Charles Kingsley also indicated his admiration for Darwin's work in his desire to have a photograph of the man:

> I say honestly, there is no living man whom I have *not* seen, whom I wish to see as much as I do you: and failing that, to have some Eidolon of you for myself and mine. If you have any photograph of yourself – or if you can tell me where I can procure one, you will do me a real favour. The more I think, & see, the more I find to make me thankful for your great book.[139]

Kingsley's request for a photographic portrait of Darwin served to proclaim his approbation for his controversial theory of natural selection. For Darwin, therefore, this simple social formality of offering or requesting *cartes de visite* was a way of not only introducing a level of familiarity into his professional intercourse and of fostering distant collaborators for his ongoing work but also a means of gauging support for his theories. In Germany, for instance, where, compared to England and especially France, there was a greater propensity among naturalists to be predisposed to Darwin's theory of natural selection, most leading scientists were also public figures attached to conservative institutions and who, for professional reasons, resisted publicly identifying

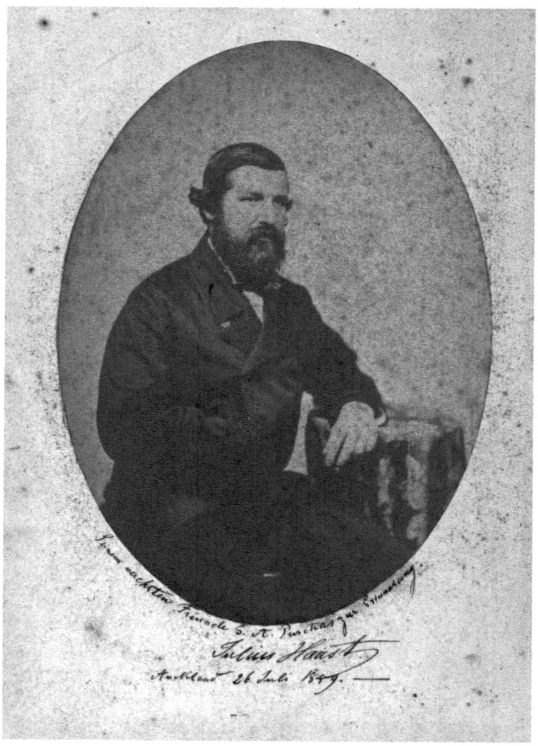

Figure 4.18 Photographer unknown, *Portrait of Julius Haast, 26 July 1859*. Collection: Sir George Grey Special Collections, Auckland Libraries (881-1).

themselves with such new and radical views. Ernst Haeckel was one of the most prominent of Darwin's interlocutors in the German scientific establishment and, like Kingsley, was an immensely effective populariser of his theories. 'With [his] lectures and various writings', no-one, in Darwin's view, could do as much as Haeckel in 'spreading and perfecting sound views on species in Germany'.[140] Thus, when Haeckel wrote to Darwin to personally express support for his views and offered to send photographs of German colleagues who also admired him, Darwin immediately replied and sent, under separate cover, half a dozen copies of his own portrait.[141]

The portrait that Haast sent to Darwin in 1863 was most likely a studio photograph taken in 1859, as he appears to have had no other formal portraits taken during this early stage of his career; furthermore, he clearly favoured this image of himself as he had sent copies of it – sometimes autographed – to other colleagues and acquaintances (Figure 4.18). Consequently, Darwin would have been well aware that the person posing with the moa skeletons in the later photographs he received was not just any human figure included as a comparison of scale but Haast himself, who clearly desired to be personally identified with new discoveries and the ongoing research into the spectacular moa.[142] Haast's deliberate amalgamation of the two genres of portrait photography and scientific documentation was clearly devised with a similar purpose of self-promotion and network building as that of Gerard Krefft's most striking portraits, in which Krefft posed variously with a snake stretched across his lap and then with a manta ray to effectively demonstrate some of his research and achievements at the Australian Museum (Figure 2.5, Plate 5). Haast's request to Darwin for a portrait was not acknowledged immediately and was in fact reiterated several times over the following year or so.[143] In contrast, Owen, who had also posed with moa on a number of occasions, employing the same device in his own cycle of portraits, was well aware of the significant value and recognition that could be gained from this simple visual conceit. As he wrote to Haast soon after receiving the same photographs as Darwin: 'I wish I could have received assurance that the "spectator" in the collective Photograph was the accomplished Geologist & Naturalist whom I have now the pleasure to address …'.[144]

Conclusion

In the nineteenth-century museum, and in science and culture in general, the significance of the moa was manifold. In one respect, the discovery, display and representation of extinct fauna advertised imperial authority and prowess: the 'annexation of fresh territory to the Empire of the Known', as Owen characterised it.[145] Yet, it was equally symbolic of the desire for a broader recognition of local or colonial natural history as it represented a significant and unique contribution to the natural sciences that was visually impressive enough to distinguish New Zealand in the cultural and scientific landscapes of the world.[146] Another, although not entirely separate, sector was non-specialist producers and consumers of natural history participating in their own transnational and intercolonial effort to communicate different experiences and new knowledge of the natural world.[147]

Accounts of the reception of the moa in European culture have often centred on the figure of Richard Owen and the privileged position he occupied in imperial science emanating from his positions within metropolitan institutions, notably London's Hunterian Museum and the British Museum. For instance, Jacob Gruber claims that Owen surrounded himself with a 'Moa cottage industry' that not only furnished him with a ready supply of specimens for his own research but that he 'studiously processed for an ever-expanding community of professionals'.[148] As a classic narrative of imperial discovery and the dominance of metropolitan science, this portrayal underestimates the significant input from provincial and colonial locations, as well as the interconnections between individuals and institutions across a range of sites.

This chapter has highlighted the significant role played by photographs and other images representing extinct fauna in the institutionalisation of natural history – both in promoting natural history as public spectacle and as a potent symbol of colonial as well as imperial aspirations. According to the Royal Colonial Institute, the extinct animals of Great Britain's colonies represented a window into the 'dark backward or abysm of time' and also offered a new category of antiquities that could give 'the rising generation in Australia and New Zealand ... an interest in the land which they inhabit, and ... cause patriotic feeling to arise in their minds'.[149] From this perspective, the depiction of extinct nature from the colonies represented a unique amalgam of temporal and spatial distance and, like the Victorian museum, was 'capable of articulating two temporal narratives'.[150] It was a glimpse back at the prehistoric world that had considerable significance for the prospects of colonial society far into the future. The extensive circulation of images and fossil specimens and the manner in which they were presented in public museums and popular publications had extraordinary immediacy for audiences across the globe. By tracing the trajectories of individual specimens and images, I have charted the cultural biography and popular reception of the moa as a mode of commoditisation, which, in turn, can be seen reflected in a range of relationships between individuals, institutions and related objects. With the boundary between objects and people having recently been conceptualised as more fluid than previously acknowledged, it is now clearer than ever that the interpretation of museum objects is contingent on a range of factors and processes, both cognitive and cultural.[151] Like other artefacts, the meanings and status of objects from nature routinely shifted over space and time and were affected by changing social and intellectual circumstances both within and outside the museum framework.

Reflecting the typology of mid-century colonial museums, the formation and early progress of the Canterbury Museum in Christchurch from the early 1860s provides the primary focus for this case study. Having been annexed by the British in 1840, the fledgling New Zealand colony contrasts with Britain's older territories for being a distinctively Victorian creation and, as a consequence, its early values, ambitions, initiatives and institutions offer special insights into that period.[152] Museum-based naturalists such as Julius von Haast, while situated at the periphery of imperial science, were able to operate at a level that transformed the practice of local natural history into a global enterprise and brought their museums into an international network. A crucial factor in this process was the fascination that attached to the moa itself and the kudos that could be derived from that interest.

The moa attained something of a mythical status in the Victorian imagination. The monstrous size of its fossilised bones fuelled exaggeration and miscalculation, with some speculating that the tallest of the species might stand as much as 5 metres in height. Furthermore, for many decades, the possibility of finding live moa seemed to be a legitimate expectation. As London's *New Zealand Journal* observed, there was an anticipation that moas would eventually be 'seen striding among the emus and ostriches in the Regent's Park'.[153] Not surprisingly, throughout this period, moa skeletons were displayed prominently among the most prized specimens in public museums and featured in popular and scientific publications. For disinterested consumers as well as for ardent imperialists, the moa and other discoveries of extinct fauna came to be viewed as a justification for Britain's colonial possessions. For a growing number of naturalists, the addition of the moa to the fossil record revealed New Zealand to be a testing ground for evolution.[154] As Blanchard Jerrold advised in his 1852 guide to the British Museum, it was not just the extreme rarity of the giant bird that the visitor should note – 'now almost extinct', he surmised – but also its great importance to men of science and to contemporary theories of nature.[155] Such vestiges of deep time proved to be extremely adaptable and able to be modified for a variety of theoretical messages, and they could also be circulated in the popular arena or used outside the realm of pure scientific inquiry for social or political purposes.[156] In this respect, the role of the museum, photography and mass media in the production, display and circulation of images and specimens from natural history were often much more influential, and concomitantly less predictable, in their effect on the broader culture than is generally acknowledged.

Notes

1 Alexander von Humboldt, *Personal Narrative of Travels to the Equinoctial Regions of the New Continent*, vol. 1, Henry G. Bohn, London, 1852, p. xxi.
2 William Cowper, *The Diverting History of John Gilpin: Shewing that He Went Further than Intended, and Came Safe Home Again*, 1782.
3 Dinornis Sumnerensis (Alexander McKay), *The Canterbury Gilpin; or, The Capture and Flight of the Moa: A Poem*, James Hughes, Wellington, 1880, pp. 3–4.
4 P. H. Gosse, *The Romance of Natural History*, Gould & Lincoln, Boston, 1861; Ferdinand Hochstetter, *New Zealand: Its Physical Geography, Geology and Natural History*, J. G. Cotta, Stuttgart, 1867.
5 Tim Barringer, 'Victorian Culture and the Museum: Before and After the White Cube', *Journal of Victorian Culture*, vol. 11, no. 1, 2006, p. 134; Elizabeth Prettejohn, *The Art of the Pre-Raphaelites*, Princeton University Press, Princeton, 2000, pp. 89–96.
6 Martin Rudwick, *Scenes from Deep Time*, University of Chicago Press, Chicago and London, 1992, p. 1; Michael Freeman, *Victorians and the Prehistoric*, Yale University Press, New Haven and London, 2004, p. 227.
7 Kate Flint, *The Victorians and the Visual Imagination*, Cambridge University Press, Cambridge, 2002, p. 139.
8 Lynda Nead, *Victorian Babylon: People, Streets and Images in Nineteenth-Century London*, Yale University Press, New Haven and London, 2005, p. 8.
9 Freeman, *Victorians and the Prehistoric*, pp. 227–8.
10 E. Bulwer Lytton, *England and the English*, vol. 2, J. J. Harper, New York, 1833, p. 165 (emphasis in original).
11 Barbara Black, *On Exhibit: Victorians and their Museums*, University of Virginia Press, Charlottesville, 2000, p. 99.
12 Rudwick, *Scenes from Deep Time*, p. 220.
13 Georges Cuvier, *Recherches sur les Ossemens Fossiles de Quadrupèdes*, Déterville, Paris, 1812; Dorinda Outram, *Georges Cuvier: Vocation, Science and Authority in Post-Revolutionary France*, Manchester University Press, Manchester, 1984; Nicolaas Rupke, *Richard Owen: Victorian Naturalist*, Yale University Press, New Haven, 1994.
14 Rudwick, *Scenes from Deep Time*, pp. 222–3.
15 For 'virtual witnessing', see Steven Shapin and Simon Schaffer, *Leviathan and the Air-Pump: Hobbes, Boyle, and the Experimental Life*, Princeton University Press, Princeton, 1985.
16 Martin Rudwick, *Scenes from Deep Time*, p. 223.
17 Joseph Hooker to James Hector, 21 July 1864 – John Yaldwyn and Juliet Hobbs (eds.), *My Dear Hector: Letters from Joseph Dalton Hooker to James Hector 1862–1893*, Technical Report 31, Museum of New Zealand Te Papa Tongarewa, Wellington, 1998, p. 50.

18 Thomas Allis to Richard Owen, 23 June 1864: f.113; 14 November 1864: f.115–116; 16 November 1864: f.117–118, Owen correspondence, Natural History Museum, London; Thomas Allis, 'Notice of a Nearly Complete Skeleton of a *Dinornis*, Presented by Dr Gibson to the Yorkshire Philosophical Society', *Journal of the Proceeding of the Linnean Society – Zoological*, vol. 8, no. 29, 1864, pp. 50–2; Thomas Allis, 'Further Note on a Skeleton of a *Dinornis* Owen, in the York Museum', *Journal of the Proceeding of the Linnean Society – Zoological*, vol. 8, no. 31, 1865, pp. 140–1.

19 W. B. Clarke, '*Dinornis* an Australian Genus', *Geological Magazine*, vol. 6. no. 26, 1869, pp. 383–4; W. B. Clarke, 'On Dromornis australis (Owen): a new fossil bird of Australia', *Journal and Proceedings of the Royal Society of New South Wales*, no. 11, 1878, pp. 41–49.

20 W. B. Clarke, 'Dinornis – To the Editor of the Herald', *Sydney Morning Herald*, 20 May 1869, p. 3; Gerard Krefft, 'To the Editor of the Herald', *Sydney Morning Herald*, 19 May 1869, p. 13.

21 W. B. Clarke to Owen, 9 June 1876: f.150–1, Owen correspondence, Natural History Museum, London; see also Gerard Krefft to Owen, 25 March 1870: f.33–8, Owen correspondence, Natural History Museum, London.

22 Johann Franz Julius Haast was born in Bonn, 1 May 1822. From 1835 to 1838, he attended Höhere Bürgerschule in Köln, a secondary institution aimed primarily at training prospective merchants and engineers, where his studies included natural history, geography and English. He then returned to Bonn and undertook a commercial apprenticeship, possibly as a mining technician, and attended lectures on geology, chemistry and mineralogy at Bonn University as part of that training. From 1841, he travelled extensively in Europe before taking up a position in the book trade in Frankfurt. In 1857, he was commissioned by the English ship owners Willis, Gann & Co to travel to New Zealand to investigate the prospects for large-scale emigration by Germans; he arrived in Auckland on 21 December 1858, the day before the Austrian frigate *Novara* sailed into port as part of its global scientific expedition sponsored by Emperor Franz Josef, which led him to join Ferdinand Hochstetter's geological survey of the Auckland and Nelson regions. Hochstetter was temporarily enticed away from the *Novara* expedition to undertake this nine-month exploration, which was the first to describe and interpret the geology of New Zealand – Colin J. Burrows, *Julius Haast in the Southern Alps*, Canterbury University Press, Christchurch, 2005; Heinrich F. von Haast, *The Life and Times of Julius von Haast*, H. F. von Haast, Wellington, 1848; Peter B. Maling, 'Haast, Johann Franz Julius von', from the *Dictionary of New Zealand Biography: Te Ara - the Encyclopedia of New Zealand* – http://www.TeAra.govt.nz/en/biographies/1h1/haast-johann-franz-julius-von (accessed 2 April 2011).

23 Julius von Haast, 'On the Origin and Early History of the Canterbury Museum: Being the Annual Address', *Transactions of the Philosophical Institute of Canterbury*, 1881, pp. 503–16.

24 Haast, 'On the Origin and Early History of the Canterbury Museum', p. 504.

25 Sally Burrage, 'Exchange of Native Bird Skins, Skeletons and Eggs by the Canterbury Museum from 1869–1913', *Records of the Canterbury Museum*, vol. 15, 2001, pp. 1–7.

26 Haast, 'On the Origin and Early History of the Canterbury Museum', p. 209.

27 Jacob W. Gruber, 'The Moa and the Professionalizing of New Zealand Science', *Turnbull Library Record*, vol. 20, no. 2, 1987, pp. 61–100.

28 Susan Sheets-Pyenson, *Cathedrals of Science: The Development of Colonial Natural History Museums during the Late Nineteenth Century*, McGill-Queens's University Press, Kingston and Montreal, 1988; Ruth Barton, 'Haast and the Moa: Reversing the Tyranny of Distance', *Pacific Science*, vol. 54, no. 3, pp. 251–63.

29 Jacob W. Gruber, 'From Myth to Reality: The Case of the Moa', *Archives of Natural History*, vol. 14, no. 3, 1987, p. 339.

30 Haast, 'On the Origin and Early History of the Canterbury Museum', p. 504.

31 H. F. von Haast, *The Life and Times of Sir Julius von Haast*, p. 220.

32 Canterbury Museum Records: Files kept by Directors previous to 1948; Reports to the Provincial Council, 4/1; Box 1, Folder 1 – *Report on the Geological Survey of the Province of Canterbury. Journal of Proceedings, Session XXII, 1864,* p. 11.

33 Haast, 'On the Origin and Early History of the Canterbury Museum', p. 505.

34 Gustav Jaeger, *Bericht Uper Ein Fast Vollstandiges Skelet Von Palapteryx Ingens Uber Dessen Restauration Und Die Davon Angefertigten Gypsabgusse; Mit Einigen Bemerkungen Uber Die Aufstellung Der Vogelskelete Uberhaupt*, Wilhelm Braumuller, Wien, 1863.
35 Sheets-Pyenson, *Cathedrals of Science*, p. 25.
36 Julius von Haast, *Report of a Topographical and Geological Exploration of the Western Districts of the Nelson Province, New Zealand*, C. & J. Elliott, Nelson, 1861. The list of 40 or so individuals and organisations to whom Haast sent his Nelson report included Hochstetter; Murchison; Hooker; Darwin; the Royal, Zoological and Linnean societies; the Geological and Geographical societies of London and Paris; the Bodleian and Trinity College libraries; and the editors of London's *Times* and *Saturday Review* – Mark Caudel, *Julius Haast: Towards a New Appreciation of His Life and Work*, MA thesis, University of Canterbury, 2007, p. 97.
37 Haast to Owen, 27 August 1861: f.113–4, Owen correspondence, Natural History Museum, London.
38 Lyell to Haast, 25 December 1861: MS Papers 0037116, Haast family Papers, Alexander Turnbull Library, Wellington.
39 Julius Haast, 'Notes on the Mountains and Glaciers of the Canterbury Province, New Zealand', *Journal of the Royal Geographical Society*, vol. 34, 1864, pp. 87–96; Haast, *Sir Julius von Haast*, pp. 975–1022.
40 Haast to W. J. Hooker, 9 June 1862 – quoted in Haast , *Sir Julius von Haast*, p. 213.
41 Idem.
42 Haast, 'Notes on the Mountains and Glaciers of the Canterbury Province'; Julius Haast, 'Altitude Sections of the Principal Routes between the East and West Coasts of the Province of Canterbury, New Zealand: Across the Southern Alps', *Journal of the Royal Geographical Society*, vol. 37, 1867, pp. 328–38.
43 Editor's note, *Journal of the Royal Geographical Society*, vol. 34, 1864, p. 92; J. S. Gully 'Gully, John' from the *Dictionary of New Zealand Biography: Te Ara - the Encyclopedia of New Zealand* – http://www.TeAra.govt.nz/en/biographies/1g25/gully-john1 (accessed 2 April 2011).
44 Gully to Haast, 4 May 1863: Haast papers, MS Papers 37/68, Alexander Turnbull Library.
45 Roderick Murchison, 'President's Address', *Journal of the Royal Geographical Society*, 1864, p. cliv.
46 Idem.
47 Julius Haast, *Report on the Formation of the Canterbury Plain: With a Geological Sketch-map and Five Geological Sections*, Press Office, Christchurch, 1864; Julius Haast, *Report on the Headwaters of the River Rakaia: With Twenty Illustrations,* James Edward Fitzgerald, Govt. Printer, Christchurch, 1866; Julius Haast, *Reports of Geological Explorations during 1871–2: With Maps and Sections*, George Didsbury, Wellington, 1872; Julius Haast, *Reports of Geological Explorations during 1873–4: With Maps and Sections*, James Hughes, Wellington, 1877.
48 Darwin to Haast, 22 January 1863: letter no. 3935, Darwin Correspondence Project Database, http://www.darwinproject.ac.uk/entry-3935 (accessed 9 April 2011).
49 M. P. K. Sorrenson, 'Mantell, Walter Baldock Durrant, 1820–1895', *Dictionary of New Zealand Biography* – http://www.teara.govt.nz/en/biographies/1m11/1 (accessed 14 March 2011).
50 Walter Mantell to Gideon Mantell, 3 February–7 April 1847: ms papers 0083-419, Mantell family collection, Alexander Turnbull Library, Wellington.
51 Walter Mantell to Gideon Mantell, 12 September 1849: ms papers 0083-419, Mantell family collection, Alexander Turnbull Library, Wellington.
52 Sorrenson, 'Mantell, Walter Baldock Durrant, 1820–1895'.
53 James Brown, *Mr. W. B. D. Mantell, An amplification of his well known discovery of Moa remains!* c.1851, pencil on paper. Hocken Collections, Uare Taoka o Hakena, University of Otago.
54 Burrage, 'Exchange of Native Bird Skins, Skeletons and Eggs', p. 1.
55 Canterbury Museum Records: Donations and Exchanges 1869–1913, Canterbury Museum.
56 Sharon Macdonald, 'Museums, National, Postnational and Transcultural Identities', *Museum and Society*, vol. 1, no. 1, 2003, p. 3.

57 Haast, 'On the Origin and Early History of the Canterbury Museum', p. 50.
58 Idem.
59 CP 656b/3 Report for Canterbury Museum by Trustees, 1873.
60 Haast, 'On the Origin and Early History of the Canterbury Museum', p. 509.
61 Susan L. Star and James R. Griesemer, 'Institutional Ecology, "Translations" and Boundary Objects: Amateurs and Professionals in Berkeley's Museum of Vertebrate Zoology, 1907–39', *Social Studies of Science*, vol. 19, 1989, pp. 387–420.
62 Francis Lucian Reid, 'The Democratic Politician does not Trouble himself with Science': Class and Professionalism in the New Zealand Institute, 1867–1903', *Tuhinga*, no. 16, pp. 21–23.
63 Barbara Black, *On Exhibit*, p. 106.
64 Sue Ann Prince, 'Stuffing Birds, Pressing Plants, Shaping Knowledge: Natural History in North America, 1730–1860', *Transactions of the American Philosophical Society*, vol. 93, no. 4, 2003, p. 1.
65 Rudwick, *Scenes from Deep Time*, p. 148.
66 Owen, *On the Extent and Aims of a National Museum of Natural History*, pp. 10–11 and 13.
67 David Murray, *Museums: Their History and their Use*, vol. 1, James Maclehose, Glasgow, 1904, p. 1 and Chapter 14.
68 Ibid., pp. 186 and 208.
69 Edward Forbes, 'On the Educational Uses of Museums' in Hugh H. Genoways and Mary Andrei (eds.), *Museum Origins: Readings in Early Museum History and Philosophy*, Left Coast Press, Walnut Creek, 2008.
70 W. Stanley Jevons, 'The Use and Abuse of Museums' in *Methods of Social Reform and other Papers*, Macmillan & Co, London, 1883, pp. 61–2.
71 Tony Bennett, *The Birth of the Museum*, Routledge, London and New York, 1995, pp. 39–43; Carol Duncan and Alan Wallach, 'The Universal Survey Museum', *Art History*, vol. 3, no. 4, 1980, pp. 448–69.
72 Alfred Russel Wallace, 'Museums for the People', *Macmillan's Magazine*, vol. 19, no. 111, pp. 244–55, in Genoways and Andrei, *Museum Origins*, p. 206.
73 Susan Stewart, *On Longing: Narratives of the Miniature, the Gigantic, the Souvenir, the Collection*, Duke University Press, Durham and London, 1993, p. 153.
74 Owen, On *the Extent and Aims of a National Museum of Natural History*, p. 67.
75 Ibid., p. 68.
76 C. G. Carus, *The King of Saxony's Journey through England and Scotland in the Year 1844*, Chapman & Hall, London, 1846, p. 60.
77 Owen, *On the Extent and Aims of a National Museum of Natural History*, p. 68.
78 Henry A. Ward, *Catalogue of Fossils: From the Principal Museums of Europe and America*, Benton and Andrews. Rochester, 1866, p. viii (emphasis in original).
79 Ibid., p. i; Haast to the Secretary for Public Works, Christchurch, 19 September 1865–30 June 1868, CP 349a-b Canterbury Museum 1864–7/9, Archives New Zealand.
80 Ward, *Catalogue of Fossils*, pp. i–ii.
81 Ibid., pp. iii–iv.
82 Ibid., p. vii.
83 Ibid., pp. v and 47–9; W. Blanchard Jerrold, *How to See the British Museum in Four Visits*, Bradbury & Evans, London, 1852, p. 3.
84 D. L. Mundy, 'Photographic Experiences in New Zealand', *The Photographic News*, 18 December 1874, pp. 602–3.
85 Haast to the secretary for public works, Christchurch, 15 November 1867, CP 349b Canterbury Museum 1864–9, 1617/19.12.65, mss, Archives New Zealand.
86 Idem.
87 Alexander Agassiz to Haast, 16 November 1868, Agassiz letter books, F890: vol. 5, f.36, Museum of Comparative Zoology, Harvard University.
88 Museum of Comparative Zoology, Harvard University – Agassiz letter books, F890: vol. 2, f.201 – Louis Agassiz to Owen, 27 April 1863.
89 Owen, *On the Extent and Aims of a National Museum of Natural History*, p. 108.
90 Elizabeth Edwards, *Raw Histories, Anthropology and Museums*, Berg, Oxford, 2001, pp. 30–32.

91 Charles Darwin, *On the Origin of Species by Means of Natural Selection*, John Murray, London, 1859, p. 339.
92 Darwin, *On the Origin of Species*, p. 391.
93 Darwin to Haast, 28 January 1868: Haast family papers, MS-Papers-0037-051, Alexander Turnbull Library, Wellington.
94 Haast to Darwin, 9 December 1862, Darwin Correspondence Project Database – http://www.darwinproject.ac.uk/entry-3851 (accessed 9 February 2010).
95 Darwin to Haast, 18 July (1863), Darwin Correspondence Project Database – http://www.darwinproject.ac.uk/entry-4245 (accessed 9 February 2010).
96 Charles Darwin, *On the Origin of Species by Means of Natural Selection*, John Murray, London, 1866, pp. 442–3.
97 Haast to Darwin, 4 December 1867, Darwin Correspondence Project Database – http://www.darwinproject.ac.uk/entry-5705 (accessed 9 February 2010).
98 Haast to Darwin, 4 December 1867.
99 Idem.
100 Julius Haast, 'On the Measurements of Dinornis Bones, Obtained from Excavations in a Swamp, Situated at Glenmark, on the Property of Messrs Kermode & Co, up to 15, February 1868', *Transactions and Proceedings of the Royal Society of New Zealand*, vol. 1, 1868, pp. 80–9.
101 Idem.
102 'The Canterbury Museum, New Zealand', *Illustrated London News*, 6 February 1868, p. 144.
103 Idem.
104 Idem.
105 'The New Zealand Industrial Exhibition', *Illustrated London News*, 13 May 1865, p. 462.
106 'The Dinornis, or Gigantic Bird of New Zealand', *Illustrated London News*, 10 June 1865, p. 54.
107 Gyan Pakrash, 'Science "Gone Native" in Colonial India', *Representations*, no. 40, 1992, p. 155.
108 Edgar Thurston, *Administration Report of the Government Central Museum for 1894–95*, p. 1 – quoted in Pakrash, 'Science "Gone Native" in Colonial India', pp. 170–1.
109 *Centenary Souvenir: Madras Government Museum (1851-1951)*, Principal Commissioner of Museums, Chennai, 1999; *Hand Book on Government Museum, Chennai*, S. S. Jawahar, Chennai, 2011.
110 J. R. Asokan, Curator, Government Museum, Chennai – correspondence with author, 22 January 2010.
111 Douglas M. Peers, 'Colonial Knowledge and the Military in India, 1780–1860', *The Journal of Imperial and Commonwealth History*, vol. 33, no. 2, 2005, pp. 157–80.
112 John McCosh, *Advice to Officers in India*, W. H. Allen, London, 1856, p. 6 – quoted in Peers, 'Colonial Knowledge and the Military in India, 1780–1860', p. 158.
113 Captain J. Mitchell, *Catalogue of the Mollusca in the Collection of the Government Central Museum, Madras*, William Thomas, Madras, 1867; 'Madras Government Proceedings', *Quarterly Journal of Microscopical Science*, vol. S2–4, no. 15, 1864, pp. 219–32.
114 *Annual Report of the Superintendant of the GSI and Museum of Economic Geology, 1859* – quoted in Savithri Preetha Nair, 'Science and the Politics of Colonial Collecting: The Case of Indian Meteorites, 1856–70', *British Journal for the History of Science*, vol. 39, no. 1, 2006, p. 101.
115 Idem.
116 Ibid., pp. 99-101; Andrew Grout, 'Oldham, Thomas (1816–1878)', *Oxford Dictionary of National Biography*, Oxford University Press, 2004 – http://www.oxforddnb.com/view/article/20691 (accessed 10 April 2011).
117 Jesse Mitchell [c/- John Colebrooke, (Late surgeon) Madras Army] to Owen, 26 December 1865: f.328–9, vol. viii, Owen correspondence, Natural History Museum, London (emphases in original).
118 Richard Owen, 'On *Dinornis*, Part XII: containing a description of the femur, tibia, and metatarsus of *Dinornis maximus* Owen', *Transactions of the Zoological Society of London*, vol. 6, 1869, pp. 497–500; the same images were also later reproduced as plates

79 and 80 in Richard Owen, *Memoirs of the Extinct Wingless Birds of New Zealand*, John Van Voorst, London, 1879.

119 Haast to Owen, 26 October 1869: f.125–31 & f.33–40, vol. xiv, Owen Correspondence, Natural History Museum, London.

120 Capt. J. Michael to Owen, 27 March (1867): f.125a-6, vol. xix, Owen Correspondence, Natural History Museum, London.

121 Nelson Goodman, *Ways of Worldmaking*, Hackett, Indianapolis, 1978.

122 Haast, *Sir Julius von Haast*, p. 536.

123 Haast, Provincial Geologist, to the Secretary for Public Works, Christchurch, 20 February 1868: 266/20.2.68: CP 349b Canterbury Museum 1864–69, Archives New Zealand; 'Correspondence Relative to the Arrangement of the Collections, Canterbury Museum by Dr. J. Haast, Provincial Geologist', 30 June 1868, 11 Session XXX No. 18: CP 349b Canterbury Museum 1864–69, Archives New Zealand.

124 W. H. Hooker to J. D. Hooker, 19 February–[8 March] 1868 [1869]: MS 4023, National Library of Australia, Canberra.

125 Alfred Barker to Matthias Barker, 10 August 1858; 4 January 1861; 12 January 1864: Folders 1–6, Dr. A. C. Barker letters, Canterbury Museum; C. C. Burdon, *Dr. A. C. Barker, 1819–1873, A Photographer, Farmer, Physician*, McIndoe, Dunedin, 1972; Joan Woodward, *A Canterbury Album*, Te Waihora Press, Lincoln, 1987.

126 Alfred Barker to Matthias Barker, 5 March 1862; 13 October 1863; 12 July 1864: Dr. A. C. Barker letters, Canterbury Museum, Christchurch.

127 Alfred Barker to Matthias Barker, 3 October 1867: Folder 9, Item 101, Dr. A. C. Barker letters, Canterbury Museum, Christchurch (emphasis in original).

128 Alfred Barker to Matthias Barker, 10 May 1868: Folder 10, Item 110, Dr. A. C. Barker letters, Canterbury Museum, Christchurch.

129 Hardwicke Knight, *Photography in New Zealand: A Social and Technical History*, John McIndoe, Dunedin, 1971, pp. 32–35.

130 Haast, *Sir Julius von Haast*, pp. 463 and 626 passim.

131 Mundy to Haast, 5 May 1875 and 27 July 1875: Daniel Mundy correspondence, Ref no: MS-Papers-0037-085, Alexander Turnbull Library; Daniel Mundy, *Rotomohana; and the Boiling Springs of New Zealand. A Photographic Series of Sixteen Views by D. L. Mundy with Descriptive Notes by Ferdinand Hochstetter*, Sampson, Low, Marston, Low & Searle, London, 1875.

132 Knight, *Photography in New Zealand*, pp. 32–35.

133 Clippings: Folder 18, 1871–72, Item 201, Dr. A. C. Barker letters, Canterbury Museum, Christchurch.

134 Haast, *Sir Julius von Haast*, p. 142.

135 Haast to Darwin, 9 December 1862: Darwin Correspondence Project Database – http://www.darwinproject.ac.uk/entry-3851 (accessed 9 February 2010); see also Darwin to J. D. Hooker, 21 September 1862: Darwin Correspondence Project Database – http://www.darwinproject.ac.uk/entry-3735 (accessed 9 February 2010).

136 Haast to Darwin, 21 July [– 7? August] 1863: Darwin Correspondence Project Database – http://www.darwinproject.ac.uk/entry-4249 (accessed 9 February 2010).

137 See Geoffrey Belknap and Sophie Defrance, 'Photographs as Scientific and Social Objects in the Correspondence of Charles Darwin' in Elizabeth Edwards and Christopher Morton (eds.), *Photographs, Museums, Collections: Between Art and Information*, Bloomsbury Academic, London, 2015.

138 Darwin to Fritz Muller, 11 January 1866: Darwin Correspondence Project, 'Letter no. 4972' – http://www.darwinproject.ac.uk/DCP-LETT-4972 (accessed 29 November 2016).

139 Charles Kingsley to Darwin, 30 May 1865: Darwin Correspondence Project Database – http://www.darwinproject.ac.uk/entry-4843 (accessed 6 July 2011).

140 Darwin to Haeckel, 6 December [1865]: Darwin Correspondence Project Database – http://www.darwinproject.ac.uk/entry-4947 (accessed 11 January 2011).

141 Haeckel to Darwin, 11 January 1866; Darwin to Haeckel, 20 January 1866: Darwin Correspondence Project Database – http://www.darwinproject.ac.uk/entry-4973; http://www.darwinproject.ac.uk/entry-4980 (accessed 11 January 2011).

142 Haast to Darwin, 4 December 1867: Darwin Correspondence Project Database – http://www.darwinproject.ac.uk/entry-5705 (accessed 9 February 2010).

143 Haast to Darwin, 27 September 1865: Darwin Correspondence Project Database – http://www.darwinproject.ac.uk/entry-4900 (accessed 9 February 2010); Darwin to Haast, 26 December 1865: Darwin Correspondence Project Database – http://www.darwinproject.ac.uk/entry-4956 (accessed 9 February 2010).

144 Owen to Haast, 23 January 1868: Alexander Turnbull Library (Originals at MS-Papers – 0037, Folders 120–130, Copy Micro 0717, Reel 10).

145 Richard Owen, Hunterian anniversary dinner speech: Royal College of Surgeons, mss Misc./Hunt. D (1) – quoted in Rupke, *Richard Owen: Victorian Naturalist*, p. 81.

146 Joyce E. Chaplin, 'Nature and Nation: Natural History in Context', *Transactions of the American Philosophical Society*, vol. 93, no. 4, 2003, pp. 75–95.

147 Ibid., pp. 78–81.

148 Gruber, 'From Myth to Reality', p. 349.

149 John Dennistoun Wood in Richard Owen, 'On the Extinct Animals of the Colonies of Great Britain', *Proceedings of the Royal Colonial Institute*, vol. 10, 1879–79, Sampson, Low, Marston, Searle & Rivington, London, 1879, p. 293.

150 Macdonald, 'Museums, National, Postnational and Transcultural Identities', p. 3; see also Nicolaas Rupke, 'Metonymies of Empire: Visual Representations of Prehistoric Times, 1830–90' in Renato Mazzolini (ed.), *Non-Verbal Communication in Science prior to 1900*, Florence, 1993, pp. 513–28.

151 Igor Kopytoff, 'The Cultural Biography of Things: Commoditization as Process' in Arjun Appadurai (ed.), *The Social Life of Things: Commodities in Cultural Perspective*, Cambridge University Press, Cambridge, 1986, p. 64; Janet Hoskins, 'Agency, Biography and Objects' in Christopher Tilley, Webb Keane, Susanne Kuechler, Mike Rowlands and Patricia Spyer (eds.), *Handbook of Material Culture*, Routledge, London, 2006, pp. 74–84.

152 Michael King, *The Penguin History of New Zealand*, Penguin, Auckland, 2003, pp. 151–67.

153 *The New Zealand Journal*, 30 March 1844 – quoted in Atholl Anderson, *Prodigious Birds: Moas and Moa-Hunting in Pre-Historic New Zealand*, Cambridge University Press, Cambridge, 2003, p. 2.

154 Owen, *On the Extinct Animals of the Colonies of Great Britain*, p. 273.

155 Jerrold, *How to See the British Museum in Four Visits*, p. 32.

156 Rudwick, *Scenes from Deep Time*, pp. 170 and 253.

Conclusion

On the evening of 14 August 1872, the distinguished naturalist William Carpenter delivered the presidential address at the annual assembly of the British Association for the Advancement of Science. The association was, by this time, more than 40 years old, and its annual meetings had addressed almost every aspect of Victorian scientific inquiry and had provided the occasion for various scientific controversies and announcements. As the 'principal public scientific pronouncement of the year', the presidential address aimed to set the tone for the conference and to reflect upon the state of contemporary scientific activity and its place in British life and culture.[1] Accordingly, Carpenter's speech addressed a subject of the most fundamental importance to Victorian science, namely changes in the way nature was being interpreted and the different methodologies this change entailed. The title of his lecture, *Man the Interpreter of Nature*, drew directly upon Francis Bacon's legacy of empirical observation and the inductive approach to scientific method.[2] The Baconian process of describing and analysing only one's own observations so as to identify visible similarities between phenomena was in fact rapidly yielding to other approaches that allowed for more speculation and imaginative thinking.[3] However, in his speech, Carpenter argued that, by applying Bacon's methodology within the context of a more interpretative framework, there was still much scope for perpetuating and building upon this time-honoured approach. As Charles Kingsley declared, Baconianism was the most British of scientific methodologies and one that was sanctioned by God and generations of English scientists.[4]

Carpenter had also appealed to British national pride in an 1871 lecture delivered at the Royal Institution, in which he had warned that Britain's pre-eminence in the natural sciences was at risk of being overshadowed by the considerable advances being made by other nations.[5] While his 1872 presidential address championed English scientific traditions, it would be wrong to conclude that the British Association or Carpenter himself were concerned exclusively with promoting science as a nationalist enterprise or, for that matter, were only attentive to the interests and procedures of elite metropolitan practitioners. The endeavours of Britain's colonies, as well as its counties; the vital role played by the empire in the nation's economic, intellectual and creative activities; and its wider connections within the global scientific community were a constant feature of the British Association's annual meetings, as was the concern to bring science to a non-specialist audience. Many participants either travelled across the world to attend these meetings in person or, as corresponding members, sent papers to be read on their behalf.

The meetings routinely included lecturers who promoted new procedures, technologies and institutional reforms that would enhance the ways in which scientists produced and communicated new knowledge. Carpenter himself exemplified this tenet, making it his duty throughout his career, as Roger Smith notes, 'to mediate between scientific and lay opinion through books, articles and lectures'.[6] At an earlier meeting in Norwich in 1868, the then president Joseph Hooker, who was also director of the Royal Botanic Gardens at Kew, reported on the administration of the natural history collections at the British Museum, which he lauded as including the finest zoological collections in the world, and geological and palaeontological collections of 'prodigious extent and value'.[7] While acknowledging that much had been written upon the uses of museums, he declared that the subject was far from being exhausted – particularly, he suggested, as museums remained the most efficient way of teaching zoology and physiology to schools. For this task, Hooker argued, provincial and local museums were indispensable, but, while their natural history collections 'should not contain a single specimen more than is wanted', they needed to be complemented by 'many pictorial illustrations [and] magnified views of the smaller organs and their structure'.[8] Similarly, when John Edward Gray – who was president of the association's zoological and botanical section as well as keeper of zoology at the British Museum – presented the opening address for his section at the 1864 meeting in Bath, he reflected on the subject of public museums and their uses by describing the different ways of arranging and disseminating natural history collections for the general public and for naturalists.[9]

Photography was one of the more prominent new technologies promoted by the association for the purpose of advancing the study and appreciation of natural history. For example, Samuel Highley's presentation to the 1864 gathering in Bath focussed on the uses of photography and the magic lantern for demonstrating natural history and microscopical subjects. The advantages of these technologies over ordinary diagrams, he advised, were their 'truthfulness to nature, even to the minutest details; the impressive character of the image projected, being often stereoscopic in aspect; the lecturer being able to fix the attention of the student upon *one* object at a time; and their portability and cheapness'.[10] As a new mode for interpreting aspects of nature – a fundamental theme in Carpenter's 1872 speech – Highley emphasised photography's important role in the natural sciences, especially as naturalists had begun to use it to execute and publicise some of the more exploratory aspects of their work. Quickly embraced as more than simply a tool to maintain the hegemony of British scientific methods, photography's expanding role in natural history epitomised the way Victorian science itself was being transformed. This was exemplified in the British Association's desire to move science beyond the confines of London's elite societies and into the public realm of museums and other public institutions, a change they achieved by enhancing the social standing and authority of professional scientists, promoting international exchange and encouraging public participation.

Whereas early proponents advocated photography solely as a reliable means of recording observations from nature, a new generation of scientists were becoming attuned to photography's broader influence in contemporary society. As a new mode of scientific documentation, photography was not only expedient but became increasingly indispensable for a much wider range of purposes. This coincided with the move towards professionalisation in the sciences, mass popularisation of the medium and increased public access to the realm of natural history. The rapidly developing field of photography and its diversity of uses, therefore, meant that naturalists' modes of visual production were much broader than has been previously recognised.

The impetus for these alternative uses of photography cannot be fully appreciated by reference to its status in visual culture alone but must be seen as a response to changes in the practice and presentation of natural history itself. These developments include the rise of the modern museum, the professionalisation of the sciences, and the popularisation and increased dissemination of natural science to an expanding middle class. Such issues were aired widely in journals at the time. For instance, in a series of articles published in the *Photographic News* between January 1863 and September 1864, Highley took great pains to demonstrate how photography not only delineated the detail of forms in nature and recorded changes in phenomena over time but also opened up hitherto unseen worlds so as to make whole aspects of nature visible to the naked eye.[11] Moreover, he suggested that the potential uses of photography and the magic lantern in natural history were not confined to specialist application or scholarly instruction but could also extend into the public domain by being 'placed in the open cases of museums ... beside allied objects' or projected in popular presentations on a large enough scale 'to arrest attention'.[12] As any experienced showman might appreciate, the value in appealing to the eye was not simply to present information but also to stir and tantalise the imagination. Highley advocated photography as a means to inform the scientific community about new fields of natural history and also to expand the networks of science ever wider by introducing the wonders of nature to new and diverse audiences.

The various activities featured in the British Association's week-long annual meetings were typical of the array of sights and cultural experiences available to the Victorians. These events, and the founding of the association itself, had coincided with the rise of the public museum and other modes of rational entertainment that showcased scientific discoveries from around the globe. Together, these activities and institutions demonstrated that, in Victorian Britain, the intersecting influences of science, popular culture and the empire could be encountered in any number of different forms and at a variety of locations. The British Association's 1872 conference, for instance, was convened in Brighton, occupying the magnificent and eclectic Royal Pavilion, an orientalist fantasy designed by John Nash for the Prince Regent, later crowned King George IV. With its series of marvellous Mughal-inspired domes, delicate latticework and enchanting Chinoiserie interiors, the choice of the Royal Pavilion as the venue for the association's convention was, without doubt, an unambiguous visual reminder to participants of the considerable influence of Britain's formal and informal empire on the imagination and accomplishments of the nation. It also signalled the important role the association performed in its stated objective 'to promote the intercourse of those who cultivate Science in different parts of the British Empire with one another and with foreign philosophers'.[13]

The meeting was also timed to coincide with the unveiling of the impressive Brighton Aquarium, highlighting the association's objective to bring science to the general public by making it engaging and accessible. The *Illustrated London News* devoted not only its front page but also a long article to publicising the concurrence of these two events and predicted that in its first week – during the association's annual congress – the new aquarium would undoubtedly be visited by thousands of people: 'this establishment, which promises to be a great help to the popular study of an interesting branch of zoology, [will also] supply the seaside loungers with constant means of rational entertainment'.[14] As with previous gatherings of the British Association, Carpenter's presidential speech launched a lively programme encompassing lectures, exhibitions and magic lantern presentations; scientific displays and demonstrations; excursions; and elegant soirées that drew the ordinary public from the provinces and cities as well as members and visitors from all parts of the empire and across the globe, as detailed in the series of pictorial vignettes in *The Graphic*. The newspapers were brimming with reports of visiting local and international celebrities who made appearances during the week, including the Japanese ambassador extraordinary, his Excellency Iwakura Tomomi, who, with a 'dozen other members of his embassy ... had travelled twelve-thousand miles simply to learn'; the recently exiled Emperor Napoleon III and Empress Eugenie, who attended the opening soirée; and the so-described 'Americo-African lion, Mr. Stanley' who, accompanied by his African protégé Kalulu, was ogled, interrogated and applauded throughout the evening's proceedings as one of the 'great travellers, who at the peril of their lives, have added to the stock of human knowledge'.[15] As Carpenter observed, one of the most valuable functions of these yearly meetings was the chance for any and all attendees to make the personal acquaintance of 'those distinguished men whose names are to every cultivator of science, "household words"' and whose 'brilliant example' was the most 'precious influence [on one's] own aspirations'.[16]

Outside these structured and very public events, the ideals of scientific internationalism, public involvement and due recognition of knowledge transmitted between parts of the empire, while pressing issues, were rarely addressed so openly. The most sustained and productive exchanges often came from less public and more informal connections, which were seldom acknowledged beyond those immediate contacts. As well as possessing the unique attributes that Highley was so keen to emphasise, photographs were also endlessly reproducible and eminently suitable for placing in an envelope to complement a letter or sketch as a de facto specimen, to fulfil a request for one's portrait or to verify an observation in the field. They could thus be used to enhance the professional and personal exchanges that took place between naturalists as part of their ever-expanding imperial and global networks, numerous examples of which have been addressed in this book. This was especially the case as the field of natural history and its consumers became an increasingly diverse community during the Victorian era. With a burgeoning middle class, industrialisation and the expansion of communications, transport, trade and immigration during the nineteenth century, naturalists found themselves less and less constrained by geographical, national or social boundaries and, concomitantly, less distinct as an identifiable or cohesive group.

This movement of people, ideas and attitudes effecting the trajectory and expression of empire was not confined to Britain and her colonies but also encompassed other cultures. This was nowhere more apparent than in the domain of natural history, which developed through negotiations between far-flung individuals, museums and other institutions and contributed to a culture of interdependency between centre and periphery. While most naturalists offered to this expanding web their first-hand information and expertise, they were also contributing to a body of knowledge and areas of endeavour that were being transformed in a variety of ways. Their exchanges were very often informal communications transacted directly with other individuals rather than through official government channels. These more informal interactions produced many new and diverse connections involving savants and institutions in other centres of knowledge and power as well as those in Britain.

The presence of naturalists from various backgrounds in this global network presented a complex cultural terrain or, at the very least, introduced another set of interests and loyalties that existed in parallel with those of the British Empire. These non-British naturalists included many Germans who, with their experience of state-supported sciences and science education, travelled extensively and took up positions of authority in significant numbers. Julius von Haast's first foray into the natural history of New Zealand, for example, was as part of fellow German-speaking Ferdinand Hochstetter's government-commissioned geological exploration of the North Island in 1859. Their relationship continued and strengthened over the following decades as a result of the many exchanges of specimens between the Canterbury Museum and the Imperial Museum in Vienna. Similarly, prior to his appointment at the Australian Museum, Gerard Krefft already had an excellent reputation among his German confreres as a first-rate supplier of scientific specimens, having been praised by the German Museum Commission in 1862 for the quality and importance of the Australian fauna he had gathered for their public institutions.[17] In addition to these transnational exchanges, greater intercolonial communication and coordination between naturalists and museums from the 1860s onwards meant that the work and reputations of figures such as Haast and Krefft added materially to the wealth of knowledge being developed in other parts of the periphery.[18] These exchanges were fundamental not only to the research of naturalists but also to the appreciation of natural history by the public at large. To regard the nature and extent of these relationships solely in terms of the transfer of actual specimens means neglecting other significant, although possibly less prominent, factors that also defined photography's role within the practice of colonial natural history.

Such a vast and complex network would in fact be inconceivable without a range of visual and reproductive technologies, new modes of image processing and transmission, and elaborate information systems. In the course of the nineteenth century, the volume of raw data in circulation expanded exponentially. For the Victorians, therefore, the matter of where scientific knowledge originated or was interpreted and deployed – and by whom – became at least as important, and sometimes even more so, than the methods by which it was obtained. The private exchange of images between the centre of imperial science and the periphery sometimes presented an invaluable opportunity for scientists working in the field, often on behalf of museums, to gain a foothold that could help propel them onto the international natural history stage or at least publicise their work through more popular avenues.

Issues of the trustworthiness, reliability and credibility of scientific information were closely related to what was known about the character and circumstances of those individuals and groups who were generating new knowledge: their moral standing, social or cultural background and affiliation with professional institutions and associations or eminent individuals. Even those two great lions of Victorian natural history, Richard Owen and Charles Darwin, sought broader approval for their work to provide momentum for their respective career trajectories. This desire for wider recognition coincided with a growing fascination with celebrity, which was fuelled by the proliferation of photographic portraits of contemporary 'worthies' in commercial publications. Motifs representing naturalists' vocation and achievements were customary features in their portraits, and commercial portraits increasingly became vehicles through which naturalists could communicate specific aspects of their scientific practice to a broader audience. For those working in distant, unfamiliar or even local but specialised environments, information about their surroundings also contributed to their profiles. Thus, the burgeoning trade in studio portraits and commercially produced views, far from being seen as inconsequential to the core practice of nineteenth-century natural history – as has been commonly presumed – should in fact be viewed as an integral part of the process of ascertaining the respectability and authority of naturalists and thus establishing a foundation upon which the merit of their scientific observations could be considered. Their importance is very evident in correspondence between naturalists in which portraits, as well as views of local museums, were as common as other types of information. For example, in order to launch his extensive publicity campaign promoting the new Canterbury Museum, Julius von Haast commissioned commercial and amateur photographers and sought to combine several photographic genres by posing in the museum's public gallery surrounded by an awe-inspiring display of rare, gigantic moa skeletons. The resulting photographs were circulated widely within scientific and museum circles and provided Haast with a pretext for initiating regular correspondence with some of the most prominent scientific figures of the day. Through this process, portraits and views formed part of a complementary web of information that was exchanged so as to establish and promote networks of trust and facilitate the transmission of more conventional scientific data.[19]

While the unidirectional movement of official directives and despatches between the periphery and centre of empire – the 'weighty, broad and permanent conversations of pen and paper', as Alan Atkinson describes them – represents the way imperial power was constituted and impressed upon colonial society, informal exchanges reveal a more nuanced picture comprising alternative and complex flows of information, debate and speculation between individuals and the multifarious uses of photography these prompted.[20] When Gerard Krefft wrote to his colleague John Edward Gray at the British Museum, for example, soon after his appointment as curator at the Australian Museum, he not only attached to his letter 12 small photographs of whale vertebrae and several sketches showing the whale's Atlas bone and another whale specimen obtained from Port Curtis in Queensland but also added his own *carte-de-visite* portrait in anticipation that Gray 'would favour [him] with [his] own [portrait]'. In addition, Krefft enclosed a sketch of a fish for Gray 'to hand to [Albert] Gunther' and offered to send specimens to this German compatriot at the British Museum as he had 'several to spare'.[21] Far from being a reluctant or unseen contributor to the empire's expanding body of scientific knowledge, Krefft sought to put a face to himself and his correspondent Gray in their written exchanges and to be recognised as a co-worker and collaborator among natural history researchers rather than an anonymous, colonial collector. As a self-taught naturalist, he also cultivated his German contacts who were influential in imperial and colonial science, which enabled him to move into the museum world and to advance his standing in professional science.

To gain recognition in the more prestigious aspects of Victorian science, colonial naturalists needed to retain control over discoveries made on behalf of their own museums, especially their most spectacular finds. They achieved this by sending photographs and sketches of some of the most treasured specimens to London and offering the specimens themselves only if duplicates became available. Consequently, although exchanges of innumerable observations and images suggest a seemingly unlimited number and range of specimens available to Gray, Owen and others at the hub of the empire, the production and circulation of an expanding range of photographs and other visual media increasingly became a means for colonial naturalists to receive acknowledgement for the crucial role of the periphery in imperial science. These media allowed naturalists who were building their own local institutions in different parts of the globe and actively developing a culture of interdependence and collaboration to advance their own unique work rather than to merely defer to the centre of imperial science.

As the natural sciences were transformed and multiplied during the Victorian era and came to encompass a much wider field of producers and consumers, the ways in which natural history was represented became correspondingly diverse, leading to the incorporation of a combination of not only new media but also different visual genres. The resulting collections of images might be regarded as unconventional if one considers the visual work of naturalists only as a means of illustrating natural history specimens. Viewed in the context of the broader social and economic imperatives of nineteenth-century natural history, however, they are in fact indicative of the expanding scope and changing purpose of a field that, more than any other science, was venturing increasingly into the public domain.

Once photography was introduced into museums in the 1850s – a period during which the sciences were becoming increasingly institutionalised – it quickly came to play a role in the substantial museum reforms taking place at the time, such as improving public access. In the early years, photography's future in the museum context was far from assured and relied not only on the foresight of particular individuals to recognise its significance but also their influence to promote its use. Nevertheless, photography's increasing accessibility and broad popularity due to its commercialisation meant that the medium was readily adapted to a variety of genres and rapidly became a desirable and practical means for naturalists – both professional and amateur – to circulate an unprecedented quantity and range of information. Rather than being perceived merely as a tool to simplify or supplant the visual processes of naturalists or to verify their intellectual work, photography introduced a productive complexity that facilitated the contributions of a wider range of practitioners and addressed the changing interests of consumers. Photography and its allied technologies, such as the stereoscope, allowed specialists and the general public alike to experience natural history in unprecedented ways. With photography being rapidly embraced by a growing middle class located in all corners of the empire, the medium soon began to record not only nature itself but the changing institutions and social realm of natural history, a process in which photography, in its multifarious applications, was itself an agent of change.

Notes

1 George Basalla, William Coleman and Robert H. Kargon (eds.), *Victorian Science: A Self-Portrait from the Presidential Addresses to the British Association for the Advancement of Science*, Doubleday Anchor, Garden City, 1970, p. 4.
2 Francis Bacon, *The Novum Organum*, Book 1, 1620.
3 Jim Endersby, *Imperial Nature: Joseph Hooker and the Practices of Victorian Science*, University of Chicago Press, Chicago and London, 2008, p. 363 n. 50.
4 Charles Kingsley, 'How to Study Natural History', lecture delivered at Reading, 1849; Richard Yeo, 'Scientific Method and the Image of Science' in Roy MacLeod and Peter Collins (eds.), *Parliament of Science*, Science Reviews Ltd., Northwood, 1981, p. 65.
5 Margaret Deacon, *Scientists and the Sea, 1650–1900*, Ashgate, London, 1991, p. 333.
6 Roger Smith, 'The Human Significance of Biology: Carpenter, Darwin and the *vera causa*' in U. C. Knoepflmacher and G. B. Tennyson (eds.), *Nature and the Victorian Imagination*, University of California Press, Berkeley and Los Angeles, 1977, p. 217.

7 'Address of Joseph Hooker, F. R. S., President', *Report of the British Association for the Advancement of Science*, 1868, p. lxii.
8 Idem.
9 J. E. Gray, 'On Museums, their Use and Improvement', *Annals and Magazine of Natural History*, vol. 14, 1864, pp. 283–97.
10 Samuel Highley, 'On the Application of Photography and the Magic Lantern to Class Demonstrations in Microscopic Science and Natural History', *Report of the Thirty-Fourth Meeting of the British Association for the Advancement of Science Held at Bath in September 1864,* John Murray, London, 1865, p. 98 (emphasis in original).
11 Samuel Highley, 'The Application of Photography to the Magic Lantern Educationally Considered', *The Photographic News*, vol. 7, 6 March 1863, pp. 113–14.
12 Ibid., vol. 7, 23 January 1863, pp. 43–44; 6 February 1863, pp. 68–69; 6 March 1863, pp. 113–14.
13 'Objects and Rules of the Association', *Report of the Forty-Second Meeting of the British Association for the Advancement of Science Held at Brighton in August 1872,* John Murray, London, 1873, p. xvii.
14 'The Brighton Aquarium', *Illustrated London News*, 10 August 1872, p. 123.
15 'British Association at Brighton', *Leicester Chronicle and the Leicester Mercury*, 24 August 1872, p. 2; 'The British Association at Brighton', *The Graphic*, 24 August 1872, p. 166; 'The British Association at Brighton', *Brighton Daily News* 15 August, 1872, p. 2.
16 'Address of William B. Carpenter, M.D., L.L.D., F.R.S., President' in *Report of the Forty-Second Meeting of the British Association for the Advancement of Science Held at Brighton in August 1872,* John Murray, London, 1873, p. lxix.
17 A. Panning, *Mitteilungen aus dem Hamburgischen Zoologischen Museum und Institut*, vol. 54, 1958, p. 19 – cited in Gilbert Whitley, 'The Life and Work of Gerard Krefft', *Proceedings of the Royal Zoological Society of New South Wales for the Year 1958–59*, 1961, p. 23.
18 See Gerard Krefft, *Notes on the Fauna of Tasmania*, F. White, Sydney, 1868.
19 Idem.
20 Alan Atkinson, *The Europeans in Australia: A History*, vol. 1, Oxford University Press, Melbourne, 1997–2004, p. x.
21 Krefft to Gray, 19 June 1865: J. E. Gray, *Collection of Illustrations of* Cetaceans *and* Sirenia *Comprising Prints, Original Drawings, Tracings and Photographs Dating from the 18th Century*, vol. 1, pp. 25–28, Natural History Museum, London.

References

Archival sources

Alexander Turnbull Library, Wellington

Haast family papers, MS-Papers-0037-292.
Mantell family papers, MS-Group-305.
Mundy Albums 1–4, 1860–1870.
Mundy, Daniel, correspondence, MS-Papers-0037-122.
Ramsay, E. P., correspondence, MS-Papers-0942.

Archives New Zealand, Christchurch

CP 349a Canterbury Museum 1864–7; CP 349b Canterbury Museum 1864–9; CP 349c Canterbury Museum 1868–70; CP 349d Canterbury Museum 1874–5.
Canterbury Museum CH287 – Annual Reports, Progress Reports; Exchanges; Donations.

Australian Museum, Sydney

Albert Gunther papers, correspondence of Gerard Krefft, 1861–78, AMS48.
Annual Reports of the Trustees of the Australian Museum, for the years 1858–99.
Krefft family letters – translated by Bodo Matziche from originals held by the Krefft family, Hamburg.
Papers Relating to the Exploration of Caves and Rivers (Fossil Remains at Wellington Caves, NSW), 1869–1882, Series 438.
Photograph albums, AMS 421/1–19.
Photographs by the Staff of the Australian Museum, Sydney, NSW, 1880–90: album VA232.
Series 6, Outward Letterbooks, 1837–1923.
Series 7, Letters Received, 1853–1883.
Whitley, G. P., 'History of the Australian Museum', 1963, unpublished ms.

Bodleian Library, University of Oxford

Papers of the British Association for the Advancement of Science: records relating to annual meetings, minute books of 'Sections' A–M and X, plus press cuttings – 1832–1962, (Dep. BAAS 11–12); papers of the council, 1850–1900, (Dep. BAAS 18–20, 32–3).

British Library, London

A collection of plans and views of portions of the buildings and contents of old Montague House and the present British Museum, comprising 32 engraved plans and 1 plan drawn in pencil on tracing paper and 86 engraved views, 7 watercolour views, 9 photographic views and 1 pen-and-ink view drawn on tracing paper, together with 3 engraved portraits, including one of Sir Robert Smirke, the architect, various descriptive cuttings from the *Illustrated London News*, *Builder*, *New London Magazine* – assembled by John Edward and Maria Emma Gray: Maps C.26.f.7.

Cooper, Thompson, *Men of Mark: A Gallery of Contemporary Portraits of Men Distinguished in the Senate, the Church, in Science, Literature, and Art, the Army, Navy, Law, Medicine, Etc. Photographed from life by Lock and Whitfield, with brief biographical notices by Thompson Cooper*, Sampson Low, Marston, Searle and Rivington, London, 1876–83.

British Museum, London

Indexes to trustees' general and committee meetings, 1754–1939: CE 26.

Letterbooks (1st series), 1829–1857: CE27.

Minutes of meetings of Trustees' Standing Committee, 1754–1963: CE3.

Officers' Reports, 1805–1869: CE5.

Original Papers, 1743–1963: CE4.

The 'York' Album, c.1875.

Canterbury Museum, Christchurch

Burrage, Sally, 'History of the Canterbury Museum. Chronological Summary of Events 1850–2003', unpublished ms, 2004.

Canterbury Museum Records – Donations & Exchanges (Duplicates) 1872–1887 (1869–1883)

Director's Correspondence – 3/1; Box 1; Folders 1–9.

Dr A. C. Barker letters, 1858–73: items 1–199.

Dr A. C. Barker Albums: albums 1–11.

Files Kept by Directors Prior to 1948: Reports to the Provincial Council – 4/1; Box 1; Folders 1–13.

Hutton Album, no. 1, *carte-de-visite* album: album 13, 18XX.2.1.

Museum of Comparative Zoology Library Archives, Harvard University

Agassiz, Alexander, correspondence: Museum Directors' Papers, 1859–1927.

Geological Society, London

Murchison Collection: letters from Richard Owen, M/07/1–35.

The Archives, Royal Botanic Gardens, Kew

Lovell Reeve Papers – LRP/1/23 Indexed Letter Book Nov 1847–Sep 1872; LRP/6/1 Press Cuttings 1851–1910; LRP/8/1: Catalogues, prospectuses and circulars c 19th C–1910s.

Linnean Society, London

Gunther, A. C. L. G., 'A Series of Albums of Printed Illustrations, Drawings, Photographs and Printed Ephemera', 39 volumes, mss. case 22.

R. C. A Prior Album: 'Literary & Scientific Portrait Club'.

Macleay Museum, Sydney

W. S. Macleay correspondence.

Mitchell Library, Sydney

A Few Letters and Testimonials from Distinguished Men of Science Addressed to Mr Gerard Krefft, Curator and Secretary of the Australian Museum, From 1858–1874, Privately Printed, 1874: A26; P1, Gerard Krefft Papers, Mitchell Library; unpublished ms, Gerard Krefft papers, Mitchell Library, microfilm no. 930, A265 and A267.

E. P. Ramsay Correspondence: Ramsay family – papers 1815–1970, MLMSS 7429.

Gerard Krefft letters from Charles Darwin 1872–76, ML mss 5828.

Krefft papers, Mitchell Library, microfilm no. 930, A265 and A267.

Mundy, Daniel Louis, *Rotomohana; and the boiling springs of New Zealand. A photographic series of sixteen views by D.L. Mundy with descriptive notes by Ferdinand Hochstetter*, 1875.

Reeve, Lovell, *Portraits of Men of Eminence in literature, science, and art: the photographs from life by Ernest Edwards*, Lovell Reeve & Co, London, 1863.

Thompson, Stephen, *Catalogue of a series of photographs by Stephen Thompson, from the collection in the British Museum, introduction by Charles Harrison*, W. A. Mansell, London, 1872.

Museum of New Zealand Te Papa Tongarewa, Wellington

Sir James Hector, Colonial Museum – outward letters 1865–1904: MU000013.

National Gallery of Australia, Canberra

Maull and Polyblank, *Photographic Portraits of Living Celebrities; executed by Maull and Polyblank, with biographical notices by E. Walford*, W. Kent & Co., 1856–59.

National Library of Australia, Canberra

Sir William Henry Flower, 1831–99, correspondence, DF 932 semi-official papers, Australian and New Zealand papers.

W. H. Hooker to J. D. Hooker correspondence: MS 4023.

Natural History Museum, London

Gray, J. E., *Collection of Illustrations of* Cetaceans *and* Sirenia *Comprising Prints, Original Drawings, Tracings and Photographs Dating from the 18th Century*, 2 volumes.

Gruber, Jacob W., 'The Richard Owen Correspondence: An Introductory Essay', unpublished ms, Natural History Museum, London – Owen Collection 87/1.

Richard Owen Collection of Palaeontological and Zoological Drawings

Correspondence of Sir Richard Owen.

Rare Books and Special Collections Library, University of Sydney

Correspondence Relative to Exploration of Wellington Caves, Presented to the Houses of Parliament, New South Wales, 1870.

Royal Society, London

Wharton Jones, J., 'Observations on Single Vision with Two Eyes', communicated by Richard Owen to the Royal Society, London, 30 November 1839: ms. AP/23/33.

State Library of Queensland

Walford, Edward, *Representative Men in Literature, Science and Art*, Alfred William Bennett, London, 1868.

State Library of Victoria, Melbourne

The Stereoscopic Magazine: A Gallery of Landscape Scenery, Architecture, Antiquities, and Natural History, Lovell Reeve, London, 1858–65.

Newspapers and periodicals

The Athenaeum; *The Art-Journal*; *Brighton Daily News*; *British Journal of Photography*; *Canterbury Times*; *Dunedin Punch*; *Fraser's Magazine for Town and Country*; *The Graphic*; *Illustrated London News*; *Krefft's Nature in Australia*; *Leicester Chronicle and the Leicester Mercury*; *Literary Gazette and Journal of Belle Lettres, and Art*; *Lyttelton Times*; *Natural Science*; *Nature*; *Notes and Queries*; *Photographic News*; *The Press*; *Proceedings of the Geological Society of London*; *The Sydney Mail*; *Sydney Morning Herald*; *Sydney Punch*; *The Literary Gazette: A Weekly Journal of Literature, Science, and the Fine Arts*; *The Photographic News*; *The Stereoscopic Magazine*; *The Times*.

Secondary sources

— *Catalogue of the Natural and Industrial Products of New South Wales Forwarded to the Paris Universal Exhibition Commissioners*, Thomas Richards, Government Printer, Sydney, 1867.

— *Centenary Souvenir: Madras Government Museum (1851–1951)*, Principal Commissioner of Museums, Chennai, 1999; *Hand Book on Government Museum, Chennai*, S. S. Jawahar, Chennai, 2011.

— *L' Exposition Universelle de 1867 Illustrée: Publication International Autorisée par la Commission Impériale*, Paris, 1867.

— 'Madras Government Proceedings', *Quarterly Journal of Microscopical Science*, vol. S2-4, no. 15, 1864, pp. 219–32.

— 'Memorial of George Browne Goode', *Annual Report of the Board of Regents of the Smithsonian Institution*, vol. 575, 1980, p. 56.

— 'Portrait Gallery at Drayton Manor', *The Gentleman's Magazine*, vol. 27, 1847, pp. 289–291.

— *Report from the Select Committee on the British Museum*, House of Commons Parliamentary Papers, London, 1860.

— *Report of the Forty-Second Meeting of the British Association for the Advancement of Science Held at Brighton in August 1872*, John Murray, London, 1873.

— *Report of the Thirtieth Meeting of the British Association for the Advancement of Science Held at Oxford in June and July 1860*, John Murray, London, 1861.

— 'The Canterbury Museum, New Zealand', *Illustrated London News*, 6 February 1868.

— *The Fifty-First Annual Exhibition of the Royal Photographic Society of Great Britain: Illustrated Catalogue*, The New Gallery, Regent Street, London, 20 Sept–27 Oct 1906.

— 'The Walhalla at Tamworth', *The Living Age*, vol. 36, 1853, pp. 444–5.

— *York & Son Collection*, National Monuments Records, Swindon, (n.d.).
Acland, L. G. D., *The Early Canterbury Runs: Comprising the First, Second and Third (New) Series*, Whitcombe & Tombs Ltd, Christchurch, 1946.
Adams, James Eli, *Dandies and Desert Saints: Styles of Victorian Manhood*, Cornell University Press, Ithaca and London, 1995.
Agassiz, G. R. (ed.), *Letters and Recollections of Alexander Agassiz*, Constable & Co Ltd, London, 1913.
Alberti, Samuel, 'Amateurs and Professionals in One County: Biology and Natural History in Late-Victorian Yorkshire', *Journal of the History of Biology*, vol. 34, 2001, pp. 115–47.
Alberti, Samuel J. M. M., 'Conversaziones and the Experience of Science in Victorian England', *Journal of Victorian England*, vol. 8, 2008, pp. 208–30.
Allana, Rahaab, 'Performance for Camera: Shapoor N. Bhedwar and the Dimensions of Studio Photography in Bombay' in Partha Mitter *et al.*, *The Artful Pose: Early Studio Photography in Mumbai, c.1855–1940*, Mapin Publishing, Ahmedabad, 2010, pp. 56–73.
Allen, D. E., 'On Parallel Lines: Natural History and Biology from the Late Victorian Period', *Archives of Natural History*, vol. 25, no. 3, 1998, pp. 361–71.
Allis, Thomas, 'Further Note on a Skeleton of a *Dinornis Robustus* Owen, in the York Museum' (Read 17 November 1864), *Journal of the Proceeding of the Linnean Society – Zoological*, vol. 8, no. 31, 1865, pp. 140–1.
Allis, Thomas, 'Notice of a nearly Complete Skeleton of a *Dinornis*, presented by Dr Gibson to the Museum of the Yorkshire Philosophical Society' (Read 16 June 1864), *Journal of the Proceeding of the Linnean Society – Zoological*, vol. 8, no. 29, 1864, pp. 50–2.
Alpers, Svetlana, *The Art of Describing: Dutch Art in the Seventeenth Century*, University of Chicago Press, Chicago, 1983.
Altick, Richard, 'National Monuments' in David Boswell and Jessica Evans (eds.), *Representing the Nation*, Routledge, London, 1999, pp. 240–57.
Anderson, Atholl, *Prodigious Birds: Moas and Moa-hunting in Pre-historic New Zealand*, Cambridge University Press, Cambridge, 2003.
Anderson, Warwick, 'Postcolonial Technoscience', *Social Studies of Science*, vol. 32, nos.5–6, 2002, pp. 643–58.
Annear, Judy (ed.), *Portraits of Oceania*, Art Gallery of New South Wales, Sydney, 1997.
Appadurai, Arjun (ed.), *The Social Life of Things: Commodities in Cultural Perspective*, Cambridge University Press, Cambridge, 1986.
Armstrong, Carol, *Scenes in a Library: Reading the Photograph in the Book, 1843–1875*, MIT Press, Cambridge (Mass.), 1998.
Ashcroft, Bill, Gareth Griffiths and Helen Tiffen, *Post-Colonial Studies: The Key Concepts*, Routledge, London and New York, 2000.
Asma, Stephen T., *Stuffed Animals and Pickled Heads: The Culture and Evolution of Natural History Museums*, Oxford University Press, Oxford, 2001.
Atkinson, Alan, *The Europeans in Australia: A History*, vol. 1, Oxford University Press, Melbourne, 1997–2004.
Bacon, Francis, *The Novum Organum*, Book 1, 1620.
Bala, Arun, *The Dialogue of Civilisations in the Birth of Modern Science*, Palgrave Macmillan, New York, 2006.
Baldwin, Gordon, Malcolm Daniel and Sarah Greenough (eds.), *All the Mighty World: The Photographs of Roger Fenton, 1852–1860*, Metropolitan Museum of Art, New York, 2004.
Ballerini, Julia, 'Recasting Ancestry: Statuettes as Imaged by Three Inventors of Photography' in Anne W. Lowenthal (ed.), *The Object as Subject: Studies in the Interpretation of Still Life*, Princeton University Press, Princeton, 1996, pp. 41–58.
Ballerini, Julia, 'The In-visibility of Hadji-Ishmael: Maxime Du Camp's 1850 Photographs of Egypt' in Kathleen Adler and Marcia Pointon (eds.), *The Body Imaged: The Human Form and Visual Culture since the Renaissance*, Cambridge University Press, Cambridge, 1993, pp. 147–60.

Bann, Stephen, *Parallel Lines: Printmakers, Painters, and Photographers in Nineteenth-Century France*, Yale University Press, New Haven, 2001.

Barker, Thomas, *Photographs of Eminent Medical Men*, John Churchill & Son, London, 1867–68.

Barrell, John, *The Idea of Landscape and the Sense of Place, 1730–1840*, Cambridge University Press, Cambridge, 1972.

Barrett, Ann, 'Nineteenth-Century German Science in London', *International Conference on Archives of Science*, Deutsches Museum, Munich, 2005 – http://www.sac.cat/ficha_notes.php?sitelang=en&idnoti=364&pw=barret (accessed 29 November 2016).

Barringer, Tim, *Men at Work: Art and Labour in Victorian Britain*, Yale University Press, New Haven and London, 2005.

Barringer, Tim, *The Pre-Raphaelites*, Calmann & King, London, 1998.

Barringer, Tim, 'Victorian Culture and the Museum: Before and After the White Cube', *Journal of Victorian Culture*, vol. 11, no. 1. 2006, pp. 133–45.

Barringer, Tim, Geoff Quilley and Douglas Fordham (eds.), *Art and the British Empire*, Manchester University Press, Manchester and New York, 2007.

Barton, Ruth, 'Haast and the Moa: Reversing the Tyranny of Distance', *Pacific Science*, vol. 54, no. 3, 2000, pp. 251–63.

Barton, Ruth, '"Men of Science": Language, Identity and Professionalization in the Mid-Victorian Scientific Community', *History of Science*, vol. 41, no. 1, 2003 pp. 73–119.

Basalla, George, 'The Spread of Western Science' in William K. Storey (ed.), *Scientific Aspects of European Expansion*, Variorum, Brookfield, Vt., 1996, pp. 1–21.

Basalla, George, William Coleman and Robert H. Kargon (eds.), *Victorian Science: A Self-Portrait from the Presidential Addresses to the British Association for the Advancement of Science*, Doubleday Anchor, Garden City, N.Y., 1970.

Batchen, Geoffrey, *Burning with Desire: The Conception of Photography*, MIT Press, Cambridge (Mass.), 1997.

Bates, H. W., 'Hints on the Collection of Objects of Natural History', *Proceedings of the Royal Geographical Society*, vol. 16, no. 1, 1871–72, pp. 67–78.

Bather, F. A., 'Some Colonial Museums', *Report of the Museums Association*, 1894, pp. 193–239.

Beer, Gillian, *Darwin's Plots: Evolution Narrative in Darwin, George Eliot and Nineteenth-century Fiction*, Cambridge University Press, Cambridge, 2000.

Beilharz, Peter, *Imagining the Antipodes: Culture, Theory and the Visual in the Work of Bernard Smith*, Cambridge University Press, Cambridge, 1997.

Beinart, William and Lotte Hughes, *Environment and Empire*, Oxford University Press, Oxford, 2007.

Bennett, Brett M. and Joseph M. Hodge (eds.), *Science and Empire: Knowledge and Networks of Science across the British Empire, 1800–1970*, Palgrave Macmillan, Basingstoke and New York, 2011.

Bellon, Richard, 'Joseph Dalton Hooker's Ideals for a Professional Man of Science', *Journal of the History of Biology*, vol. 34, 2001, pp. 51–82.

Bennett, George, *Gatherings of a Naturalist in Australasia*, John Van Voorst, London, 1860.

Bennett, Tony, *Pasts Beyond Memory: Evolution, Museums, Colonialism*, Routledge, London and New York, 2004.

Bennett, Tony, *The Birth of the Museum*, Routledge, London and New York, 1995.

Berman, Morris,'"Hegemony" and the Amateur Tradition in British Science', *Journal of Social History*, vol. 8, no. 2, 1975, pp. 30–50.

Bewell, Alan, 'William Jones and Cosmopolitan Natural History', *European Romantic Review*, vol. 16, no. 2, 2005, pp. 167–80.

Black, Barbara, *On Exhibit: Victorians and their Museums*, University of Virginia Press, Charlottesville, 2000.

Boas Hall, Marie, *All Scientists Now: The Royal Society in the Nineteenth Century*, Cambridge University Press, London, 1984.

Boesky, Amy, '"Outlandish- Fruits": Commissioning Nature for the Museum of Man', *ELH*, vol. 58, no. 2, 1991, pp. 305–30.

Bohrer, Frederick N., *Orientalism and Visual Culture: Imagining Mesopotamia in Nineteenth-Century Europe*, Cambridge University Press, Cambridge, 2003.

Bolger, Peter, 'William Bedford (1809–1876)', *Australian Dictionary of Biography*, vol. 3, Melbourne University Press, Melbourne, 1969.

Bonyhady, Tim, *The Colonial Earth*, Miegunyah Press, Melbourne, 2000.

Born, Georgina, 'Public Museums, Museum Photography, and the Limits of Reflexivity', *Journal of Material Culture*, vol. 3, no. 2, 1998, pp. 223–54.

Bowler, Peter J. and John V. Pickstone (eds.), *The Cambridge History of Science Volume 6: The Modern Biological and Earth Sciences*, Cambridge University Press, Cambridge, 2009.

Brain, Robert M., 'Exhibitions', *Oxford Companion to the History of Science*, Oxford University Press, Oxford, 2002, pp. 283–86.

Braudy, Leo, *The Frenzy of Renown: Fame and its History*, Oxford University Press, New York, 1986.

Braund, James, 'German-Speaking Scientists in New Zealand, 1773–1951: Research Past, Present and Future', *European Studies* 21, 2005, pp. 179–81.

Briggs, Asa, *Victorian Things*, Sutton, Thrupp, 2003.

Browne, Janet, 'Charles Darwin as a Celebrity', *Science in Context*, vol. 16, no. 1/2, 2003, pp. 175–94.

Browne, Janet, *Charles Darwin: The Power of Place*, Jonathan Cape, London, 2002.

Browne, Janet, 'Commemorating Darwin', *British Journal for the History of Science*, vol. 38, no. 3, 2005, pp. 251–74.

Browne, Janet, 'I Could Have Retched All Night: Charles Darwin and His Body' in Christopher Lawrence and Steven Shapin (eds.), *Science Incarnate: Historical Embodiments of Natural Knowledge*, University of Chicago Press, Chicago and London, 1998, pp. 240–87.

Browne, Janet, 'Looking at Darwin: Portraits and the Making of an Icon', *Isis*, vol. 100, no. 3, 2009, pp. 542–70.

Brusati, Celeste, 'Natural Artifice and Material Values in Dutch Still Life' in Wayne E. Frantis, *Looking at Seventeenth-Century Dutch Art: Realism Reconsidered*, Cambridge University Press, Cambridge, 1997, pp. 144–57.

Brusius, Mirjam, Katrina Dean and Chitra Ramalingam (eds.), *William Henry Fox Talbot: Beyond Photography*, Yale Center for British Art and The Paul Mellon Centre for Studies in British Art, New Haven and London, 2013.

Bull, Marjorie and Joseph Denfield, *Secure the Shadow: The Story of Cape Photography from Its Beginnings to the End of 1870*, Terence McNally, Cape Town, 1970.

Bulwer Lytton, E., *England and the English*, vol. 2, J. J. Harper, New York, 1833.

Burdon, C. C., *Dr A. C. Barker, 1819–1873: A Photographer, Farmer, Physician*, McIndoe, Dunedin, 1972.

Burkhardt, Frederick and Sydney Smith (eds.), *The Correspondence of Charles Darwin*, Cambridge University Press, Cambridge, 1985–2004.

Burrage, Sally, 'Exchange of Native Bird Skins, Skeletons and Eggs by the Canterbury Museum from 1869–1913', *Records of the Canterbury Museum*, vol. 15, 2001, pp. 1–7.

Burrows, Colin J., *Julius Haast in the Southern Alps*, Canterbury University Press, Christchurch, 2005.

Butcher, Barry, 'Darwin's Australian Correspondents: Deference and Collaboration in Colonial Science' in Roy McLeod and Philip F. Rehbock (eds.), *Nature in Its Greatest Extent: Western Science in the Pacific*, University of Hawaii, Honolulu, 1988, pp. 139–57.

Cannadine, David, *Ornamentalism: How the British Saw Their Empire*, Oxford University Press, Oxford and New York, 2002.

Cannon, W. Faye, 'Charles Lyell, Radical Actualism, and Theory', *British Journal for the History of Science*, vol. 9, no. 2, 1975, pp. 104–20.

Carter, Paul, 'Invisible Journeys' in Paul Foss (ed), *Island in the Stream: Myths of Place in Australian Culture*, Pluto Press, Leichhardt, 1988, pp. 47–60.
Carus, C. G., *The King of Saxony's Journey through England and Scotland in the Year 1844*, Chapman & Hall, London, 1846.
Castree, Noel, 'Place: Connections and Boundaries in an Interdependent World', in Gill Valentine, Sarah Holloway and Stephen Rice (eds.), *Key Concepts in Geography*, Sage, London, 2003, pp. 165–86.
Cato, Jack, *The Story of the Camera in Australia*, Institute of Australian Photography, Melbourne, 1955.
Caudel, Mark, *Julius Haast: Towards a New Appreciation of his Life and Work*, MA thesis, University of Canterbury, 2007.
Chalmers-Hunt, J. M. (ed.), *Natural History Auctions 1700–1972: A Register of Sales in the British Isles*, Sotheby Parke Bernet, London, 1976.
Chambers, David Wade and Richard Gillespie, 'Locality in the History of Science: Colonial Science, Technoscience and Indigenous Knowledge', *Osiris*, series 2, vol. 15, 2000, pp. 221–40.
Chaplin, Joyce E., 'Nature and Nation: Natural History in Context', *Transactions of the American Philosophical Society*, vol. 93, no. 4, 2003, pp. 75–95.
Christ, Carol T. and John O. Jordan (eds.), *Victorian Literature and the Victorian Visual Imagination*, University of California Press, Berkeley, 1995.
Clarke, W. B., '*Dinornis* an Australian Genus', *Geological Magazine*, vol. 6. no. 26, 1869, pp. 383–4.
Clarke, W. B., 'On Dromornis australis (Owen): a new fossil bird of Australia', *Journal and Proceedings of the Royal Society of New South Wales*, no. 11, 1878, pp. 41–9.
Cleevely, R. J., 'Gray, John Edward (1800–75)', *Oxford Dictionary of National Biography* – http://www.oxforddnb.com/view/article/11344 (viewed 22 August 2007).
Codell, Julie F. (ed.), *Transculturation in British Art, 1770–1930*, Ashgate, Farnham, 2012.
Cohen, Claudine, *The Fate of the Mammoth: Fossils, Myth and History*, University of Chicago Press, Chicago and London, 2002.
Colnaghi & Co, P & D, *Photography: The First Eighty Years*, P & D Colnaghi, London, 1976.
Comolli, Jean-Louis, 'Machines of the Visible' in Teresa de Lauretis and Stephen Heath (eds.), *The Cinematic Apparatus*, Macmillan, London, 1980, pp. 121–50.
Connell, R. W., *Masculinities*, Allen & Unwin, St. Leonards, 1995.
Crary, Jonathan, *Techniques of the Observer: On Vision and Modernity in the Nineteenth Century*, MIT Press, Cambridge (Mass.), 1990.
Cunningham, Peter, *Two Years in New South Wales*, Angus & Roberson, Sydney, 1966 (1827).
Currey, C. H., 'Denison, Sir William Thomas (1804–1871)' in *Australian Dictionary of Biography*, Online Edition, Australian National University, 2006, http://www.adb.online.anu.edu.au/biogs/A04008b.htm (accessed 11 December 2010).
Cuvier, Georges, *Recherches sur les Ossemens Fossiles de Quadrupèdes*, Déterville, Paris, 1812.
Dance, S. Peter, *A History of Shell Collecting*, E. J. Brill, Leiden, 1986.
Dance, S. Peter, 'Reeve, Lovell Augustus (1814–1865)', *Oxford Dictionary of National Biography*, Oxford University Press, 2004, http://www.oxforddnb.com/view /article/23300 (accessed 27 November 2010).
Dance, S. Peter, 'Ruskin the Reluctant Conchologist', *Journal of the History of Collections*, vol. 16, no. 1, 2004.
Darwin, Charles, *On the Origin of Species by Means of Natural Selection*, (ed.) Joseph Carroll, Broadview, Peterborough, 2003 (1859).
Darwin, Charles, *Voyage of the* Beagle, Penguin, London, 1989 (1839).
Darwin, Francis (ed.), *The Life and Letters of Charles Darwin, including an Autobiographical Chapter*, John Murray, 1887.
Darwin, John, 'Imperialism and the Victorians: The Dynamics of Territorial Expansion', *The English Historical Review*, vol. 112, no. 447, 1987, pp. 614–42.

Daston, Lorraine, 'The Ideal and Reality of the Republic of Letters', *Science in Context*, vol. 4, no, 2, 1991, pp. 367–86.
Daston, Lorraine and Peter Galison, *Objectivity*, Zone Books, Cambridge (Mass.), 2010.
Date, Christopher, 'Photographer on the Roof', *British Museum Society Bulletin*, July 1989, pp. 10–12.
Date, Christopher and Anthony Hamber, 'The Origins of Photography at the British Museum, 1839–1860', *History of Photography*, vol. 14, no. 4, 1990, pp. 309–25.
David, Andrew, *The Voyage of HMS Herald to Australia and the South-west Pacific 1852–1861 under the Command of Captain Henry Mangles Denham*, Melbourne University Press, Melbourne, 1995.
Davidson, Jane, *A History of Palaeontology Illustration*, Indiana University Press, Bloomington and Indianapolis, 2008.
Davies, Alan and Peter Stanbury, *The Mechanical Eye in Australia: Photography 1841–1900*, Oxford University Press, Melbourne, 1985.
Dawson, Lyndall, 'Marsupial Fossils from Wellington Caves, New South Wales: The Historic and Scientific Significance of the Collections in the Australian Museum, Sydney', *Records of the Australian Museum*, vol. 37, no. 2, 1985, pp. 55–69.
Deacon, Desley, Penny Russell and Angela Woollacott (eds.), *Transnational Ties: Australian Lives in the World*, ANU E-press, Canberra, 2008.
Deacon, Margaret, *Scientists and the Sea, 1650–1900*, Ashgate, Aldershot, 1997.
Desmond, Adrian, 'Redefining the X Axis: "Professionals", "Amateurs" and the Making of Victorian Biology – A Progress Report', *Journal of the History of Biology*, 34, 2001, pp. 3–50.
Desmond, Adrian J., 'Designing the Dinosaur: Richard Owen's Response to Robert Edmond Grant', *Isis*, vol. 70, no. 2, 1979, pp. 224–34.
Donald, Diana and Jane Munro (eds.), *Endless Forms: Charles Darwin, Natural Science and the Visual Arts*, Yale Center for British Art, New Haven, 2009.
Driver, Felix, 'Distance and Disturbance: Travel, Exploration and Knowledge in the Nineteenth Century', *Transactions of the Royal Society*, sixth series, vol. 14, 2004, pp. 73–92.
Du Chaillu, Paul and Dr J. E. Gray, 'Notice of a Skull of a New Species of Bush-goat (Cephalophus longiceps), Sent from the Gaboon', *Proceedings of the Zoological Society of London*, vol. 33, issue 1, 1865, pp.204–6.
Du Chaillu, Paul, *The Country of the Dwarfs*, Harpers & Brothers, New York, 1872.
Duffin, Christopher J., 'William Buckland (1784–1856)', *Geology Today*, vol. 22, no. 3, 2006, pp. 104–8.
Duncan, Carol and Alan Wallach, 'The Universal Survey Museum', *Art History*, vol. 3, no. 4, 1980, pp. 448–69.
Dunmore, Edward. 'On Backgrounds and Portraits Combined', *British Journal of Photography*, 1875, p. 487.
Edwards, Elizabeth, *Anthropology and Photography 1860–1920*, Yale University Press, New Haven and London, 1992.
Edwards, Elizabeth, *Raw Histories, Anthropology and Museums*, Berg, Oxford, 2001.
Edwards, Elizabeth and Christopher Morton (eds.), *Photographs, Museums, Collections: Between Art and Information*, Bloomsbury, London, 2015.
Eggleton, David, *Into the Light*, Craig Potton, Nelson, 2006.
Eisenstein, Elizabeth, *The Printing Revolution in Early Modern Europe*, Cambridge University Press, Cambridge, 2005.
Ellegård, Alvar, *Darwin and the General Reader: The Reception of Darwin's Theory of Evolution*, University of Chicago Press, Chicago, 1990.
Endersby, Jim, *Imperial Nature: Joseph Hooker and the Practices of Victorian Science*, University of Chicago Press, Chicago and London, 2008.
Ennis, Helen, *Photography and Australia*, Reaktion, London, 2007.
Falconer, John, 'Encountering India' in *Encountering India: Colonial Photography 1850–1911*, Exhibition Catalogue, Museum of Sydney, Sydney, 1997.

Fara, Patricia, 'Faces of Genius: Images of Isaac Newton in Eighteenth-century England' in Geoffrey Cubitt and Allen Warren (eds.), *Heroic Reputations and Exemplary Lives*, Manchester University Press, Manchester and New York, 2000, pp. 57–81.
Fara, Patricia, 'Michael Faraday: Media Man', *Endeavour*, vol. 30, no. 1, 2006, pp. 10–12.
Fay, Elizabeth, 'Portrait Galleries, Representative History, and Hazlitt's *Spirit of the Age*', *Nineteenth-Century Contexts*, vol. 24, no. 2, 2002, pp. 151–75.
Ferguson, Niall, *Empire: How Britain Made the Modern World*, Allen Lane, London, 2003.
Field, Barron, *Geographical Memoirs on New South Wales*, John Murray, London, 1825.
Finney, Colin, *Paradise Revealed: Natural History in Nineteenth-century Australia*, Museum of Victoria, Melbourne, 1993.
Fletcher, Pamela, 'Creating the French Gallery: Ernest Gambart and the Rise of the Commercial Art Gallery in Mid-Victorian London', *Nineteenth-century Art Worldwide*, vol. 6, no. 1, 2007 – http://www.19thc-artworldwide.org/spring07/143-creating-the-french-gallery-ernest-gambart-and-the-rise-of-the-commercial-art-gallery-in-mid-victorian-london (accessed 27 October 2010).
Flint, Kate, *The Victorians and the Visual Imagination*, Cambridge University Press, Cambridge, 2002.
Forbes, Edward, 'On the Educational Uses of Museums' in Hugh H. Genoways and Mary Andrei (eds.), *Museum Origins: Readings in Early Museum History and Philosophy*, Left Coast Press, Walnut Creek, 2008, pp. 251–6.
Forbes, James, *The Danger of Superficial Knowledge*, London, 1849.
Ford, Colin, *Julia Margaret Cameron: A Critical Biography*, The J. Paul Getty Museum, Los Angeles, 2003.
Forgan, Sophie, 'The Architecture of Display: Museums, Universities and Objects in Nineteenth-Century Britain', *History of Science*, vol. 32, no. 96, 1994, pp. 139–62.
Foucault, Michel, *The Order of Things*, Routledge, London, 2002.
Freeman, James, 'On the Progress of Photography and its Application to the Arts and Sciences', *Sydney Magazine of Science and Art*, vol. II, January–February 1859, pp. 136–41.
Freeman, Michael, *Victorians and the Prehistoric*, Yale University Press, New Haven and London, 2004
Freund, Gisèle, *Photography and Society*, Fraser, London, 1980.
Fried, Michael, *Courbet's Realism*, University of Chicago Press, Chicago and London, 1992.
Fyfe, Aileen and Bernard Lightman (eds.), *Science in the Marketplace: Nineteenth-Century Sites and Experiences*, University of Chicago Press, Chicago, 2007.
Gagnier, Regenia, *Subjectivities: A History of Self-Representation in Britain, 1832–1920*, Oxford University Press, New York and Oxford, 1991.
Gallagher, John and Ronald Robinson, 'The Imperialism of Free Trade', *The Economic History Review*, series 2, vol. 6. no. 1, 1953, pp. 1–15.
Gates, Barbara T., 'Why Victorian Natural History?' *Victorian Literature and Culture*, vol. 35, 2007, pp. 539–49.
Gellatly, Anne F., 'Historical Records of Glacier Fluctuations in Mount Cook National Park: A Century of Change', *The Geographical Journal*, vol. 151, no. 1, 1985, pp. 86–99.
Ginzburg, Carlo, 'Clues: Morelli, Freud and Sherlock Holmes' in Umberto Eco and Thomas A. Sebeok (eds.), *The Sign of Three*, Indiana University Press, Bloomington and Indiana, 1988, pp. 81–118.
Goldberg, Vicki (ed.), *Photography in Print*, Simon & Schuster, New York, 1981.
Golinski, Jan, *Making Natural Knowledge: Constructivism and the History of Science*, Chicago University Press, Chicago and London, 2005.
Goodall, Jane R., *Performance and Evolution in the Age of Darwin*, Routledge, London, 2002.
Goodman, David, 'Fear of Circuses: Founding the National Museum of Victoria', *Continuum*, vol. 3, no. 1, 1990, pp. 18–34.
Goodman, Nelson, *Ways of Worldmaking*, Hackett, Indianapolis, 1978.
Gosse, P. H., *The Romance of Natural History*, Gould & Lincoln, Boston, 1861.

Gould, Stephen Jay, 'An Awful, Terrible Dinosaurian Irony', *Natural History*, vol. 107, no. 1, 1998, pp. 24–26.

Gray, J. E., *Catalogue of Shield Reptiles in the Collection of the British Museum*, Pt. II, British Museum, London, 1872.

Gray, J. E., 'On Museums, their Use and Improvement', *Annals and Magazine of Natural History*, vol. 14, 1864, pp. 283–97.

Green-Lewis, Jennifer, *Framing the Victorians: Photography and the Culture of Realism*, Cornell University Press, Ithaca, N.Y., 1996.

Grout, Andrew, 'Oldham, Thomas (1816–1878)', *Oxford Dictionary of National Biography*, Oxford University Press, 2004 – http://www.oxforddnb.com/view/article/20691 (accessed 20 December 2016).

Gruber, Jacob W., 'From Myth to Reality: The Case of the Moa', *Archives of Natural History*, vol. 14, no. 3, 1987, pp. 339–52.

Gruber, Jacob W., 'The Moa and the Professionalising of New Zealand Science', *Turnbull Library Record*, vol. 20, no. 2, 1987, pp. 61–100.

Gruber, Jacob W. and John C. Thackray, *Richard Owen Commemoration: Three Studies*, Natural History Museum, London, 1992.

Gully, J. S., 'Gully, John' from the *Dictionary of New Zealand Biography: Te Ara – the Encyclopedia of New Zealand* – http://www.TeAra.govt.nz/en/biographies/1g25/1 (accessed 2 April 2011).

Gunther, A. E., *The Founders of Science at the British Museum 1753–1900*, Halesworth Press, Halesworth, 1980.

Gunther, Albert E., *A Century of Zoology at the British Museum: Through the Lives of Two Keepers, 1815–1914*, Dawsons, London, 1975.

Haast, H. F. von, *The Life and Times of Julius von Haast*, Wellington, 1848.

Haast, Julius, 'Altitude sections of the principal routes between the east and west coasts of the province of Canterbury, New Zealand, across the Southern Alps', *Journal of the Royal Geographical Society*, vol. 37, 1867, pp. 328–38.

Haast, Julius, 'Notes on the Mountains and Glaciers of the Canterbury Province, New Zealand', *Journal of the Royal Geographical Society*, vol. 34, 1864, pp. 87–96.

Haast, Julius, 'On the Measurements of Dinornis Bones, Obtained from Excavations in a Swamp, Situated at Glenmark, on the Property of Messrs Kermode & Co, up to Feb 15, 1868', *Transactions and Proceedings of the Royal Society of New Zealand*, vol. 1, 1868, pp. 80–9.

Haast, Julius, *Report on the Formation of the Canterbury Plain: With a Geological Sketch-map and Five Geological Sections*, Press Office, Christchurch, 1864.

Haast, Julius, *Report on the Headwaters of the River Rakaia: With Twenty Illustrations*, James Edward Fitzgerald, Govt. Printer, Christchurch, 1866.

Haast, Julius, *Reports of Geological Explorations during 1871–2: With Maps and Sections*, George Didsbury, Wellington, 1872.

Haast, Julius, *Reports of Geological Explorations during 1873–4: With Maps and Sections*, James Hughes, Wellington, 1877.

Haast, Julius von, 'On the Origin and Early History of the Canterbury Museum: Being the Annual Address', *Transactions of the Philosophical Institute of Canterbury*, 1881, pp. 503–16.

Haast, Julius von, *Report of a Topographical and Geological Exploration of the Western Districts of the Nelson Province, New Zealand*, C. & J. Elliott, Nelson, 1861.

Hackforth-Jones, Jocelyn, *Augustus Earle, Travel Artist: Paintings and Drawings in the Rex Nan Kivell Collection*, National Library of Australia, Canberra, 1980.

Hackmann, W. D., *Apples to Atoms: Portraits of Scientists from Newton to Rutherford*, National Portrait Gallery, London, 1986.

Hall, Catherine, Keith McClelland and Jane Rendall, *Defining the Victorian Nation: Class, Race, Gender and the Reform Act of 1867*, Cambridge University Press, Cambridge, 2000.

Hamber, Anthony, '*A Higher Branch of Art': Photographing the Fine Arts in England, 1839–1880*, Gordon and Breach, Amsterdam, 1996.

Hamilton, Peter and Roger Hargreaves, *The Beautiful and the Damned: The Creation of Identity in Nineteenth-century Photography*, Lund Humphries and National Portrait Gallery, London, 2001.

Hannavy, John (ed.), *Encyclopedia of Nineteenth-Century Photography*, Routledge, London, 2007.

Hannavy, John, 'Roger Fenton and the British Museum', *History of Photography*, vol. 12, no. 3, 1988, pp. 193–204.

Hannavy, John, *Roger Fenton of Crimble Hall*, Gordon Fraser Gallery, London, 1975.

Harley, Alexis, 'Autobiography, Authority, Atheism: The Evolving Charles Darwin', unpublished ms., Fisher Rare Books, University of Sydney: PrEs.

Harrison, Carol E., 'Citizens and Scientists: Toward a Gendered History of Scientific Practice in Post-revolutionary France', *Gender & History*, vol. 13, no. 3, 2001, pp. 444–80.

Heckenberg, Kerry, 'Photography and Exploration: Issues Related to the Uptake of Photography in Australian Exploration', *Journal of Australian Colonial History*, vol. 9, 2007, pp. 155–188.

Hedley, Charles, 'A Query as to the Synonymy of *Rhysota Armiti*, Smith', *The Annals and Magazine of Natural History*, series 6, vol. 16, no. 92, 1895, p. 201.

Hedley, Charles, 'The Imputed Jealousy of European Workers on Australasian Faunas by Local Writers', *The Annals and Magazine of Natural History*, series 6, vol. 17, no, 99, 1896, pp. 258–60.

Henry, David, 'York & Son', *New Magic Lantern Journal*, vol. 3, no. 1, 1984, p.12 and vol. 3, no. 2, 1984, p. 13.

Heyck, T. W., *The Transformation of Intellectual Life in Victorian England*, Croom Helm, London and Canberra, 1982.

Higgitt, Rebekah, *Recreating Newton: Newtonian Biography and the Making of the Nineteenth-century History of Science*, Pickering & Chatto, London, 2007.

Highley, Samuel, 'On the Application of Photography and the Magic Lantern to Class Demonstrations in Microscopic Science and Natural History', *Report of the Thirty-Fourth Meeting of the British Association for the Advancement of Science Held at Bath in September 1864*, John Murray, London, 1865, p.98.

Highley, Samuel, 'The Application of Photography to the Magic Lantern Educationally Considered', *The Photographic News*, vol. 7, 23 January 1863, pp. 43–4; 6 February 1863, pp. 68–9; 6 March 1863, pp. 113–4.

Hight, Eleanor and Gary Sampson (eds.), *Colonialist Photography: Imag(in)ing Race and Place*, Routledge, London, 2002.

Hochstetter, Ferdinand von, *New Zealand: Its Physical Geography, Geology and Natural History*, J. G. Cotta, Stuttgart, 1867.

Hoffenberg, Peter H., 'Nineteenth-Century Australian Scientists and the Unholy Australian Trinity: Overcoming Distance, Exile and Wandering at Exhibitions', *British Scholar*, vol. II, issue 2, 2010, pp. 227–53.

Hogg, Robert, 'Performing Manliness: "Unmanly" Men on British Frontiers in the Mid-Nineteenth-Century', *Journal of Australian Studies*, vol. 35, no. 3, 2011, pp. 355–372.

Home, R. W. and Sally Kohlstedt (eds.), *International Science and National Scientific Identity: Australia between Britain and America*, Kluwer Academic Publishers, Dordrecht, 1991.

Hooker, Joseph, 'Address of Joseph Hooker, F.R.S., President', *Report of the British Association for the Advancement of Science*, 1868, pp. lviii–lxxv.

Hoskins, Janet, 'Agency, Biography and Objects' in Christopher Tilley, Webb Keane, Susanne Kuechler, Mike Rowlands and Patricia Spyer (eds.), *Handbook of Material Culture*, Routledge, London, 2006, pp. 74–84.

Hoskins, Janet, *Biographical Objects: How Things Tell Stories about People's Lives*, Routledge, London, 1998.

Humboldt, Alexander von, *Personal Narrative of Travels to the Equinoctial Regions of the New Continent*, vol. 1, Henry G. Bohn, London, 1852.

Hunt, Robert, 'The Stereoscope', *The Art-Journal*, 1856, p. 118.

Ingles, Jean M. and Frederick C. Sawyer, 'A Catalogue of the Richard Owen Collection of Palaeontological and Zoological Drawings in the British Museum (Natural History)', *Bulletin of the British Museum (Natural History)* (historical series), vol. 6, no. 5, 1979, pp. 109–97.

Inkster, Ian, 'Scientific Enterprise and the Colonial "Model": Observations on Australian Experience in Historical Context', *Social Studies of Science*, vol. 15, 1985, pp. 677–704.

Jaeger, Gustav, *Bericht Uper Ein Fast Vollstandiges Skelet Von Palapteryx Ingens Uber Dessen Restauration Und Die Davon Angefertigten Gypsabgusse; Mit Einigen Bemerkungen Uber Die Aufstellung Der Vogelskelete Uberhaupt*, Wilhelm Braumuller, Wien, 1863.

James, Frank A. J. L., *Guides to the Royal Institution of Great Britain: History*, Royal Institution, London, 2000.

Jardine, N., J. A. Secord, and E. C. Spary (eds.), *Cultures of Natural History*, Cambridge University Press, Cambridge, 1996.

Jerrold, W. Blanchard, *How to See the British Museum in Four Visits*, Bradbury & Evans, London, 1852

Jevons, W. Stanley, 'The Use and Abuse of Museums' in *Methods of Social Reform and other Papers*, Macmillan & Co, London, 1883.

Johnson, Samuel, *The Works of Samuel Johnson, with an Essay on his Life and Genius by Arthur Murphy*, vol. 1, George Dearborn, New York, 1837.

Jordan, Caroline, *Picturesque Pursuits: Colonial Women Artists and the Amateur Tradition*, Melbourne University Press, Melbourne, 2005.

Jordanova, Ludmilla, 'Medical Men 1780–1820' in Joanna Woodall, *Portraiture: Facing the Subject*, Manchester University Press, Manchester, 1997, pp. 101–15.

Jordanova, Ludmilla, 'Museums: Representing the Real?' in George Levine (ed.), *Realism and Representation: Essays on the Problem of Realism in Relation to Science, Literature and Culture*, University of Wisconsin Press, Madison,1999, pp. 225–278.

Kean, John, 'Observing Mondellimin, Or When Gerard Krefft "Saved Once More the Honour of the Exploring Expedition"', *Proceedings and Transactions of the Royal Society of Victoria*, vol. 121, no. 1, 2008, pp. 109–28.

Kean, John, *The Art of Science*, Museum of Victoria, Melbourne, 2013.

Keller, Susanne B., 'Sections and Views: Visual Representations in Eighteenth-century Earthquake Studies', *British Journal for the History of Science*, vol. 31, no. 2, pp. 129–59.

Kelly, Alfred, *The Descent of Darwin: The Popularization of Darwin in Germany, 1860–1914*, University of North Carolina Press, Chapel Hill, 1981.

Kemp, Martin and Craig Ashley Hanson, 'Science and art', *Grove Art Online – Oxford Art Online*, Oxford University Press – http://www.oxfordartonline.com/subscriber/article/grove/art/T076962 (accessed 28 November 2016).

Kerr, Joan (ed.), 'Johann Ludwig Gerard (Louis) Krefft (1830–1881)', *The Dictionary of Australian Artists: Painters, Sketchers, Photographers and Engravers to 1870*, Oxford University Press, Melbourne, 1992, p. 437.

King, Michael, *The Penguin History of New Zealand*, Penguin, Auckland, 2003.

Kingsley, Charles, *Works*, Macmillan, London, 1880–5.

Kingston, Beverly, *The Oxford History of Australia, vol. 3: Glad, Confident Morning, 1860–1900*, Oxford University Press, Melbourne and Oxford, 2001.

Knight, Hardwicke, *Photography in New Zealand: A Social and Technical History*, John McIndoe, Dunedin, 1971.

Knoepflmacher, U. C. and G. B. Tennyson (eds.), *Nature and the Victorian Imagination*, University of California Press, Berkeley and Los Angeles, 1977.

Kohler, Robert E., 'Finders, Keepers: Collecting Sciences and Collecting Practice', *History of Science*, vol. 45, 2007, pp. 428–54.

Kohlstedt, Sally Gregory, 'Australian Museums of Natural History: Public Priorities and Scientific Initiatives in the 19th Century', *Historical Records of Australian Science*, vol. 5. 1983, pp. 1–29.

Krefft, Gerard, 'A Biographical Sketch of the Youth and Early Manhood of Dr Ludwig Leichardt', *The Maitland Mercury and Hunter River Advertiser*, 3 February 1866, p. 3.

Krefft, Gerard, 'A Few Remarks on the Habit and Economy of the Brown-capped Potamorhinus (P. ruficeps, Hartlaub)', *Proceedings of the Zoological Society of London*, vol. 23, pp. 352–3.

Krefft, Gerard, 'Description of a Gigantic Amphibian Allied to the Genus *Lepidosiren*, from Wide-Bay District, Queensland', *Proceedings of the Zoological Society of London*, 28 April 1870, pp. 221–4.

Krefft, Gerard, *Notes on the Fauna of Tasmania*, F. White, Sydney, 1868.

Krefft, Gerard, 'The Improvements Effected in Modern Museums in Europe and Australia', *Transactions of the Royal Society of New South Wales for the Year 1868*, Sydney, 1869, pp. 15–25.

Krefft, Gerard, *The Mammals of Australia*, Thomas Richards, Sydney, 1871.

Krefft, Gerard, *The Snakes of Australia: An Illustrated and Descriptive Catalogue of all Known Species*, Thomas Richards, Govt. Printer, Sydney, 1869.

Kuhn, Thomas, *The Structure of Scientific Revolutions*, University of Chicago Press, Chicago and London, 1962.

Lalvani, Suren, *Photography, Vision, and the Production of Modern Bodies*, State University of New York Press, Albany, 1996.

Lamping, H. and M. Linke (eds.), *Australia: Studies on the History of Discovery and Exploration*, Institutes für Wirtschafts und Sozialgeographische der Johann Wolfgang Goethe-Univeristät, Frankfurt am Main, 1994.

Larson, Barbara and Fay Brauer (eds.), *The Art of Evolution: Darwin, Darwinisms, and Visual Culture*, Dartmouth College Press, Hanover, 2009.

Larson, Barbara and Sabine Flach (eds.), *Darwin and Theories of Aesthetics and Cultural History*, Ashgate, Farnham, 2013.

Latour, Bruno, *Science in Action: How to Follow Scientists and Engineers through Society*, Harvard University Press, Cambridge (Mass.), 1987.

Latour, Bruno, 'Visualisation and Cognition: Drawing Things Together' – http://www.bruno-latour.fr/ (accessed 8 June 2009).

Latour, Bruno and Steve Woolgar, *Laboratory Life: The Construction of Scientific Facts*, Princeton University Press, Princeton, 1986.

Lewes, G. H., 'Professor Owen and the Science of Life', *Fraser's Magazine for Town and Country*, vol. 53, 1856, pp. 79–92.

Lightman, Bernard and Bennett Zon (eds.), *Evolution and Victorian Culture*, Cambridge University Press, Cambridge, 2014.

Livingstone, David N., *Putting Science in its Place: Geographies of Scientific Knowledge*, University of Chicago Press, Chicago and London, 2003.

Livingstone, David N., and Charles W. J. Withers (eds.), *Geographies of Nineteenth-Century Science*, University of Chicago Press, Chicago and London, 2011.

Lloyd, Valerie, *Roger Fenton: Photographer of the 1850s*, Hayward Gallery, London, 1988.

Lochhead, Ian, *A Dream of Spires: Benjamin Mountfort and the Gothic Revival*, Canterbury University Press, Christchurch, 1999.

Lydon, Jane, *Eye Contact: Photographing Indigenous Australians*, Duke University Press, Durham and London, 2005.

Lyell, Charles, *Principles of Geology*, John Murray, London, vol. 3, 1833.

Mabberley, D. J., *Jupiter Botanicus: Robert Brown of the British Museum*, J. Cramer, Braunshweig, 1985.

Macdonald, Sharon, 'Museums, National, Postnational and Transcultural Identities', *Museum and Society*, vol. 1, no. 1, 2003, pp. 1–16.

Mackay, Mary, *The Geological Sublime: A New Paradigm*, PhD dissertation, University of Sydney, 1990.

MacKenzie, John M., *Museums and Empire: Natural History, Human Cultures and Colonial Identities*, Manchester University Press, Manchester and New York, 2009.

MacLeod, Roy, 'Introduction. Nature and Empire: Science and the Colonial Enterprise', *Osiris*, 2nd series, vol. 15, 2000, pp. 1–13.

MacLeod, Roy, 'On Visiting the "Moving Metropolis": Reflections on the Architecture of Imperial Science', *Historical Records of Australian Science*, vol. 5, no. 3, 1982, pp. 1–16.

MacLeod, Roy, 'The "Practical Man" Myth and Metaphor in Anglo-Australian Science', *Australian Cultural History*, vol. 8, 1989, pp. 24–49.

MacLeod, Roy and Peter Collins (eds.), *The Parliament of Science*, Science Reviews Ltd., Northwood, 1981.

Maddock, W., *Visitors' Guide to Sydney*, Sydney, 1872.

Madox Brown, Ford, *The Exhibition of* WORK *and other Paintings, by Ford Madox Brown, at the Gallery, 191, Piccadilly*, London, 1865.

Main, William and John B. Turner, *New Zealand Photography from the 1840s to the Present*, Photoforum Inc., Auckland, 1993.

Maling, Peter B., 'Haast, Johann Franz Julius von, from the *Dictionary of New Zealand Biography: Te Ara – the Encyclopedia of New Zealand* – http://www.TeAra.govt.nz/en/biographies/1h1/haast-johann-franz-julius-von (accessed 2 April 2011).

Malraux, André, *Museum without Walls*, Secker & Warburg, London, 1965.

Mandelstam, Joel, 'Du Chaillu's Stuffed Gorillas and the Savants from the British Museum', *Notes and Records of the Royal Society of London*, vol. 48, no. 2, 1994, pp. 227–45.

Marien, Mary Warner, *Photography: A Cultural Reader*, Laurence King, London, 2002.

Massey, Doreen, *Space, Place, and Gender*, Polity, Cambridge, 1994.

Mazzolini, Renato (ed.), *Non-verbal Communication in Science Prior to 1900*, L.S. Olschki, Florence, 1993.

McCarthy, Colin, 'Paper Snakes', *Nature*, vol. 306, 1 December 1983, p. 514.

McCauley, Elizabeth Anne, *A. A .E. Disdéri and the Carte-de-Visite Portrait Photograph*, Yale University Press, New Haven, 1985.

McCook, Stuart, 'Paul du Chaillu and the Legitimation of Evidence in the Field Sciences', *Osiris*, 2nd ser., vol. 11, 1996, pp. 177–97.

McCoy, Frederick, *On the Formation of Museums in Victoria*, Goodhugh & Hough, Melbourne, 1857.

Michael Shortland and Richard Yeo (eds.), *Telling Lives in Science: Essays on Scientific Biography*, Cambridge University Press, Cambridge, 1996.

Miller, Edward, *That Noble Cabinet: A History of the British Museum*, Andre Deutsch, 1973.

Mills, Christopher, *Images from Nature: Drawings and Paintings from the Library of the Natural History Museum, London*, Natural History Museum, London, 1998.

Mitchell, Captain J., *Catalogue of the Mollusca in the Collection of the Government Central Museum, Madras*, William Thomas, Madras, 1867.

Mitchell, Thomas L., *Three Expeditions into the Interior of Eastern Australia: With Descriptions of the Recently Explored Region of Australia Felix and of the Present Colony of New South Wales*, T. W. Boone, London, 1839.

Mitchell, W. J. T. (ed.), *Landscape and Power*, Chicago University Press, Chicago and London, 1994.

Mitchell, W. J. T., *Picture Theory*, University of Chicago Press, Chicago and London, 1994.

Mitchell, W. J. T., *What Do Pictures Want?*, University of Chicago Press, Chicago and London, 2005.

Moore, Clive, 'Australian Masculinities', *Journal of Australian Studies*, no. 56, 1998, pp. 1–16.

Morrell, Jack, 'Professionalization' in Robert Olby *et al.* (eds.), *Companion to the History of Modern Science*, Routledge, London and New York, 1990, pp. 980–9.

Morris, John, *John Dillwyn Llewelyn, 1810–1882: The First Photographer in Wales*, Welsh Arts Council, 1980.

Morus, Iwan, Simon Shaeffer and Jim Secord, 'Scientific London' in Celina Fox (ed.) *London: World City 1800–40*, Yale University Press and Museum of London, New Haven, 1992, pp. 129–42.

Moyal, Ann, 'Sir Richard Owen and his Influence on Australian Zoological and Palaeontological Science', *Historical Records of Australian Science*, vol. 3, no. 2, 1975, pp. 41–56.

Moyal, Ann, *The Web of Science: The Scientific Correspondence of Rev. W. B. Clarke, Australia's Pioneer Geologist*, Australian Scholarly Publishing, Kew, 2003.

Mundy, D. L., 'Photographic Experiences in New Zealand', *The Photographic News*, 18 December 1874, pp. 602–3.

Mundy, Daniel, *Rotomohana; and the Boiling Springs of New Zealand. A Photographic Series of Sixteen Views by D.L. Mundy with Descriptive Notes by Ferdinand Hochstetter*, Sampson, Low, Marston, Low & Searle, London, 1875.

Murchison, Roderick, 'President's Address', *Journal of the Royal Geographical Society*, vol. 34, 1864, pp. cvi–cxciii.

Murray, David, *Museums: Their History and their Use*, vol. 1, James Maclehose, Glasgow, 1904.

Nair, Savithri Preetha, 'Science and the Politics of Colonial Collecting: The Case of Indian Meteorites, 1856–70', *British Journal for the History of Science*, vol. 39, no. 1, 2006, pp. 97–119.

Nancarrow, Jenny, 'Gerard Krefft: A Singular Man', *Proceedings and Transactions of the Royal Society of Victoria*, vol. 121, no. 1, 2008, pp. 146–54.

Nead, Lynda, *Victorian Babylon: People, Streets and Images in Nineteenth-Century London*, Yale University Press, New Haven and London, 2005.

Neville, Richard, *A Rage for Curiosity*, State Library of New South Wales, Sydney, 1997.

Newton, Gael, *Shades of Light: Photography and Australia 1839–1988*, Australian National Gallery and Collins, Canberra and Sydney, 1988.

Nochlin, Linda, *Realism*, Penguin, Harmondsworth, 1971.

Outram, Dorinda, *Georges Cuvier: Vocation, Science and Authority in Post-revolutionary France*, Manchester University Press, Manchester, 1984.

Owen, Professor, 'On a Collection of Australian Fossils in the Museum of the Natural History Society of Worcester; with Descriptions of the Lower Jaw and Teeth of the Nototherium Inferme and Nototherium Mitchelli, Owen; Demonstrating the Identity of the Latter Species with the Zygomaturus of Macleay', *Quarterly Journal of the Geological Society of London*, vol. 15, no. 1–2, 1859, pp. 176–86.

Owen, Professor, 'On some Outline-Drawings and Photographs of the Skull of the Zygomaturus trilobus, *Macleay* (Nototherium, *Owen?*)', *Quarterly Journal of the Geological Society of London*, vol. 15, no. 1–2, February 1859, pp. 168–76.

Owen, Rev Richard, *The Life of Richard Owen*, 2 volumes, John Murray, London, 1894.

Owen, Richard, 'Exhibition of a Bone of an Unknown Struthious Bird of Large Size from New Zealand', *Proceedings of the Zoological Society of London*, vol. 7, 1839, pp. 169–71.

Owen, Richard, *Memoir on the Pearly Nautilus (Nautilus Pompilius Linn.): With Illustrations of its External Form and Internal Structure by Richard Owen*, W. Wood, London, 1832.

Owen, Richard, 'On *Dinornis*, Part XII: Containing a description of the femur, tibia, and metatarsus of *Dinornis maximus* Owen', *Transactions of the Zoological Society of London*, vol. 6, 1869, pp. 497–500.

Owen, Richard, 'On *Dinornis*, Part XXV: Containing a Description of the Sternum of *Dinornis elephantopus*', *Transactions of the Zoological Society of London* vol. 12, no. 1, 1886, pp. 1–3.

Owen, Richard, *On the Extent and Aims of a National Museum of National History*, Saunders & Oatley, London, 1862.

Owen, Richard, 'On the Extinct Animals of the Colonies of Great Britain', *Proceedings of the Royal Colonial Institute*, vol. 10, 1878–9, Sampson, Low, Marston, Searle & Rivington, London, 1879, pp. 267–97.

Owen, Richard, 'Osteological Contributions to the Natural History of the Anthropoid Apes – No. 7', *Transactions of the Zoological Society of London*, vol. 5, pt.1, 1866, pp. 1–32.

Owen, Richard, *Palaeontology or, A Systematic Study of Extinct Animals and their Geographical Relations*, A & C Black, Edinburgh, 1861.

Pakrash, Gyan, 'Science "Gone Native" in Colonial India', *Representations*, no. 40, 1992, pp. 153–78.

Paterson, K. David, 'Paul B. Chaillu and the Exploration of Gabon, 1851–1865', *The International Journal of African Historical Studies*, vol. 7, no. 4, 1974, pp. 647–67.

Paul, Janet, 'Twelve Water Colours of Glaciers in the Province of Canterbury', *Turnbull Library Record*, no. 7, 1974, pp. 4–10.

Pearce, Susan (ed.), *Exploring Science in Museums*, Athlone, London and Atlantic Highlands N. J., 1996.

Pearson, Karl, *The Grammar of Science*, A & C Black, London, 1911.

Peel, George (ed.), *The Private Letters of Sir Robert Peel*, John Murray, London, 1920.

Peers, Douglas M., 'Colonial Knowledge and the Military in India, 1780–1860', *The Journal of Imperial and Commonwealth History*, vol. 33, no. 2, 2005, pp. 157–80.

Pelizzari, Maria Antonella (ed.), *Traces of India: Photography, Architecture, and the Politics of Representation, 1850–1900*, Canadian Center for Architecture and Yale University Press, Montreal and New Haven, 2003.

Pelzer, Ruth, 'Technical Reproduction and its Significance' in Matthew Rampley (ed.), *Exploring Visual Culture*, Edinburgh University Press, Edinburgh, 2005, pp. 197–213.

Peponis, John and Jenny Hedin, 'The Layout of Theories in the Natural History Museum', *9H*, no. 3, 1982, pp. 21–5.

Pickstone, John, 'Museological Science? The Place of the Analytical/Comparative in Nineteenth-Century Science, Technology and Medicine', *History of Science*, no. 32, 1994, pp. 111–38.

Plunkett, John, 'Celebrity and Community: The Poetics of the Carte-de-Visite', *Journal of Victorian Culture*, vol. 8, no. 1, 2003, pp. 55–79.

Pole, Dr, 'Photography for Travellers and Tourists', *Proceedings of the Royal Geographical Society*, 1871, pp. 47–50.

Potts, Timothy, *Civilisation: Ancient Treasures from the British Museum*, Australian National Gallery, Canberra, 1990.

Pratt, Mary Louise, *Imperial Eyes: Travel Writing and Transculturation*, Routledge, New York, 2008.

Prelli, Lawrence J. (ed.), *Rhetorics of Display*, University of South Carolina, Colombia, 2006.

Prescott, Gertrude Mae, 'Faraday: Image of the Man and the Collector' in David Gooding and Frank A. J. L. James (eds.), *Faraday Rediscovered: Essays on the Life and Work of Michael Faraday, 1791–1867*, Macmillan, Basingstoke, 1985, p. 15–32.

Prescott, Gertrude Mae, *Fame and Photography: Portrait Publications in Great Britain, 1856–1900*, PhD dissertation, University of Texas, 1985.

Prettejohn, Elizabeth, *The Art of the Pre-Raphaelites*, Princeton University Press, Princeton, 2000

Prince, Sue Ann, 'Stuffing Birds, Pressing Plants, Shaping Knowledge: Natural History in North America, 1730–1860', *Transactions of the American Philosophical Society*, vol. 93, no. 4, 2003, pp. 1–9.

Prodger, Philip, *Darwin's Camera: Art and Photography in the Theory of Evolution*, Oxford University Press, New York, 2009.

Public Works Department of NSW, *The Australian Museum: A Conservation Analysis of the Complex and its Site with a Statement of its Significance*, 1984.

Ramamurthy, Anandi, 'Spectacles and Illusions: Photography and Commodity Culture' in Liz Wells (ed.), *Photography: A Critical Introduction*, Routledge, London, 2004, pp. 193–244.

Reeder, Warwick, 'The Stereograph and the Album Portrait in Colonial Sydney 1859–62', *History of Photography*, vol. 23, no. 2, 1999, pp. 181–6.

Reeve, Lovell, *Conchologia Iconica, or, Illustrations of the Shells of Molluscous Animals*, 20 volumes, Reeve Brothers, London, 1843–78.

Reid, Francis Lucian, 'The Democratic Politician does not Trouble himself with Science': Class and Professionalism in the New Zealand Institute, 1867–1903', *Tuhinga*, no. 16, pp. 21–31.

Richards, Robert J., 'The Impact of German Romanticism on Biology in the Nineteenth Century' in Karl Ameriks (ed.), *The Impact of Idealism: The Legacy of Post-Kantian German Thought*, Cambridge University Press, Cambridge, 2013, pp. 235–42.

Richards, Thomas, *The Imperial Archive*, Verso, London, 1991.

Robin, Libby, 'The Platypus Frontier: Eggs, Aborigines and Empire in 19th Century Queensland' in Deborah Bird Rose and Richard Davis (eds.), *Dislocating the Frontier: Essaying the Mystique of Outback*, ANU E-Press, Canberra, 2005, pp. 99–120.

Rosen, Charles and Henri Zerner, *Romanticism and Realism: The Mythology of Nineteenth-Century Art*, Faber and Faber, London, 1984.

Rudwick, Martin, *Georges Cuvier, Fossil Bones and Geological Catastrophes*, University of Chicago Press, Chicago, 1997.

Rudwick, Martin, 'Georges Cuvier's Paper Museum of Fossil Bones', *Archives of Natural History*, vol. 27, no. 1, 2000, pp. 51–68.

Rudwick, Martin, *Scenes from Deep Time*, University of Chicago Press, Chicago and London, 1992.

Rudwick, Martin, 'The Emergence of a Visual Language for Geology', *History of Science*, vol. 14, no. 3, 1976, pp. 149–195.

Rudwick, Martin, *The Great Devonian Controversy*, University of Chicago Press, Chicago, 1985.

Rudwick, Martin, *The Meaning of Fossils: Episodes in the History of Palaeontology*, Macdonald & Co., London, 1972.

Rudwick, Martin, *Worlds Before Adam: The Reconstruction of Geohistory in the Age of Reform*, Chicago University Press, Chicago, 2008.

Rupke, Nicolaas, *Richard Owen: Biology without Darwin*, University of Chicago Press, Chicago and London, 2009.

Rupke, Nicolaas, *Richard Owen: Victorian Naturalist*, Yale University Press, New Haven and London, 1994.

Ryan, James, *Picturing Empire: Photography and the Visualization of the British Empire*, Reaktion Books, London, 1997.

Ryan, Simon, *The Cartographic Eye: How Explorers Saw Australia*, Cambridge University Press, Cambridge, 1996.

Said, Edward, *Culture and Imperialism*, Vintage, London, 1993.

Sapir, Michael, 'The Impossible Photograph: Hippolyte Bayard's "Self portrait as a Drowned Man"', *Modern Fiction Studies*, vol. 40, no. 3, 1994, pp. 619–29.

Schor, Naomi, *Reading in Detail: Aesthetics and the Feminine*, Routledge, New York and London, 2007.

Sear, Martha, '"Curious and Peculiar"?: Women Taxidermists in Colonial Australia' in Joan Kerr and Jo Holder (eds.), *Past Present: The National Women's Art Anthology*, Craftsman House, Sydney, 1999, pp. 85–99.

Secord, Anne, 'Corresponding Interests: Artisans and Gentlemen in Nineteenth-Century Natural History', *British Journal for the History of Science*, no. 27, December 1994, pp. 383–408.

Secord, Anne, 'Science in the Pub: Artisan Botanists in Early Nineteenth-Century Lancashire', *History of Science*, no. 32, 1994, pp. 269–315.

Seiberling, Grace, *Amateurs, Photography and the Mid-Victorian Imagination*, University of Chicago Press, Chicago, 1986.

Serres, Michel with Bruno Latour, *Conversations on Science, Culture, and Time*, University of Michigan Press, Ann Arbor, 1995.

Shadbolt, George, 'On the Mode of Producing Extremely Minute Photographs for Microscopic Examination', *Journal of the Photographic Society*, no. 4, 1858, pp. 75–81.

Shapin, Steven, 'The Invisible Technician', *American Scientist*, vol. 77, no. 66, 1989, pp. 554–63.

Shapin, Steven and Arnold Thackray, 'Prosopography as a Research Tool in History of Science: The British Scientific Community 1700–1900', *History of Science*, no. xii, 1974, pp. 1–28.

Shapin, Steven and Simon Schaffer, *Leviathan and the Air-Pump: Hobbes, Boyle, and the Experimental Life*, Princeton University Press, Princeton, 1985.

Shaw, A. G. L., 'Ludwig Becker', *Dictionary of Australian Artists Online* – http://www.daao.org.au/bio/ludwig-becker/biography/ (accessed 1 October 2010, updated 2012).

Sheets-Pyenson, Susan, 'Civilizing by Nature's Example: The Development of Colonial Museums of Natural History, 1850–1900' in Nathan Reingold and Marc Rothenbeg (eds.), *Scientific Colonialism: A Cross Cultural Comparison* Smithsonian Institution Press, Washington D.C., 1987, pp. 351–78.

Sheets-Pyenson, Susan, *Cathedrals of Science: The Development of Colonial Natural History Museums during the Late Nineteenth Century*, McGill-Queens University Press, Kingston and Montreal, 1988.

Sheppard, Eric, 'David Harvey and Dialectical Space-time' in Noel Castree and Derek Gregory (eds.), *David Harvey: A Critical Reader*, Wiley-Blackwell, Malden and Oxford, 2006, pp. 121–41.

Sherman, Daniel and Irit Rogoff (eds.), *Museum Culture: Histories, Discourses, Spectacles*, Routledge, London, 1994.

Shinn, Terry and Richard Whitley (eds.), *Expository Science: Forms and Functions of Popularisation*, D. Reidel, Dordrecht, Netherlands, 1985.

Shteir, Ann B. and Bernard Lightman (eds.), *Figuring It Out: Science, Gender, and Visual Culture*, Dartmouth College Press, Hanover and London, 2006.

Shweber, S. S., 'Scientists as Intellectuals: The Early Victorians' in James Paradis and James Postlewait (eds.), *Victorian Science and Victorian Values*, Annals of the New York Academy of Science, vol. 360, 1981, pp. 1–37.

Silverman, Robert J., 'The Stereoscope and Photographic Depiction in the 19th Century', *Technology and Culture*, vol. 34, no. 4, 1993, pp. 729–56.

Sinha, Mrinalini, *Colonial Masculinity: The Manly Englishman and the Effeminate Bengali*, 1995.

Smiles, Samuel, *Self Help*, Ward & Lock, London, 1859.

Smith, Andrew, *Illustrations of the Zoology of South Africa*, Smith, Elder & Co, London, 1838.

Smith, Bernard, *European Vision and the South Pacific*, Yale University Press, New Haven and London, 1988.

Smith, Bernard, 'Evolution and Australian Nature', *Meanjin*, April 1959, pp. 83–7.

Smith, Bernard, *Imagining the Pacific: In the Wake of the Cook Voyages*, Melbourne University Press, Melbourne, 1992.

Smith, Bernard, *Place, Taste and Tradition*, Ure Smith, Sydney, 1945.

Smith, Edgar A., 'Notes on *Rhysota Armiti*', *The Annals and Magazine of Natural History*, series 6, vol. 16, no, 94, 1895, pp. 351–2.

Smith, Jonathan, *Charles Darwin and Victorian Visual Culture*, Cambridge University Press, Cambridge, 2006.

Smyth, Charles Piazzi, *Teneriffe, An Astronomer's Experiment: or, Specialities of a Residence Above the Clouds*, Lovell Reeve, London, 1858.

Snyder, Joel, 'Nineteenth-Century Photography of Sculpture and the Rhetoric of Substitution' in Geraldine A Johnson (ed.), *Sculpture and Photography: Envisioning the Third Dimension*, Cambridge University Press, Cambridge, 1998, pp. 21–34.

Sorrenson, M. P. K., 'Mantell, Walter Baldock Durrant, 1820–1895', *Dictionary of New Zealand Biography* – http://www.teara.govt.nz/en/biographies/1m11/1 (accessed 14 March 2011).

Spary, E. C., 'Rococo Readings of the Book of Nature' in Marina Frasca-Spada and Nick Jardine (eds.), *Books and the Sciences in History*, Cambridge University Press, Cambridge, 2000, pp. 255–75.

Spenser, Thomas, 'An Account of some Experiments Made as to How Voltaic Electricity may be Usefully Applied for the Purpose of Working in Metal', *Westminster Review*, vol. 34, 1840, p. 459.

Stafford, Barbara Maria, *Artful Science: Enlightenment, Entertainment, and the Eclipse of Visual Education,* MIT Press, Cambridge (Mass.), 1994.

Stafford, Barbara Maria, *Voyage into Substance: Art, Science, Nature, and the Illustrated Travel Account, 1760–1840,* MIT Press, Cambridge (Mass.), 1984.

Stafford, Robert, *Scientist of Empire: Sir Roderick Murchison, Scientific Exploration and Victorian Imperialism*, Cambridge University Press, Cambridge, 1989.

Stanbury, Peter and Julian Holland, *Mr. Macleay's Celebrated Cabinet*, Macleay Museum, University of Sydney, Sydney, 1988.

Star, Susan L. and James R. Griesemer, 'Institutional Ecology, "Translations" and Boundary Objects: Amateurs and Professionals in Berkeley's Museum of Vertebrate Zoology, 1907–39', *Social Studies of Science*, vol. 19, 1989, pp. 387–420.

Stark, Amy, 'Lovell Reeve (1814–1865), Publisher and Patron of the Stereograph', *History of Photography*, vol. 5, no. 1, 1981, pp. 3–15.

Starr, Joan and Doug McMillan, *The Wellington Caves*, Macquarie Publications, Dubbo, 1985.

Stewart, Susan, *On Longing: Narratives of the Miniature, the Gigantic, the Souvenir, the Collection*, Duke University Press, Durham and London, 1993.

Stoler, Laura, 'Rethinking Colonial Categories: European Communities and the Boundaries of Rule', *Comparative Studies in Society and History*, vol. 31, no. 1, 1989, pp. 134–61.

Strachey, General R., 'On the Scope of Scientific Geography', *Proceedings of the Royal Geographical Society*, vol. XVI, no. 5, 1871–2, pp. 443–50.

Strahan, Ronald, *Rare and Curious Specimens: An Illustrated History of the Australian Museum 1827–1979*, Australian Museum, Sydney, 1979.

Sumnerensis, Dinornis (Alexander McKay), *The Canterbury Gilpin; or, The Capture and Flight of the Moa: A Poem*, James Hughes, Wellington, 1880.

Tagg, John, *The Burden of Representation: Essays on Photographies and Histories*, Macmillan, Basingstoke, 1988.

Tee, Garry J., 'Darwin's Correspondents in New Zealand', *Proceedings of the Australasian Association for the History, Philosophy and Social Studies of Science*, vol. 12, no. 2, 1980–1, pp. 367–79.

Terry, Martin, 'Curious Caverns: Picturesque Geology in Australian Art', *The Australian Antique Collector*, no. 33, 1986, pp. 46–7.

Thackray, John, 'Separately-published Prints of Fossils in Nineteenth-century Britain', *Archives of Natural History*, vol. 12, no. 2, September 1985, pp. 175–99.

Thackray, John C., *A Guide to the Official Archives of the Natural History Museum, London*, Natural History Museum, London, 1998.

Thomas, Ann (ed.), *Beauty of Another Order: Photography in Science*, Yale University Press, New Haven and London, 1997.

Thomas, Nicholas, *Entangled Objects: Exchange, Material Culture, and Colonialism in the Pacific*, Harvard University Press, Cambridge (Mass.) and London, 1991.

Thompson, Stephen, *Catalogue of a Series of Photographs by Stephen Thompson, from the Collection in the British Museum, Introduction by Charles Harrison*, W. A. Mansell, London, 1872.

Thompson, Stephen, 'On Portraits and Portraiture', *British Journal of Photography*, 1862, pp. 21–2.

Thomson, Keith W., *Art Galleries and Museums of New Zealand*, Reed, Wellington, 1981.

Todhunter, Isaac, *William Whewell: An Account of his Writings, with Selections from his Correspondence*, vol. 2, Macmillan & Co, London, 1876.

Trachtenberg, Alan, 'Likeness and Identity: Reflections on the Daguerrean Mystique' in Graham Clarke (ed.), *The Portrait in Photography*, Reaktion Books, London, 1992, pp. 173–92.

Tucker, Jennifer, *Nature Exposed: Photography as Eyewitness to Science*, John Hopkins University Press, Baltimore, 2005.

Tucker, Jennifer, 'The Historian, the Picture, and the Archive', *Isis*, vol. 97, 2006, pp. 111–20.

Tucker, Jennifer, 'Voyages of Discovery on Oceans of Air: Scientific Observation in the Age of "Balloonacy"', *Osiris*, 2nd series, vol. 11, 1996, pp. 144–76.

Turnbull, Paul, '"Savages, Fossils and Recent": Gerard Krefft and the Production of Racial Knowledge, 1869–1873' in Martin Crotty, John Germov and Grant Rodwell (eds), *A Race for Place: Eugenics, Darwinism and Social Thought and Practice in Australia*, Proceedings of the History of Sociology Eugenics Conference, University Newcastle, 2000, pp. 133–40.

Turner, Frank M., *Contesting Cultural Authority: Essays in Victorian Intellectual Life*, Cambridge University Press, Cambridge, 1993.

Tyndall, John, 'Michael Faraday' in Leslie Stephen (ed.), *Dictionary of National Biography*, vol. 18, Smith, Elder & Co., London, 1889, p. 201.

Vogeler, Martha S., 'Harrison, Charles (1835–1897)', *Oxford Dictionary of National Biography*, Oxford University Press, 2004 – http://www.oxforddnb.com/view/article/38891 (accessed 6 May 2011).

Wakefield, N. A, 'Mammals of the Blandowski Expedition to North-Western Victoria, 1856–57', *Proceedings of the Society of Victoria*, vol. 79, 1966, pp. 371–91.

Walker, David and Jürgen Tampke (eds.), *From Berlin to the Burdekin: The German Contribution to the Development of Australian Science, Exploration and the Arts*, New South Wales University Press, Sydney, 1991.

Wallace, Alfred Russel, 'Museums for the People', *Macmillan's Magazine*, January 1869.

Wanhalla, Angela and Erika Wolf (eds.), *Early New Zealand Photography: Images and Essays*, Otago University Press, Dunedin, 2011.

Ward, Henry A., *Catalogue of Fossils, From the Principal Museums of Europe and America*, Benton and Andrews, Rochester, 1866.

Warren, James H., 'Contesting Colonial Masculinity/Constituting Imperial Authority: Ceylon in Mid-nineteenth-century British Public Debate', *New Zealand Journal of Asian Studies*, vol. 6, no. 2, 2004, pp. 39–62.

Weiss, Marta, *Julia Margaret Cameron: Photographs to Electrify You with Delight and Startle the World*, MACK and V&A Publishing, London, 2015.

Wheatstone, Charles, 'Contributions to the Physiology of Vision – Part the First: On Some Remarkable, and Hitherto Unobserved, Phenomena of Binocular Vision', *Philosophical Transactions of the Royal Society of London*, vol. 128, 1838, pp. 371–94.

Whewell, William, 'On the Connexion of the Physical Sciences, by Mrs Somerville', *Quarterly Review*, vol. 51, 1834, pp. 54–68.

White, Paul, 'Darwin's Emotions: The Scientific Self and the Sentiment of Objectivity', *Isis*, vol. 100, no. 4, 2009, pp. 811–26.

White, Paul, 'Darwin Wept: Science and the Sentimental Subject', *Journal of Victorian Culture*, vol. 16, no. 2, 2011, pp. 195–21.

White, Paul, *Thomas Huxley: Making the 'Man of Science'*, Cambridge University Press, Cambridge and New York, 2003.

Whitley, G. P. and Martha Rutledge, 'Krefft, Johann Ludwig Gerard (Louis) (1830–1881)', *Australian Dictionary of Biography*, Volume 5, Melbourne University Press, 1974, pp. 42–44.

Whitley, Gilbert,'The Life and Work of Gerard Krefft', *Proceedings of the Royal Zoological Society of New South Wales for the Year 1958–59*, 1961, pp. 21–34.

Wilder, Kelley, *Photography and Science*, Reaktion, London, 2009.

Wiley, Peter Booth and Frances Chaves, *John Wiley & Sons: 200 Years of Publishing*, John Wiley & Sons, New York, 2010.

Willis, Anne-Marie, *Picturing Australia: A History of Photography*, Angus & Robertson, Sydney, 1988.

Wilson, David, *The British Museum: A History*, British Museum Press, London, 2002.

Wilson, David M., *The British Museum: Purpose and Politics*, British Museum, London, 1989.

Wind, Edgar, *Art and Anarchy*, Duckworth, London, 1985.
Woodward, Horace B., *The History of the Geological Society*, Arno Press, New York, 1978 (1907).
Woodward, Joan, *A Canterbury Album*, Te Waihora Press, Lincoln, 1987.
Yaldwyn, John and Juliet Hobbs (eds.), *My Dear Hector: Letters from Joseph Dalton Hooker to James Hector 1862–1893*, Technical Report 31, Museum of New Zealand Te Papa Tongarewa, Wellington, 1998.
Yanni, Carla, *Nature's Museums: Victorian Science and the Architecture of Display*, John Hopkins University Press, Baltimore, 1999.
Yeo, Richard, *Defining Science: William Whewell, Natural Knowledge and Public Debate in Early Victorian Britain*, Cambridge University Press, Cambridge, 1993.

Index

'actor-network theory' 10
Agassiz, Alexander 174–5
Agassiz, Louis 175
age of transition 156
Alfred, Prince *see* Duke of Edinburgh
Alfred Charles Barker at the Canterbury Museum Skeleton Hall 183
Allis, Thomas 159
amateur practitioners 5, 40, 44, 79, 99, 109–10, 112–13, 120, 124, 130, 138, 145, 168, 203
Anderson, John 97
Antiquarian Photographic Club 33
The Antiquity of Man 120
Appadurai, Arjun 68
Art Journal 46, 51, 108
Assyrian collections (British Museum) 22–8, 30, 34–5, 54
The Assyrian Gallery, British Museum 53–4
Athenaeum 82, 131
Atkinson, Alan 202
Australian Museum 1, 9, 12–13, 38, 68
Australian Museum, Sydney New South Wales 74

Bacon, Francis 197
Baconianism 70, 126, 133, 158, 197
Band of Hope Gold Mine 75
Barker, A. C. 182–4
Barnes, Henry 44, 74, 80
Baron Macaulay 131
Barton, Ruth 112
Bayard, Hippolyte 130
Bayertz, Kurt 70
Becker, Ludwig 40
Bedford, William 40
Beer, Gillian 14
Bennett, George 1–2, 73, 79
Blandowski, William 72, 87
Bohrer, Frederick 27
Bones of a New Zealand Moa (Dinornis robustus Owen*)* 157
'boundary objects' 10, 168
Bowerbank, James Scott 140
Bowler, Peter 4
'brainworkers' 70
Braudy, Leo 144
British Association for the Advancement of Science 125, 134, 197–200
British Journal of Photography 118
British Museum, photography and natural history at *see* paper museum
Brown, Ford Madox 70
Brown, James 166–7
Brown, Robert 29
Browne, Janet 122
Buckland, William 126
Byrne, Charles 117

Cameron, Julia Margaret 123
Campbell, George 185
The Canterbury Gilpin; or, The Capture and Flight of the Moa 154
Canterbury Museum 9, 17, 159
Carlyle, Thomas 70
Carpenter, William 29, 45–6, 197
carte de visite 185; introduction of 108; popularity of 118; portraits, men of science sitting for 138; representational power of 133
Carus, Carl Gustav 173
Catalogue of Fossils 173
Cathedral Cave 89–90
celebrity culture, rising *see* portraiture and the naturalist as celebrity
Ceratoptera alfredi 75
Charles Darwin (Edwards) 122
Charles Darwin (Maull and Polyblank) 121
Clarke, W. B. 42, 73, 159
Claudet, Antoine 108–9, 123
Clift, William 39
clyptodon clavipes 173
Cohen, Claudine 135
Cole, Henry 46

colonial masculinity 69–75
Colonial Museum (New Zealand) 141
colonial museum *see* museum traffic
commercial photography, role of 12
comparative anatomy, founder of 26
Conchologia Iconica 54
Country of the Dwarfs 84–5
Cousins, Samuel 128
Cowper, William 154
'Crown Domination' 167
Crystal Palace exhibition 30
cultivator of science 139
cultural marketplace 171
Cunningham, Peter 79
Cuvier, Georges 26, 38, 135, 158

Darwin, Charles 16, 73, 109–10, 120, 164, 175, 185, 201
Darwin, Emma 121
Darwin, Erasmus 121
Dasyurus Viverrinus 73
data assimilation 37
Denison, William 1, 9, 73
The Descent of Man 120
Desmond, Adrian 34
Dinornis 32
Dinornis crassus 177
Dinornis elephantopus 32–3, 35, 41–2
Dinornis giganteus 35, 176
Dinornis gravis 41
Dinornis maximus 137, 182
Dinornis robustus 41
Dinornis robustus (frontal view) 169
Dinornis robustus in York Museum 158
Dinornis robustus (side view) 170
Dinornis Sumnerensis see Alexander McKay
Disdéri, André-Adolphe-Eugène 118
Dromornis australis 44
Du Chaillu, Paul 68–9, 82, 85
Du Chaillu affair 68, 120
Duke of Edinburgh 74–5
Dunmore, Edward 118

Earle, Augustus 89–90, 92
Earth before the Deluge 155
East India Company Geological Survey of India 180
Edwards, Elizabeth 41
Edwards, Ernest 114, 120, 122
Elephantine Moa (Dinornis elephantopus), an Extinct Wingless Bird, in the Gallery of Fossils, British Museum 36
Ellegård, Alvar 120
Ellis, Henry 27
Empire of the Known 187
Enlightenment: intellectual 'possession' of nature during 45; philosophical inquiry epitomised by 134; quest for encyclopaedic knowledge 21; science and art, European perception and 8
European Vision and the South Pacific 7
Europe Mammals 55
exemplarity, rhetoric of *see* portraiture and the naturalist as celebrity
expeditionary photography 82–95
Explorations and Adventures in Equatorial Africa 84
The Expression of Emotions in Man and Animals 176
extinct nature, representation of 17

Faraday, Michael 108, 114, 117
Fenton, Roger 15, 25, 32, 36, 49, 51–4, 69, 83
Figuier, Louis 155
Flower, William 174
Forbes, Edward 172
Forde, Mrs Edward (Helena) 71
Fossil fish, Holoptychius nobiliformus 32
Fossil Gallery, 51, 53
Fossil Mammals of Australia 92
Freeman, James 6
Freund, Gisèle 111
Friday Evening Discourses 108
From Central Terminal Moraine of Great Godley Glacier, 5 March 1862 (Gully) 163
From Central Terminal Moraine of Great Godley Glacier, 5 March 1862 (Haast) 162, 164
Fuller, Frederick 185

Galapagos Islands reptiles 175
Gascoigne, John 59
Geller, W. O. 118
Geological Evidences of the Antiquity of Man 185
geological record, imperfection of 36
Geological Society of London 1–3
German expatriate network 8
German Museums Commission 73
German naturalists 76, 80–1, 186
The Gigantic Irish Deer, British Museum 51
Gilpin, John 154
Glaisher, James 48
Golinski, Jan 11
Gooding, David 123
'The Gorilla War' 86
Gosse, P. H. 154–5
Govett, William 91
The Graphic 200
Gray, John Edward 24, 29, 58, 95, 15, 138, 198
'Great Wizard of the North' 96

Griesemer, James 10
Group of Corals, British Museum 53
Group of Shells by Fenton 54
Group of women looking at stereoscope photographs 46
Gruber, Jacob 160, 187
Gully, John 163
Gunther, Albert 8, 15, 38–9, 52, 55, 58, 76

Haast, Julius von 8, 17, 154, 159, 162, 174, 201–2
Haeckel, Ernst 187
Harcourt, William 134
Harrison, Carol 71
Harrison, Charles 58
Harvard Museum of Comparative Zoology 174
Hawkins, Edward 27–8, 35
Hector, James 66, 141–2, 144, 159
Hedley, Charles 66–7
Heywood, J. M. 178
Highley, Samuel 198–200
HMS *Herald* 1
Hochstetter, Ferdinand 155, 160–1, 174
Hodge, Joseph 59
Holoptychius nobiliformus 32
Hooker, Joseph 66, 109, 121, 145, 159, 198
Hooker, William Henslow 182
Hooker, William (Jackson) 114, 162–3
H. R. H. the Duke of Edinburgh and Suite in Mining Costume after Descending the Band of Hope Gold Mine, Ballarat, Australia 77
Humboldt, Alexander von 88, 95
Hunt, Robert 47
Hunter, John 116
Hunterian Museum 39, 124, 158, 173–4, 176, 187
Huxley, Thomas 82, 114, 116, 120

identity: colonial 71; construction 9, 113, 145; masculine 70; national 6, 126; transnational 75
Illustrated London News 24–5, 26, 117, 119, 131, 135, 177–8, 179–80, 199
'immutable mobiles' 10
imperial–colonial relations, binary model of 59
Imperial-Royal Polytechnic Institut (Vienna) 174
imperial territories, museums in 5
inductive approach *see* Baconianism
informative entertainments *see* rational amusement 4

Jaeger, Gustav 160
James Hector 144
Jevons, W. S. 70, 172
Johann Franz Julius von Haast at the Canterbury Museum Skeleton Hall 182
Journal of Researches 120
Journal of the Photographic Society 33
Julius Haast standing beside a moa skeleton on display at Canterbury Museum 171
Julius von Haast and taxidermist Frederick Fuller 184

Kakapo 176
Kaup, Johann 40
King George IV 99
King of Saxony 173
Kingsley, Charles 186, 197
knowledge formation 94, 138
Krefft, Gerard 8–9, 38, 68–9, 71–4, 86, 91, 147, 159, 201–2
Krefft's Nature in Australia 98

landscapes, survey of 6
Latour, Bruno 10, 175
Layard, Austen Henry 26
The Leader 51
Le Noyé (Self Portrait as a Drowned Man) 130
Lewes, G. H. 125–6
Lightman, Bernard 111
The Linnean 159
lions of Victorian natural history 201
Literary and Scientific Portrait Club 120, 130–1
Literary Gazette 51
Livingstone, David 132
Llewelyn, John Dillwyn 34
looking, process of 2, 142
Lovell Reeve 143
Lyell, Charles 114, 120, 162, 185
Lytton, Edward Bulwer 156, 158

Macdonald, John 1
Macdonald, Sharon 168
Macleay, William John 73, 79
Macleay, William Sharp 1–2
Madras Museum (India) 17, 178–80
magic lantern 86, 198–200
male heroic tradition 71
Malraux, André 21
Man's Place in Nature 120
Mantell, Gideon 165
Mantell, Walter 165–7
Man the Interpreter of Nature 197
Maull, Henry 130
Maull and Fox 145
Maull and Polyblank 139–42, 145
Maurice, F. D. 70
McCosh, John 180

McCoy, Frederick 73
McCrae, George Gordon 76
McGregor, John 144
McKay, Alexander 154, 167
McLeod, Roy 70
Megaceros hibernicus 51
Megatherium 32
The Megatherium, British Museum 49
Melhuish, Arthur 48
Memoir on the Pearly Nautilus 129
Memoirs of the Extinct Wingless Birds of New Zealand 137, 181
Men of Eminence 113
Men of Mark 113
Mesozoic land reptiles 135
Michael, J. 179
Michael Faraday 109
Mill, John Stuart 133
Mitchell, Jesse 178
Mitchell, Thomas 87, 89
modernity, Victorian perspective on 156
Monkhouse and Co. 157
Moore, Clive 71
Moore, G. H. 174
Morelli, Giovanni 135
Mosman's Cave, Wellington Valley, New South Wales, no. 3 90
Mosman's Cave, Wellington Valley, New South Wales no. 5 92
motifs 147, 201
Müller, Fritz 186
Mundy, Daniel 169–70, 185
Murchison, Roderick 1, 32, 82, 114–15, 164, 180
Murray, David 172
Murray, John 82, 86
Musée d'Histoire Naturelle (Paris) 40–1
Musée National d'Histoire Naturelle 135
Museum of Practical Geology 47
Museums: Their History and their Use 172
museum traffic 66–100; authentication of scientific procedures 88; 'brainworkers' 70; colonial jealousy 67; colonial museums, increasing prominence of 81; concern for public instruction and entertainment 70; cosmopolitan 73; dismissal of curator 98; Du Chaillu affair 68, 82, 86; expeditionary photography and the 'New Traveller's Tales' 82–95; field-journal style of reporting 99; German immigrants 72; German Romantic influence in scientific inquiry 88; from German settler to cosmopolitan scholar (colonial masculinity and the rise of the self-made man) 69–75; global connections 79; 'Great Wizard of the North' 96; 'green-eyed monster' 67; impromptu performance 75; 'invisible technicians' 67; knowledge formation 94; male heroic tradition 71; museum reforms 73; naturalist correspondents, photography and the reframing of natural history 76–82; negative commentary 98; portrait series 75; 'practical man,' 70, 76, 81, 93, 94; prestigious aspects of natural history 67; proxy specimens 94; Royal Navy expeditions 78; scientific legitimation 99; Smilesian ethos of the self-made man 70; typical exchange of letters 79; visual conventions 74; Wellington Caves expeditions 87, 90, 93; Western masculinity 71
museum trustees 1, 9, 15, 22–34, 40–1, 46–7, 50, 56, 58, 69, 73, 79, 95–100
Museum without walls 21
mylodon robustus 173

Nash, John 199
natural history, as a vocation 9, 81, 109, 111–12, 131, 147, 201
natural history illustration 8, 16, 33, 36, 71, 79, 80, 94, 111, 158
naturalist correspondents *see* museum traffic
Nature 96, 99
Nead, Lynda 156
Negretti & Zambra 48
networks 3, 9, 10–13, 15, 18, 22, 38, 40, 59, 68, 75, 81, 160, 168, 174, 199, 200, 202
New South Wales Legislative Assembly 96
Newton, Isaac 125
'New Traveller's Tales' 82–95
New Zealand 39
New Zealand Fauna and Flora of Past and Present Times 156
New Zealand: Its Physical Geography, Geology and Natural History 155
Nototherium 94

Oldham, Thomas 180
Origin of Species 73, 175
Osteological Museum of the Royal College of Surgeons 47
outline drawings 2
Owen, Richard 1–2, 15–16, 22, 30, 43, 57, 78, 92, 109, 120, 154, 174, 181, 201

Palaeontology 42
palaeontology 169, 178
Palaeontology or, A Systematic Study of Extinct Animals and their Geographical Relations 41, 43
Palapteryx ingens 160
Panizzi, Antonio 30

paper museums 21–59; advent of photography, opportunity provided by 25; artefacts, meanings of 21; Assyrian tablets 25, 28, 35; Banksian Department 29; calotype photographs 24; conchological collections 55; connoisseurs 37; Coral Room 24; correspondents 40; Crystal Palace exhibition 30; data assimilation 37; elaborate paper museums 44; Enlightenment quest for encyclopaedic knowledge 21; Fossil Gallery 51, 53; fossil remains 32; from institutional priorities to imperial interests 22–33; geological record, imperfection of 36; glass-plate negatives 46; gratification derived from stereo views 53; Great Zoological Gallery 24; growing tension in modern museums 35; heuristic approach to cognitive understanding 50; imperial–colonial relations, binary model of 59; impression of tactility 49; influence of photographs in the museum context 41; institutional ambivalence 22; 'List of Suitable Objects to Be Photographed' 31; Montague House 24; naturalists, museum-based professional 38; 'noble science,' geology as 37; photographic clubs 34; private operators and entrepreneurs 45–56; 'proxy specimens' 38; Reptile Room 52; rhetoric of substitution 35, 57; sociopolitical and intellectual background 56; Standing Committee 27–8; stereoscope 47, 57; sub-genre of natural history illustration 36; venerable history 38; virtual inventory of specimens: (circulating new knowledge and filling in the gaps) 33–45; zoological galleries 24
Phillips, Thomas 127–8
Photographic Exchange Club 33
Photographic News 115, 199
Photographic Portraits of Living Celebrities 124–5, 131–2
Photographic Portraits of Men of Eminence in Literature, Science, and Art 114–15, 117
Photographic Society of London 174
photography, natural history and the nineteenth-century museum (introduction to) 1–18; 'actor-network theory' 10; colonial information 2; commercial photography, obligatory role of 12; cultural traffic 6; Darwin's theory of evolution by natural selection 4, 7; ecological approach to knowledge making 10; emerging sites of production (comparative approach) 6–7; Enlightenment 8; exchanging views of empire 7–10; extinct nature, representation of 17; formative visions of empire 7; German expatriate naturalists 8; from 'immutable mobiles' to 'boundary objects' 10–11; identity 6; imperial territories, museums in 5; informative entertainments 4; landscapes, survey of 6; making and moving images 11–12; navigating the archive 13–18; outline drawings 2; peripatetic objects of empire and collections at the periphery 12–13; photographs 2; portraiture 16; process of looking 2; raw data in circulation 11; research 14; rethinking the role of photography in Victorian natural history 4–6; stereoscopes, mass appeal of 2; transnational influences 8; visual literacy 10
physiognomy and phrenology 145
Pickersgill, Henry W. 127, 129
Pickstone, John 38
Pittard, Simon 73
Polyblank, George 130
Porter, Roy 114
Portrait of Gerard Krefft sent to Albert Gunther at the British Museum in 1864 78
Portrait of Julius Haast, 26 July 1859 186
Portrait of Richard Owen (photographer unknown) 137
Portrait of Richard Owen (Pickersgill) 129
Portrait of Richard Owen (Walker and Boutall) 136
portraiture and the naturalist as celebrity 108–47; academic robes 114; 'British Cuvier' 135; *carte de visite*, introduction of 108; circulation across public and private spheres 123; commonly used conventions 113; contemporary notions of exemplarity 133; contemporary photographic albums 108; contrast between forms of high culture and mass-produced objects 126; cultivator of science 139; defining motif 147; Du Chaillu affair 120; figure in the public gaze 123; Friday Evening Discourses 108; gentleman amateurs and professional bodies (social formation of Victorian science) 111–20; "household words" 108; identifying motifs 142; knowledge formation 138; language of exemplarity 113; natural history motifs 138; naturalist refashioned 120–45; paradox 145; philosopher 112; Platonic and Aristotelian ideas of exemplarity 133; popular criticism 123; profile portrait 144; projection of natural history to the

public 125; public identity of the scientist 109; Saurian skull 132; scientific genius as an exemplary type 126; scientific man 112; self-consciousness of the sitter, betrayal of 118; self-description, nuanced language of 112; self-display through acting 145; self-identification 110, 146; social formation of science 112; social status 111; specialisation 134; spectacle of gigantism 137; staged nature of studio portraiture 118; studio accessories 114; 'synecdoche' 135; tasks 110; Victorian laissez-faire economy 130; Victorian scientific culture, foundation of 120
Pre-Raphaelites 155
Price, William Lake 115, 117
Prince Albert 32, 134
Prince Alfred 74, 147
Proceedings of the Zoological Society 41
professionalisation 4, 12, 111–13, 134, 146, 198–9
proxy specimens 28, 35, 94
publicist, colonial naturalist as 174–87
Punch 96

Ramsay, Edward 72, 98
Rankin, George 87
rational amusement 3, 4, 22, 25, 47–8, 69, 70, 199, 200
R. C. A. Prior Album 139–40
Reeve, Lovell 15, 47–9, 50–7, 115–16, 139–141
Refers to Mr. W. B. Mantell's Moa bone discoveries. The gravel and sand bag bespeak him a Geologist 166
Regent's Park Zoological Gardens 57
Reischek, Andreas 8
reliability: of naturalists 81, 100, 111, 113, 120, 145; of observations and information 11, 67, 80, 88, 201; of photography 80, 94, 198
Representative Men in Literature, Science and Art 121
The Reptile Room, British Museum 52
retrodiction 154
Reynolds, Joshua 115, 118
Rhenish Friedrich Wilhelm University of Bonn 8
Riou, Eduard 155
Roberts Brothers 77
The Romance of Natural History 155
Rowe, George 126–7
Royal Botanic Gardens at Kew 66, 198
Royal Geographical Society 87
Royal Navy expeditions 78
Royal Pavilion 199
Royal Society of New South Wales 69
Rudwick, Martin 9, 38
Rupke, Nicolaas 125

A Saylore on Horse-backe. Ye Adelaide Troopes are reviewed by a Prussian Colonel 76
Scharf, George 22–3
scientific internationalism 200
scientific legitimation 99
Scott, Alexander Walker 72
Scott, Harriet 71–2
Seiberling, Grace 34
Self-Help 88
self-identification 110, 146
self-made man, rise of 69–75
Shadbolt, George 34
shell craze 55
Shires, Linda 143, 145
Skeleton of Man and of the Male Gorilla (Troglodytes Gorilla) I 82
Sloane, Hans 22
Smiles, Samuel 88
Smirke, Robert 24
Smith, Bernard 7
Smith, Edgar 66
Smith, Roger 198
Snakes of Australia 71
South Kensington Museum 46
spectacle, nature as 154–89; age of transition 156; attitude to colonial endeavour 165; avifauna 160; 'boundary objects' 168; Canterbury Museum, origin of natural history collections at 159–70; 'Crown Domination' 167; cynicism amongst colony-based researchers 167; difficulties associated with photographing in the colonies 183; emphasis on the sublime aspects of New Zealand's nature 163; Empire of the Known 187; expanding cultural marketplace 171; fascination with time 156; fascination with unearthing the unknown 156; fossil record, addition of the moa to 189; Galapagos Islands reptiles 175; geological surveys 163; glaciers 164; Glenmark moa discovery 168; honorific portraits 154; inventory of known fossils 174; key elements to early progress 161; modernity, Victorian perspective on 156; new era of the museum begins (debate over order versus spectacle) 170–4; New Zealand landmarks, naming of 162; palaeontology, growing complexity of 169; portraits and the press (colonial naturalist as publicist) 174–87; prehistoric fauna 158; publicity 174; public's appreciation of natural history, change

of 169; retrodiction 154; revivifications of the moa 154; stakeholders, focus of 185; temporal conflation 155; typology of mid-century colonial museums 188; variant of the genre of natural history illustration 158; world making 182
spirit collections 138
Spoel, Jacob 46
Spurgeon, Charles 86
Staircase of the Old British Museum, Montague House 23
Star, Susan 10
stereoscope 47; cultural position occupied by 57; innovations 50; mass appeal of 2; portable 48
Stereoscopic Magazine 47–56
Stevens, Samuel 141
Stoler, Laura 72
Succulent House at Kew 47
Sydney Morning Herald 1, 82, 86, 159
Sydney Punch 97
'synecdoche' 135

Talbot, William Henry Fox 27–8
Taylor, Richard 156
Thackray, John 36
theory of evolution by natural selection (Darwin) 4, 7
Thompson, Stephen 57–8, 108, 118, 122–3
Thomson, Alexander 91–2
Thomson, Thomas 139
Three Expeditions into the Interior of Eastern Australia 87, 89
Thurston, Edgar 179
Thylacoleo carnifex 91
The Times 82
Titian 115
Tomomi, Iwakura 200
Triptree and Sons 143
trustworthiness *see* reliability
'truthfulness to nature' 198
Tucker, Jennifer 88
Tyndall, John 70
The Unfortunate Moa, What Was Mired in the Slough of Despond at Waikouaiti in New Zealand. Anno Mundi??? 165

Velázquez 115
Victorian natural history: lions of 201; photography as integral aspect of 7; rethinking the role of photography in 4–6
Victorian science: culture, foundation of 120; social formation of 111–20
visual literacy 10

Walford, Edward 120–1
Walker and Boutall 136
Wallace, Alfred Russel 123
Ward, Henry A. 173–4
Warren, James 71
Wellington Caves 68, 87, 90
Westwood, J. O. 142
Whewell, William 134
William Buckland (Cousins) 128
William Buckland (Rowe) 127
Wilson, James Glen 1
Woolner, Thomas 121
Work 70
world making 182

Yeo, Richard 133
York, Frederick 57
Yorkshire Philosophical Society 158

zoological galleries (British Museum) 24
Zoological Gardens of Vienna 160
Zoological Society of London 73
Zygomaturus 1–3

FSC
www.fsc.org
MIX
Papier | Fördert
gute Waldnutzung
FSC® C083411

Kolibri 360 GmbH
Ferdinand-Jühlke-Straße 7
99095 Erfurt, Deutschland
produktsicherheit@kolibri360.de